COMBINATORIAL SET THEORY

STUDIES IN LOGIC

AND

THE FOUNDATIONS OF MATHEMATICS

VOLUME 91

NORTH-HOLLAND PUBLISHING COMPANY
AMSTERDAM · NEW YORK · OXFORD

COMBINATORIAL SET THEORY

Neil H. WILLIAMS
University of Queensland, Australia

1977

NORTH-HOLLAND PUBLISHING COMPANY
AMSTERDAM · NEW YORK · OXFORD

North-Holland ISBN: 0 7204 0722 2

Published by:
North-Holland Publishing Company – Amsterdam · New York · Oxford

Sole distributors for the U.S.A. and Canada:
Elsevier North-Holland, Inc.
52 Vanderbilt Avenue
New York, N.Y. 10017

Library of Congress Cataloging in Publication Data

Williams, Neil H.
Combinatorial set theory.
(Studies in logic and the foundations of mathematics ; v. 91)
Bibliography: p.
Includes indexes.
I. Combinatorial set theory. I. Title.
II. Series.
QA248.W56 511'.3 77-3462
ISBN 0-7204-0722-2

PRINTED IN THE NETHERLANDS

PREFACE

Combinatorial theory is largely the study of properties that a set or family of sets may have by virtue of its cardinality, although this may be widened to consider related properties held by sets carrying a simple structure, such as ordered or well ordered sets. These properties are relevant to either finite or infinite sets, although frequently the questions that pose interesting problems for finite sets are either meaningless or trivial for infinite sets, and vice-versa. There has been a recent great upsurge in the study of finite combinatorial problems, and a significant, though more manageable, increase in interest in the combinatorial properties of infinite sets.

This book deals solely with combinatorial questions pertinent to infinite sets.

Some results have arisen purely in the context of infinite sets. One of the early results in the subject is the following: given an infinite set S of power κ then there is a family of more than κ subsets of S any two of which intersect in a set of size less than κ. Chapter 1 looks at questions related to this. Problems of a different nature for families of sets are studied in Chapter 4, firstly a decomposition problem and then delta- and weak delta-systems. As a special case is the following: given any family $\mathcal{A}$ of $\aleph_2$ denumerable sets there is a subfamily $\mathcal{B}$ of $\mathcal{A}$ of size $\aleph_2$ such that the intersection of two sets from $\mathcal{B}$ is the same for all pairs from $\mathcal{B}$. Chapter 3 is devoted to the study of set mappings, that is, functions f defined on a set S such that given any x in S then $f(x)$ is a subset of S for which $x \notin f(x)$. Conditions are placed on the family $\{f(x); x \in S\}$ which ensure the existence of a large free set T (a subset of S such that $x \notin f(y)$ and $y \notin f(x)$ for all x, y from T).

Other results stem from problems that have been extensively studied for finite sets, and have been found to yield interesting questions when reformulated to apply to infinite sets. For example, in Chapter 5 we study infinite graphs, and in particular show that for any infinite cardinal κ there is a graph with chromatic number κ which contains no triangle, or indeed no pentagon; however any graph which contains no quadrilateral has chromatic number at most $\aleph_0$. Chapters 2, 4 and 7 are devoted to various extensions of Ramsey's classical theorem, which (in its finite form) states: given integers n, k, r if the

n-element subsets of a finite set S are divided into r classes then provided that S is sufficiently large there will be some k-element subset of S all the n-element subsets of which fall in the one class. Chapters 2 and 4 cover ordinary partition relations, polarized partition relations and square bracket partition relations for cardinal numbers, whilst Chapter 7 is concerned with ordinary partition relations for ordinal numbers.

Familiarity with the standard notions of set theory has been assumed throughout. An Appendix summarizes those properties of cardinal and ordinal numbers and their arithmetic which are basic to a study of this book.

The development of infinitary combinatorial theory over the past twenty years or so has been greatly stimulated by associates of the Hungarian school, under the encouragement of Paul Erdös in particular. A glance at the list of references gives some indication of how many of the results in this book show his influence.

To all those people who have created this subject, I here acknowledge my debt and record my gratitude.

Brisbane, 1976. Neil H. Williams

CONTENTS

FOREWORD ON NOTATION

The following is a brief summary of standard notation in use throughout this book. More technical notation is introduced throughout the text as the need for it becomes apparent. The Index of notation (pp. 206–208) provides a ready reference to the page on which a symbol is first defined.

Set membership is denoted by $\in$, and $\subseteq$ is the inclusion relation with $\subset$ denoting proper inclusion. The set of all subsets of a set is $\mathfrak{P}x$, so $\mathfrak{P}x = \{y; y \subseteq x\}$. The union of all the sets in x is written $\bigcup x$ and the intersection $\bigcap x$, so

$$\bigcup x = \{z;\ \exists y \in x(z \in y)\};\ \bigcap x = \{z;\ \forall y \in x(z \in y)\}\ .$$

Set difference is written $x - y$, so $x - y = \{z \in x; z \notin y\}$. The set of unordered pairs, one member from A and the other from B is $A \otimes B$, so

$$A \otimes B = \{\{x, y\}; x \in A, y \in B \text{ and } x \neq y\}.$$

Ordered pairs are written $\langle x, y\rangle$, and sequences as $\langle x_\alpha; \alpha < \beta\rangle$. The *length* of the sequence $\langle x_\alpha; \alpha < \beta\rangle$ is β, written $\ell n\langle x_\alpha; \alpha < \beta\rangle = \beta$. If $\boldsymbol{x}, \boldsymbol{y}$ are two sequences then $\boldsymbol{x}\frown\boldsymbol{y}$ is the *concatenation* of $\boldsymbol{x}$ and $\boldsymbol{y}$, that is, the sequence obtained by placing the entries from $\boldsymbol{y}$ in order after those from $\boldsymbol{x}$. If A is any set, then the *domain* and the *range* of A are defined by

$$\mathrm{dom}(A) = \{x;\ \exists y(\langle x, y\rangle \in A)\}; \qquad \mathrm{ran}(A) = \{y;\ \exists x(\langle x, y\rangle \in A)\}\ .$$

That f is a function with domain A and range contained in B is indicated briefly by writing $f\colon A \to B$. The set of all such functions is ${}^AB$, so

$${}^AB = \{f; f\colon A \to B\}\ .$$

If $f\colon A \to B$, then the value of f at x is $f(x)$. The restriction of f to a set X is written $f \restriction X$, so

$$f \restriction X = \{\langle x, y\rangle \in f; x \in X\}\ ;$$

and $f[X]$ is the range of $f \restriction X$, so

$$f[X] = \{f(x); x \in X\}\ .$$

The words *set* and *family* are used synonymously. However an *indexed family*, written $(A_i; i \in I)$, stands for that function A with domain I and

$A(i) = A_i$ for each i in I. The cartesian product of an indexed family is written $\times(A_i; i \in I)$. A *decomposition* of a set A is a family $\mathbf{\Delta}$ of sets such that $A = \bigcup \mathbf{\Delta}$; this is the same as a *partition* of A. The partition $\mathbf{\Delta}$ is *disjoint* if the sets in $\mathbf{\Delta}$ are pairwise disjoint. For elements a, b of A and a partition $\mathbf{\Delta}$ of A, the notation

$$a \equiv b (\mathrm{mod}\ \mathbf{\Delta})$$

means that there is some Δ_k in $\mathbf{\Delta}$ such that $a, b \in \Delta_k$.

The cardinality of a set X is written $|X|$, and $[X]^\kappa$, $[X]^{<\kappa}$, ... denote $\{Y \subseteq X; |Y| = \kappa\}$, $\{Y \subseteq X; |Y| < \kappa\}$, ... The operations of cardinal addition, multiplication and exponentiation are written $\eta \dotplus \theta$, $\eta \cdot \theta$ and η^θ, while the corresponding ordinal operations are written $\alpha + \beta$, $\alpha\beta$ and α^β. The infinite cardinal sum and product of an indexed family $(\eta_i; i \in I)$ of cardinals are written $\Sigma(\eta_i; i \in I)$ and $\Pi(\eta_i; i \in I)$, whereas the ordinal sum and product of a well ordered sequence $\langle \alpha_\nu; \nu < \beta \rangle$ of ordinal numbers are written $\Sigma_0(\alpha_\nu; \nu < \beta)$ and $\Pi_0(\alpha_\nu; \nu < \beta)$.

Let X be a set ordered by a relation $\prec$. By $\mathrm{tp}(X, \prec)$ or $\mathrm{tp}(X)$ is meant the order type of X under $\prec$. For subsets A, B of X, write $A \prec B$ to mean that $a \prec b$ for all a in A and b in B. The notation $\{x_1, x_2, ..., x_n\}_\prec$ refers to the set $\{x_1, ..., x_n\}$ and further indicates that $x_1 \prec x_2 \prec ... \prec x_n$. A subset A of X is *cofinal* in X if, for all x in X, there is a in A with $x \preccurlyeq a$. If α is an ordinal,

$$[X]^\alpha = \{Y \subseteq X; \mathrm{tp}(Y, \prec) = \alpha\} .$$

The *ordinal numbers* are defined so that if α is an ordinal, then

$$\alpha = \{\beta; \beta \text{ is an ordinal and } \beta < \alpha\} .$$

If A is a set of ordinals then $\sup A$ is the supremum of A, so $\sup A = \bigcup A$. *Cardinal numbers* are identified with the initial ordinals. The sequence of infinite cardinals is $\aleph_0, \aleph_1, \aleph_2, ..., \aleph_\alpha, ...$. The cardinal successor to a cardinal κ is denoted by κ^+; the iteration of this n times by $\kappa^{(n+)}$. The cofinality of a cardinal κ is written κ', so κ' is the least cardinal such that κ can be written as a sum of κ' cardinals all less than κ. The cardinal κ is *regular* if $\kappa' = \kappa$, and otherwise κ is *singular.* A cardinal of the form λ^+ for some λ is a *successor cardinal*; other cardinals are *limit cardinals.* The cardinal κ is a *strong limit cardinal* if $2^\lambda < \kappa$ whenever $\lambda < \kappa$. Regular limit cardinals are *weakly inaccessible*; regular strong limit cardinals are *strongly inaccessible.* The *cardinal beths* (starting from κ) are defined by induction:

$$\beth_0(\kappa) = \kappa , \qquad \beth_{n+1}(\kappa) = 2^{\beth_n(\kappa)}$$

The *Generalized Continuum Hypothesis* (GCH) is the statement: $2^\kappa = \kappa^+$

for all infinite cardinals κ. The GCH has been assumed throughout this book whenever it leads to a simplification in the statement or proof of a result. Theorems reached with its aid have the letters GCH appended to their result number (as in Theorem 1.6 (GCH)). In many cases the full strength of the GCH is not required, or the result could have been reformulated in a more involved form so as to avoid GCH all together. It is left to the interested reader to observe when this is so.

The Greek letters κ, λ, ι, η, θ are used throughout to stand for cardinal numbers, and usually κ, λ are infinite. Other small Greek letters denote ordinal numbers, as do k, l (except that ω is always the least infinite ordinal). The letters m, n always stand for non-negative integers.

CHAPTER 1

ALMOST DISJOINT FAMILIES OF SETS

§1. Almost disjoint families

One of the early results in combinatorial set theory was the following theorem of Sierpinski [84]. He proved that any infinite set of power κ can be decomposed into a family of more than κ infinite subsets in such a way that $|A \cap B| < |A|, |B|$ for any two different subsets A, B from the family. Two sets A and B will be called *almost disjoint* when $|A \cap B| < |A|, |B|$. Almost disjoint systems of sets were examined in more detail by Tarski [94]. It seems appropriate to start this book with an investigation of this and related problems.

Definition 1.1.1. The *degree of disjunction* $\delta(\mathcal{A})$ of a family $\mathcal{A}$ is the least cardinal θ such that $|A_1 \cap A_2| < \theta$ for all pairs A_1, A_2 in $\mathcal{A}$.

We shall be concerned with finding families $\mathcal{A}$ of subsets of a given set of infinite power κ with degree of disjunction at most θ for fixed cardinal θ. We need consider only the case when $\theta \leqslant |A|$ for all A in in $\mathcal{A}$ (and so certainly $\theta \leqslant \kappa$) since otherwise the condition $\delta(\mathcal{A}) \leqslant \theta$ imposes no restriction on $\mathcal{A}$. Clearly one can always find such a family of power κ, consisting in fact of pairwise disjoint subsets, so the problem is to find when there is a family $\mathcal{A}$ with $|\mathcal{A}| > \kappa$ and $\delta(\mathcal{A}) \leqslant \theta$. An upper bound for $|\mathcal{A}|$ is given by the following theorem.

Theorem 1.1.2. *If an infinite set of power κ can be decomposed into a family $\mathcal{A}$ with $|\mathcal{A}| = \lambda$ and $\delta(\mathcal{A}) \leqslant \theta$, then $\lambda \leqslant \kappa^\theta$.*

Proof. We may suppose that in fact it is the set κ which has been decomposed. So suppose $\kappa = \bigcup \mathcal{A}$ where $|\mathcal{A}| = \lambda$ and $\delta(\mathcal{A}) \leqslant \theta$. For each A in $\mathcal{A}$ with $|A| \geqslant \theta$, choose A^* in $[A]^\theta$. The condition $\delta(\mathcal{A}) \leqslant \theta$ implies that the mapping which sends A to A^* is one-to-one. Thus $\{A \in \mathcal{A}; |A| \geqslant \theta\}$ has cardinality at most that of $[\kappa]^\theta$, and so $|\mathcal{A}| \leqslant |[\kappa]^{\leqslant\theta}| = \kappa^\theta$.

The existence of large almost disjoint decompositions will follow from the next theorem. The method of proof is essentially that of the original Sierpinski paper.

Theorem 1.1.3. *Let κ be infinite, suppose $\eta \geqslant 2$ and let θ be the least cardinal for which $\kappa < \eta^\theta$. Let λ be a cardinal with $\lambda \leqslant \theta$ and $\lambda' = \theta'$. Then every set of power κ can be decomposed into a family of η^θ almost disjoint sets each of power λ.*

Proof. We show that such a decomposition is possible for a particular set of power κ; it then follows that this can be done for any set of power κ.

Since θ is least for which $\kappa < \eta^\theta$ we have $\theta \leqslant \kappa$ and consequently $\lambda \leqslant \kappa$. Thus $\kappa \cdot \lambda = \kappa$. The condition $\lambda \leqslant \theta$ and $\lambda' = \theta'$ means that there is a sequence $\langle \alpha(\nu); \nu < \lambda \rangle$ of ordinals $\alpha(\nu)$ with $\alpha(\nu) < \theta$ which is cofinal in θ. Put $S = \bigcup \{{}^{\alpha(\nu)}\eta; \nu < \lambda\}$. The choice of θ ensures that $|{}^{\alpha(\nu)}\eta| \leqslant \kappa$ for each ν, so $|S| \leqslant \Sigma\{|{}^{\alpha(\nu)}\kappa|; \nu < \lambda\} \leqslant \kappa \cdot \lambda = \kappa$. We shall define a decomposition of S. For each function g in ${}^\theta\eta$, put

$$S_g = \{g \restriction \alpha(\nu); \nu < \lambda\} ,$$

so $|S_g| = \lambda$. And for different f, g from ${}^\theta\eta$, let β be the least ordinal such that $f(\beta) \neq g(\beta)$. Then whenever $\alpha(\nu) > \beta$, for any h in S with $\mathrm{dom}(h) \geqslant \alpha(\nu)$ it follows that $h \notin S_f \cap S_g$. Thus $|S_f \cap S_g| \leqslant |\beta| < \lambda$. This means that if $\mathcal{S} = \{S_f; f \in {}^\theta\eta\}$, then $\mathcal{S}$ is a family of η^θ almost disjoint subsets of S each of power λ.

If $|S| = \kappa$, we have finished. If $|S| < \kappa$ we must add to S to build it up to power κ. Choose any set S_1 of power κ disjoint from S, and divide S_1 into a family $\mathcal{S}_1$ of pairwise disjoint sets each of size λ (which is possible since $\kappa \cdot \lambda = \kappa$). Then the family $\mathcal{S} \cup \mathcal{S}_1$ decomposes $S \cup S_1$ in the required manner.

As a first corollary to this theorem, we deduce the Sierpinski result mentioned above. For given the infinite cardinal κ, let θ be least such that $\kappa < 2^\theta$, so necessarily θ is also infinite. Then by Theorem 1.1.3. with $\eta = 2$ and $\lambda = \theta$, any set of power κ can be decomposed into a family of 2^θ almost disjoint subsets each of power θ, that is, into a family of more than κ almost disjoint infinite sets.

If the Generalized Continuum Hypithesis is assumed, the theorem leads to the following result. Given infinite cardinals κ and λ with $\lambda \leqslant \kappa$ and $\lambda' = \kappa'$, then every set of power κ can be decomposed into a family of κ^+ almost disjoint subsets each of power λ. This follows from Theorem 1.1.3 by putting $\eta = 2$, and noting that by the GCH, $\theta = \kappa$ is the least value of θ for which $\kappa < 2^\theta$. However, this result can be deduced without making use of the CGH, by adopting a different method of approach. When $\lambda = \kappa$, this was known by Erdös in 1934 (see Erdös, Gillman and Henriksen [34, Lemma 4.1]). A different proof is given by Sierpinski [87].

Theorem 1.1.4. *Let κ and λ be infinite cardinals with $\lambda \leqslant \kappa$ and $\lambda' = \kappa'$. Then every set of power κ can be decomposed into a family of more than κ almost disjoint subsets each of power λ.*

Proof. The argument splits into two cases, depending on whether κ is regular or not.

Case 1: κ regular. Here $\kappa = \kappa' = \lambda' \leqslant \lambda \leqslant \kappa$, so $\kappa = \lambda$. Let K be any set with $|K| = \kappa$, and let $\mathfrak{K}$ be a decomposition of K into κ disjoint sets each of power κ. We shall show that no family $\mathfrak{B}$ with $|\mathfrak{B}| < \kappa^+$ is maximal (under the subset relation) in the class $\mathfrak{F}$ of all families extending $\mathfrak{K}$ consisting of almost disjoint subsets of K each of power κ. Zorn's Lemma applies to $\mathfrak{F}$, and so there is a maximal family $\mathcal{A}$ in $\mathfrak{F}$. But then we must have $|\mathcal{A}| \geqslant \kappa^+$, and the theorem holds in this case.

So let $\mathfrak{B}$ be a member of $\mathfrak{F}$ with $|\mathfrak{B}| < \kappa^+$. Since $\mathfrak{K} \subseteq \mathfrak{B}$, in fact $|\mathfrak{B}| = \kappa$. Let $\langle B_\mu; \mu < \kappa \rangle$ be an enumeration (always without repetitions) of $\mathfrak{B}$. For each μ with $\mu < \kappa$, note that $|\bigcup \{B_\mu \cap B_\nu; \nu < \mu\}| < \kappa$ since $\delta(\mathfrak{B}) \leqslant \kappa$ and κ is regular. Hence we may choose inductively elements b_μ when $\mu < \kappa$ so that

$$b_\mu \in B_\mu - \bigcup \{B_\nu; \nu < \mu\} .$$

Then $b_\mu \neq b_\nu$ whenever $\nu < \mu$. Put $B = \{b_\mu; \mu < \kappa\}$ so $|B| = \kappa$. For each μ with $\mu < \kappa$ we have $B_\mu \cap B \subseteq \{b_\nu; \nu \leqslant \mu\}$ and so $|B_\mu \cap B| < \kappa$. Hence the family $\mathfrak{B} \cup \{B\}$ is in the class $\mathfrak{F}$, and so $\mathfrak{B}$ is not maximal.

Case 2: κ singular. Choose cardinals κ_σ for σ with $\sigma < \kappa'$ such that $\kappa' < \kappa_\sigma < \kappa_1 < ... < \kappa$ and $\Sigma(\kappa_\sigma; \sigma < \kappa') = \kappa$. Since $\lambda' = \kappa'$ we can choose cardinals λ_0 for σ with $\sigma < \kappa'$ such that always $\lambda_\sigma < \lambda$ and $\Sigma(\lambda_\sigma; \sigma < \kappa') = \lambda$. Let K be any set with $|K| = \kappa$, and let $\{K_\sigma; \sigma < \kappa'\}$ be a decomposition of K into κ' disjoint sets each of power κ. This time we show that no family $\mathfrak{B}$ with $|\mathfrak{B}| < \kappa^+$ is maximal in the class $\mathfrak{G}$ of all families $\mathcal{A}$ of almost disjoint subsets of K with always $|A \cap K_\sigma| = \lambda_\sigma$ for each A in $\mathcal{A}$. Here Zorn's Lemma applies to $\mathfrak{G}$; but any maximal member of $\mathfrak{G}$ must have power greater than κ, as desired.

So let $\mathfrak{B}$ be a member of $\mathfrak{G}$ with $|\mathfrak{B}| < \kappa^+$. Let $\langle B_\mu; \mu < \theta \rangle$ be an enumeration of $\mathfrak{B}$ with $\theta \leqslant \kappa$. For each σ with $\sigma < \kappa'$, choose X_σ with $|X_\sigma| = \lambda_\sigma$ and

$X_\sigma \subseteq K_\sigma - \bigcup \{B_\mu; \mu < \kappa_\sigma\}$. Since

$$|\bigcup\{K_\sigma \cap B_\mu; \mu < \kappa_\sigma\}| \leqslant \lambda_\sigma \cdot \kappa_\sigma < \kappa = |K_\sigma| ,$$

this is always possible. Put $B = \bigcup\{X_\sigma; \sigma < \kappa'\}$, so $|B| = \Sigma(\lambda_\sigma; \sigma < \kappa') = \lambda$. Note that given μ with $\mu < \kappa$, if τ is least for which $\mu < \kappa_\tau$ then $X_\sigma \cap B_\mu = \emptyset$ whenever $\sigma \geqslant \tau$, and so $B \cap B_\mu \subseteq \bigcup\{X_\sigma; \sigma < \tau\}$. Hence $|B \cap B_\mu| \leqslant \Sigma(\lambda_\sigma; \sigma < \tau) < \lambda$. It follows that the family $\mathcal{B} \cup \{B\}$ is in $\mathfrak{G}$, and so $\mathcal{B}$ is not maximal.

If we assume the Generalized Continuum Hypothesis, then we can show that the condition $\lambda \leqslant \kappa$ and $\lambda' = \kappa'$ of Theorem 1.1.4 is also necessary for the existence of such a decomposition. This result goes back to Tarski [94]. Unfortunately there appears no way to avoid the GCH for this proof. We need first the following easy lemma.

Lemma 1.1.5. *Let λ be a cardinal, let β be an ordinal with λ not cofinal in β. For each set A with $A \in [\beta]^{\geqslant\lambda}$ there is an ordinal α with $\alpha < \beta$ such that $|\alpha \cap A| \geqslant \lambda$.*

Proof. Given A with $A \in [\beta]^{\geqslant\lambda}$, take a subset A_1 of A with order type λ. Since λ is not cofinal in β, there is α with $\alpha < \beta$ such that $A_1 \subseteq \alpha$. Then $A_1 \subseteq \alpha \cap A$ so that $|\alpha \cap A| \geqslant |A_1| = \lambda$.

Theorem 1.1.6 (GCH). *Let κ and λ be cardinals with $\lambda' \neq \kappa'$. Then any decomposition $\mathcal{A}$ of a set of power κ into subsets each of power at least λ with $\delta(\mathcal{A}) \leqslant \lambda$, has power at most κ.*

Proof. We may suppose that $\mathcal{A}$ is a decomposition of the set κ, so suppose $\mathcal{A} \subseteq [\kappa]^{\geqslant\lambda}$ with $\delta(\mathcal{A}) \leqslant \lambda$, and we need only consider the case $\lambda \leqslant \kappa$. So in fact $\lambda < \kappa$, because $\lambda' \neq \kappa'$. Since $\lambda' \neq \kappa'$ we know λ is not cofinal in κ. Applying Lemma 1.1.5, for each A in $\mathcal{A}$ there is an ordinal α with $\alpha < \kappa$ and $|\alpha \cap A| \geqslant \lambda$. For each α put $\mathcal{A}_\alpha = \{A \in \mathcal{A}; \alpha$ is least for which $|\alpha \cap A| \geqslant \lambda\}$. Then $\mathcal{A} = \bigcup\{\mathcal{A}_\alpha; \alpha < \kappa\}$. Thus $|\mathcal{A}| \leqslant \Sigma(|\mathcal{A}_\alpha|; \alpha < \kappa)$.

We seek an estimate of $|\mathcal{A}_\alpha|$. Since $\delta(\mathcal{A}) \leqslant \lambda$, for distinct A_1, A_2 from $\mathcal{A}_\alpha$ we have

$$|(\alpha \cap A_1) \cap (\alpha \cap A_2)| \leqslant |A_1 \cap A_2| < \lambda ,$$

and so $\alpha \cap A_1 \neq \alpha \cap A_2$. Thus if $\mathcal{B}_\alpha = \{\alpha \cap A; A \in \mathcal{A}_\alpha\}$ then $|\mathcal{B}_\alpha| = |\mathcal{A}_\alpha|$. Now $\delta(\mathcal{B}_\alpha) \leqslant \lambda$ and $\mathcal{B}_\alpha$ is a decomposition of the set α, so by Theorem 1.1.2 we have $|\mathcal{B}_\alpha| \leqslant |\alpha|^\lambda$. However $\lambda < \kappa$ and $|\alpha| < \kappa$ so by GCH, $|\mathcal{B}_\alpha| \leqslant \kappa$. Thus $|\mathcal{A}_\alpha| \leqslant \kappa$. Consequently $|\mathcal{A}| \leqslant \Sigma(|\mathcal{A}_\alpha|; \alpha < \kappa) \leqslant \kappa \cdot \kappa = \kappa$, and the theorem is proved.

Corollary 1.1.7 (GCH). *No set of power κ can be decomposed into a family $\mathscr{A}$ of more than κ sets each of power greater than λ such that $\delta(\mathscr{A}) \leqslant \lambda$.*

Proof. Such a decomposition is certainly impossible when $\lambda' \neq \kappa'$, by Theorem 1.1.6. So suppose $\lambda' = \kappa'$. Since $\lambda' \leqslant \lambda$ and $(\lambda^+)' = \lambda^+$, in fact $\kappa' \neq \lambda^+$. But now Theorem 1.1.6, with λ replaced by λ^+, yields the result.

It is worth remarking that if λ is finite (but κ still infinite) then Theorem 1.1.6 and Corollary 1.1.7 remain true, with the same proofs, and in fact GCH is not needed.

§2. Almost disjoint functions

In this section we shall discuss a refinement of the question considered in section 1. Rather than seeking families of almost disjoint subsets of an arbitrary set, we shall look at families of functions from a fixed set L to a fixed set K, where $|L| = \lambda$ and $|K| = \kappa$. We shall in fact from the start identify L with λ and K with κ, and so shall consider subfamilies of ${}^\lambda\kappa$, the set of all functions mapping λ into κ.

Let $\mathscr{F}$ be a subset of ${}^\lambda\kappa$. Thinking of the members of $\mathscr{F}$ as sets of ordered pairs, we can define $\delta(\mathscr{F})$, the degree of disjunction, as in Definition 1.1.1. So in fact $\delta(\mathscr{F})$ is the least cardinal η such that $|\{\alpha < \lambda; f(\alpha) = g(\alpha)\}| < \eta$ for all f, g in $\mathscr{F}$. For each value of the cardinal θ, we shall seek the largest possible cardinality of a family $\mathscr{F}$ where $\mathscr{F} \subseteq {}^\lambda\kappa$ with $\delta(\mathscr{F}) \leqslant \theta$. Clearly we need only consider the case when $\theta \leqslant \lambda$, since otherwise the condition $\delta(\mathscr{F})$ is no restriction on $\mathscr{F}$.

There is an equivalent formulation of this problem. We make the following definition.

Definition 1.2.1. A set T is said to be a *transversal* of the family $\mathscr{A}$ if $|T \cap A| = 1$ for each A in $\mathscr{A}$.

Given a family $\mathscr{A} = \{A_\nu; \nu < \lambda\}$ of λ pairwise disjoint sets each of power κ, by identifying each A_ν with κ, any transversal of $\mathscr{A}$ can be identified with a function in ${}^\lambda\kappa$. In this way, a family $\mathscr{F}$ with $\mathscr{F} \subseteq {}^\lambda\kappa$ and $\delta(\mathscr{F}) \leqslant \theta$ corresponds to a class of transversals of the family $\mathscr{A}$, any two of which meet in less than θ points. Thus the original problem is equivalent to finding the maximum size of a class $\mathscr{T}$ of transversals of $\mathscr{A}$ with $\delta(\mathscr{T}) \leqslant \theta$.

Clearly one can always find κ pairwise disjoint transversals, so we shall wish to know for which values of λ and θ is it possible that $|\mathscr{T}| > \kappa$.

We shall need to assume the Generalized Continuum Hypothesis almost throughout this section.

We start with several easy remakrs.

Lemma 1.2.2 (GCH). *Suppose either $\theta < \lambda$ and $\kappa^+ < \lambda$ or else $\theta = \lambda$ and $\kappa^+ < \lambda'$. If $\mathcal{F} \subseteq {}^\lambda\kappa$ and $\delta(\mathcal{F}) \leqslant \theta$ then $|\mathcal{F}| \leqslant \kappa$.*

Proof. For a contradiction, suppose that under these conditions there is $\mathcal{F}$ with $\mathcal{F} \subseteq {}^\lambda\kappa$ such that $\delta(\mathcal{F}) \leqslant \theta$ and $|\mathcal{F}| = \kappa^+$. For distinct f, g in $\mathcal{F}$, put $E(f, g) = \{\alpha < \lambda; f(\alpha) = g(\alpha)\}$, so $|E(f, g)| < \theta$. Put $E = \bigcup\{E(f, g); \{f, g\} \in [\mathcal{F}]^2\}$, so $|E| < \lambda$. If α is in $\lambda - E$ then $f(\alpha) \neq g(\alpha)$ for distinct f, g in $\mathcal{F}$, and so $|\mathcal{F}| \leqslant \kappa$, contradicting that $|\mathcal{F}| = \kappa^+$.

Lemma 1.2.3 (GCH). *Let $\lambda' \leqslant \kappa$. If $\mathcal{F} \subseteq {}^\lambda\kappa$ with $\delta(\mathcal{F}) \leqslant \lambda$ then $|\mathcal{F}| \leqslant \kappa^+$.*

Proof. If $\lambda \leqslant \kappa$ clearly $|\mathcal{F}| \leqslant \kappa^+$, so suppose $\kappa < \lambda$ and consequently λ must be singular. Take cardinals λ_τ where $\tau < \lambda'$ such that $\kappa^+ < \lambda_0 < \lambda_1 < \lambda_2 < \ldots < \lambda$ and $\Sigma(\lambda_\tau; \tau < \lambda') = \lambda$. For a contradiction, suppose in fact that $|\mathcal{F}| \geqslant \kappa^{++}$. Since $\delta(\mathcal{F}) \leqslant \lambda$ we have a decomposition $[\mathcal{F}]^2 = \bigcup_{\tau<\lambda'} \{\{f, g\}; |f \cap g| < \lambda_\tau\}$. At this stage, we shall appeal to the relation $\kappa^{++} \to (\kappa^+)^2_\kappa$ which follows from Theorem 2.2.4. This symbol means that whenever the class of unordered pairs from a set S of power κ^{++} is decomposed into at most κ parts there is H with $H \in [S]^{\kappa^+}$ such that all the pairs from H fall in the same part of the decomposition. In the situation here, this ensures that there is a family $\mathcal{G}$ with $\mathcal{G} \in [\mathcal{F}]^{\kappa^+}$ such that for some fixed τ with $\tau < \lambda'$ always $|f \cap g| < \lambda_\tau$ for all f, g in $\mathcal{G}$. But now if $\mathcal{H} = \{g \restriction \lambda_\tau^+; g \in \mathcal{G}\}$, it follows that $\mathcal{H}$ is a family of κ^+ functions mapping from λ_τ^+ into κ, with $\delta(\mathcal{H}) \leqslant \lambda_\tau$. This is not allowed by Lemma 1.2.2. Consequently $|\mathcal{F}| \leqslant \kappa^+$.

Lemma 1.2.4. *Let $\kappa' = \lambda'$. Then there is $\mathcal{F}$ with $\mathcal{F} \subseteq {}^\lambda\kappa$, $|\mathcal{F}| = \kappa^+$ and $\delta(\mathcal{F}) \leqslant \lambda$.*

Proof. There are ordinals γ_σ where $\sigma < \kappa'$ and δ_τ where $\tau < \lambda'$ such that $\gamma_0 < \gamma_1 < \ldots < \kappa = \sup\{\gamma_\sigma; \sigma < \kappa'\}$ and $\delta_0 < \delta_1 < \ldots < \lambda = \sup\{\delta_\tau; \tau < \lambda'\}$. We shall show that no family $\mathcal{G}$ with $\mathcal{G} \in [{}^\lambda\kappa]^{\leqslant\kappa}$ is maximal in the class of all families of almost disjoint functions in ${}^\lambda\kappa$; as in the proof of Theorem 1.1.4 this will give the result. So suppose $\mathcal{G} \in [{}^\lambda\kappa]^{\leqslant\kappa}$, and write $\mathcal{G} = \{g_\nu; \nu < \kappa\}$. Define g with $g \in {}^\lambda\kappa$ by choosing $g(\alpha)$ so that if τ is least for which $\alpha < \delta_\tau$ then $g(\alpha) \in \kappa - \{g_\nu(\alpha); \nu < \gamma_\tau\}$. Given μ with $\mu < \kappa$, if σ is least such that $\mu < \gamma_\sigma$ then $g \cap g_\mu \subseteq \{\langle\alpha, g(\alpha)\rangle; \alpha \leqslant \delta_\sigma\}$ and so $|g \cap g_\mu| \leqslant |\delta_\sigma \dotplus 1| < \lambda$. Thus g can be added to $\mathcal{G}$ and still gives an almost disjoint family.

The following lemma is a theorem of Erdös, Hajnal and Milner [36].

Lemma 1.2.5. *Let* $\kappa^+ = \lambda$. *Then there is* $\mathcal{F}$ *with* $\mathcal{F} \subseteq {}^\lambda\kappa$ *such that* $|\mathcal{F}| = \kappa^+$ *and* $\delta(\mathcal{F}) < \kappa$.

Proof. The plan for the proof is as follows. By thinking of the ordinals less than κ^+ arranged according to order type, it is clear how to define pairwise disjoint functions g_ν for ν with $\nu < \kappa^+$ mapping into κ^+, with the domain of g_ν all ordinals β with $\nu < \beta < \kappa^+$ – simply define $g_\nu(\alpha) = \alpha$.

For each α with $\alpha < \kappa^+$ we have $|\alpha| \leqslant \kappa$, and so we may take a well ordering $\langle \xi_{\alpha\nu}; \nu < \alpha \rangle$ of (a subset of) κ of order type α, and use the above construction on these well orderings to provide pairwise disjoint tail ends of functions $f_\nu : \kappa^+ \to \kappa$ where $\nu < \kappa^+$. One has only to arrange the front ends so that any two of the functions meet in a set of size less than κ, and this gives a family $\mathcal{F}$ as required.

Formally, then, we construct functions $f_\nu : \kappa^+ \to \kappa$ where $\nu < \kappa^+$ by induction as follows. Let ν with $\nu < \kappa^+$ be given, and suppose that the functions f_μ when $\mu < \nu$ have already been suitably defined. If $\alpha > \nu$, put $f_\nu(\alpha) = \xi_{\alpha\nu}$. Then when $\mu < \nu$ certainly $f_\nu(\alpha) \neq f_\mu(\alpha)$. To define $f_\nu(\alpha)$ for values of α where $\alpha \leqslant \nu$, let $\{f_\mu; \mu < \nu\} = \{f_{\nu\beta}; \beta < \min(\nu, \kappa)\}$ and choose $f_\nu(\alpha)$ in $\kappa - \{f_{\nu\beta}(\alpha); \alpha < \beta\}$. Then if $\mu < \nu$, for some β with $\beta < \kappa$ it follows that $f_\mu = f_{\nu\beta}$ and consequently $f_\mu(\alpha) \neq f_\nu(\alpha)$ whenever $\beta \leqslant \alpha \leqslant \nu$. Thus $|f_\mu \cap f_\nu| \leqslant |\beta| < \kappa$. Then if $\mathcal{F} = \{f_\nu; \nu < \kappa^+\}$, we have $|\mathcal{F}| = \kappa^+$ and $\delta(\mathcal{F}) < \kappa$ as desired.

We can now turn to the problem stated at the beginning of this section; to find the maximum cardinality of a family $\mathcal{F}$ where $\mathcal{F} \subseteq {}^\lambda\kappa$ with $\delta(\mathcal{F}) \leqslant \theta$. The case $\theta < \lambda$ is much easier than when $\theta = \lambda$, so we dismiss this case first.

Theorem 1.2.6 (GCH). *Suppose* $\theta < \lambda$. *Let* $\mathfrak{m}$ *be the maximum of the cardinalities of families* $\mathcal{F}$ *where* $\mathcal{F} \subseteq {}^\lambda\kappa$ *with* $\delta(\mathcal{F}) \leqslant \theta$. *Then* $\mathfrak{m} = \kappa$ *unless* $\lambda = \kappa^+$, *in which case* $\mathfrak{m} = \kappa^+$.

Proof. Suppose $\mathcal{F} \subseteq {}^\lambda\kappa$ with $\delta(\mathcal{F}) \leqslant \theta$.

If $\kappa < \theta$ then $\kappa^+ < \lambda$, and Lemma 1.2.2 shows $|\mathcal{F}| \leqslant \kappa$. If $\kappa > \theta$, consider $\mathcal{F} \restriction \theta^+ = \{f \restriction \theta^+; f \in \mathcal{F}\}$ and note that since $\delta(\mathcal{F}) \leqslant \theta$ the map which sends f to $f \restriction \theta^+$ is one-to-one. Now $\mathcal{F} \restriction \theta^+$ is a decomposition of $\theta^+ \times \kappa$, a set of power κ, into sets of size greater than θ with $\delta(\mathcal{F} \restriction \theta^+) \leqslant \theta$. So by Corollary 1.1.7, $|\mathcal{F} \restriction \theta^+| \leqslant \kappa$ and hence $|\mathcal{F}| \leqslant \kappa$. If $\kappa = \theta$ and $\kappa^+ < \lambda$, again Lemma 1.2.2 gives $|\mathcal{F}| \leqslant \kappa$.

Only the case $\kappa = \theta$ and $\lambda \leqslant \kappa^+$ remains. Since $\theta < \lambda$ we must have $\lambda = \kappa^+$.

Considering $\mathcal{F}$ as a decomposition of $\lambda \times \kappa$, from Theorem 1.1.2 we have $|\mathcal{F}| \leqslant (\kappa^+)^\kappa = \kappa^+$. However, Lemma 1.2.5 shows that the value $|\mathcal{F}| = \kappa^+$ can be achieved, and so in this case $\mathfrak{m} = \kappa^+$.

We are left with the situation when $\theta = \lambda$, and this is rather more involved. Indeed, the question cannot be fully answered on the basis of the usual axioms for set theory.

Theorem 1.2.7 (GCH). *Let* $\mathfrak{m}$ *be the maximum of the cardinalities of almost disjoint families* $\mathcal{F}$ *where* $\mathcal{F} \subseteq {}^\lambda\kappa$.

(i) *If* $\lambda' \neq \kappa'$ *and* $\lambda' \neq \kappa^+$, *then* $\mathfrak{m} = \kappa$.
(ii) *If* $\lambda' = \kappa'$, *then* $\mathfrak{m} = \kappa^+$.
(iii) *If* $\lambda' = \kappa^+$, *then* $\kappa^+ \leqslant \mathfrak{m} \leqslant \kappa^{++}$.

Proof. Let us do (ii) first, so suppose $\lambda' = \kappa'$. Then $\lambda' \leqslant \kappa$ and Lemma 1.2.3 shows $\mathfrak{m} \leqslant \kappa^+$. Since Lemma 1.2.4 gives $\mathfrak{m} \geqslant \kappa^+$, in fact $\mathfrak{m} = \kappa^+$.

Towards (iii), suppose $\lambda' = \kappa^+$. If λ is regular then $\lambda = \kappa^+$. We have the trivial bound $\mathfrak{m} \leqslant |{}^\lambda\kappa| = \kappa^{\kappa^+} = \kappa^{++}$, and Lemma 1.2.5 shows $\kappa^+ \leqslant \mathfrak{m}$. So suppose λ is singular, and choose cardinals λ_τ where $\tau < \lambda'$ such that $\kappa^+ < \lambda_0 < \lambda_1 < \ldots < \lambda$ and $\lambda = \Sigma(\lambda_\tau; \tau < \lambda')$. We first construct a family $\mathcal{G}$ where $\mathcal{G} \subseteq {}^\lambda\kappa$ with $|\mathcal{G}| = \kappa^+$ and $\delta(\mathcal{G}) = \kappa$, and so show $\mathfrak{m} \geqslant \kappa^+$. For τ such that $\kappa \leqslant \tau < \kappa^+$, for each ordinal α with $\lambda_\tau \leqslant \alpha < \lambda_{\tau+1}$ let $\langle \xi_{\alpha\nu}; \nu < \tau + 1\rangle$ be a well ordering of κ of order type $\tau + 1$. For ν such that $\kappa \leqslant \nu < \kappa^+$ define $g_\nu : \lambda \to \kappa$ by

$$g_\nu(\alpha) = \begin{cases} 0 \text{ if } \alpha < \lambda_\nu \\ \\ \xi_{\alpha\nu} \text{ if } \alpha \geqslant \lambda_\nu \,. \end{cases}$$

Then if $\mu \neq \nu$, we have $g_\mu(\alpha) \neq g_\nu(\alpha)$ whenever $\alpha \geqslant \max(\lambda_\mu, \lambda_\nu)$ and so $|g_\mu \cap g_\nu| \leqslant \lambda_{\max(\mu,\nu)} < \lambda$. Then $\mathcal{G} = \{g_\nu; \kappa \leqslant \nu < \kappa^+\}$ has the properties claimed.

To establish that $\mathfrak{m} \leqslant \kappa^{++}$, suppose for a contradiction that we have a family $\mathcal{F}$ of almost disjoint functions in ${}^\lambda\kappa$ with $|\mathcal{F}| \geqslant \kappa^{+++}$. The argument goes as in the proof of Lemma 1.2.3, taking the decomposition

$$[\mathcal{F}]^2 = \bigcup_{\tau < \lambda'} \{\{f, g\}; |f \cap g| < \lambda_\tau\},$$

only using now the relation $\kappa^{+++} \to (\kappa^+)^2_{\kappa^+}$ which is a consequence of Theorem 2.2.4. This ensures that there is a family $\mathcal{G}$ with $\mathcal{G} \in [\mathcal{F}]^{\kappa^+}$ such that for some fixed τ with $\tau < \lambda'$ we have $|f \cap g| < \lambda_\tau$ for all f, g in $\mathcal{G}$. But now if $\mathcal{H} = \{g \upharpoonright \lambda_\tau^+; g \in \mathcal{G}\}$, one finds $|\mathcal{H}| = \kappa^+$ and $\delta(\mathcal{H} \leqslant \lambda_\tau$, contradicting Lemma 1.2.2.

This leaves (i), so suppose henceforth that $\lambda' \neq \kappa', \kappa^+$. If $\kappa^+ < \lambda'$ then $\mathfrak{m} = \kappa$ from Lemma 1.2.2. If $\lambda < \kappa$, by considering any family $\mathcal{F}$ where $\mathcal{F} \subseteq {}^\lambda\kappa$ as a decomposition of $\lambda \times \kappa$ into sets each of power κ, when $\delta(\mathcal{F}) \leqslant \lambda$ by Corollory 1.1.7 $|\mathcal{F}| \leqslant \kappa$ and consequently $\mathfrak{m} = \kappa$. Only the situation when $\kappa < \lambda$ and $\lambda' < \kappa^+$ remains, and here λ is singular. Choose regular cardinals λ_σ for σ with $\sigma < \lambda'$ such that $\kappa^{++} < \lambda_0 < \lambda_1 < ... < \lambda$ and $\lambda = \Sigma(\lambda_\sigma; \sigma < \lambda')$. In fact $\lambda' < \kappa$, for $\lambda' = \kappa$ would mean $\lambda' = \lambda'' = \kappa'$ since always λ' is regular. We consider separately the two possible cases, $\lambda' < \kappa'$ or $\lambda' > \kappa'$. The constructions that we use come from the paper [70] of Milner.

Take first the case $\lambda' < \kappa'$. We have claimed in the statement of the theorem that $\mathfrak{m} = \kappa$. For a contradiction, suppose we have a family $\mathcal{F}$ consisting of pairwise almost disjoint functions in ${}^\lambda\kappa$ with $|\mathcal{F}| = \kappa^+$. When $\alpha < \lambda$ and $\beta < \kappa$, put $\mathcal{F}_{\alpha\beta} = \{f \in \mathcal{F}; f(\alpha) = \beta\}$. Then for each α with $\alpha < \lambda$, the sequence $\langle \mathcal{F}_{\alpha\beta}; \beta < \kappa \rangle$ gives a disjoint decomposition of $\mathcal{F}$ into κ classes. However the number of such decompositions is $\kappa^{|\mathcal{F}|}$ (since each element of $\mathcal{F}$ may be in any class), thus the number of decompositions is $\kappa^{\kappa^+} = \kappa^{++}$. Since each λ_σ is regular with $\kappa^{++} < \lambda_\sigma$, for each σ there must be a set M_σ in $[\lambda_\sigma]^{\lambda_\sigma}$ such that the decompositions $\langle \mathcal{F}_{\alpha\beta}; \beta < \kappa \rangle$ are the same for all α in M_σ. But then for α_1, α_2 in M_σ, for all f in $\mathcal{F}$ we have $f(\alpha_1) = f(\alpha_2)$. Thus each f in $\mathcal{F}$ induces a map $f^* : \lambda' \to \kappa$, where

$$f^*(\sigma) = f(\alpha) \text{ for any } \alpha \text{ in } M_\sigma \text{ .}$$

However, the number of maps from λ' to κ is $\kappa^{\lambda'}$, and this is just κ since $\lambda' < \kappa'$. We had assumed that $|\mathcal{F}| = \kappa^+$, and so there must be distinct f_1, f_2 in $\mathcal{F}$ with $f_1^* = f_2^*$. Thus $f_1(\alpha) = f_2(\alpha)$ for all α in M_σ, for any σ with $\sigma < \lambda'$. This means

$$|f_1 \cap f_2| \geqslant |\bigcup \{M_\sigma; \sigma < \lambda'\}| = \lambda \text{ ,}$$

contradicting that f_1 and f_2 are almost disjoint. The contradiction shows that in fact $|\mathcal{F}| \leqslant \kappa$.

The final case, where $\lambda' > \kappa'$, is similar but rather longer. Again the claim is that $\mathfrak{m} = \kappa$, and again we seek a contradiction from a family $\mathcal{F}$ of κ^+ almost disjoint functions in ${}^\lambda\kappa$. Since here $\kappa' < \lambda' < \kappa$, also κ is singular and we may choose regular cardinals κ_τ where $\tau < \kappa'$ such that $\kappa_0 < \kappa_1 < ... < \kappa$ and $\kappa = \Sigma(\kappa_\tau; \tau < \kappa')$. When $\alpha < \lambda$ and $\tau < \kappa'$, put $\mathcal{F}(\alpha, \tau) = \{f \in \mathcal{F}; \kappa_\tau \leqslant f(\alpha) < \kappa_{\tau+1}\}$. Then for each α with $\alpha < \lambda$, $\langle \mathcal{F}(\alpha, \tau); \tau < \kappa' \rangle$ is a disjoint decomposition of $\mathcal{F}$ into κ' classes. The number of such decompositions is $(\kappa')^{\kappa^+} = \kappa^{++}$, so as before for each σ with $\sigma < \lambda'$ there must be a set N_σ in $[\lambda_\sigma]^{\lambda_\sigma}$ such that for some fixed τ, for all α in N_σ we have $\kappa_\tau \leqslant f(\alpha) < \kappa_{\tau+1}$ for any choice of f in $\mathcal{F}$. Thus each f in $\mathcal{F}$ induces a map $f^* : \lambda' \to \kappa'$, where $f^*(\sigma)$ is that τ for which $\kappa_\tau \leqslant f(\alpha) < \kappa_{\tau+1}$ for all the α in N_σ. The number of possibilities for the map f^* is $(\kappa')^{\lambda'} = (\lambda')^+ < \kappa^+$, and so there must be $\mathcal{F}^\#$ in $[\mathcal{F}]^{\kappa^+}$ such that the maps f^* are the same for all f in $\mathcal{F}^\#$, say equal to $f^\#$. Since $f^\#$ is a map $f^\# : \lambda' \to \kappa'$ and $\kappa' < \lambda'$ with λ' regular, there must be L in $[\lambda']^{\lambda'}$ such that $f^\#$ is constant on L, say $f^\#(\sigma) = \rho$ for all σ in L. So whenever $f \in \mathcal{F}^\#$, $\sigma \in L$ and $\alpha \in N_\sigma$, we have $f^*(\sigma) = f^\#(\sigma) = \rho$ so that $\kappa_\rho \leqslant f(\alpha) < \kappa_{\rho+1}$. Put $N = \bigcup\{N_\sigma; \sigma \in L\}$; then $|N| = \lambda$ and we can consider $f \restriction N$ as a map from N into $\kappa_{\rho+1}$ whenever $f \in \mathcal{F}^\#$. So $\{f \restriction N; f \in \mathcal{F}^\#\}$ is a pairwise almost disjoint family of functions from a set of size λ into $\kappa_{\rho+1}$. Now $\lambda' < \kappa'_{\rho+1} = \kappa_{\rho+1} < \lambda$ so by the immediately preceeding case, $|\{f \restriction N; f \in \mathcal{F}^\#\}| \leqslant \kappa_{\rho+1}$. Since $|\mathcal{F}^\#| = \kappa^+$, there must be f_1, f_2 in $\mathcal{F}^\#$ for which $f_1 \restriction N = f_2 \restriction N$. But then $|f_1 \cap f_2| \geqslant |N| = \lambda$, in contradiction to the fact that $\mathcal{F}$ consists of pairwise almost disjoint functions. This completes the proof of Theorem 1.2.7.

The simplest case under (iii) in Theorem 1.2.7 where the cardinal $\mathfrak{m}$ is not explicitly determined is the conjecture that there is an almost disjoint family $\mathcal{F}$ with $\mathcal{F} \subseteq {}^{\aleph_1}\aleph_0$ and $|\mathcal{F}| = \aleph_2$. Such a family exists if there is a Kurepa tree, and consequently there is such a family in L, the constructible universe. On the other hand, Silver [91] has shown (subject to the existence of a large cardinal) that it is consistent with the axioms of set theory + GCH that there is no such family.

We shall continue this section by making a few remarks concerned with the following situation. Given a sequence $\langle \kappa_\sigma; \sigma < \lambda \rangle$ of cardinals, what is the size of the largest almost disjoint family of functions taken from the cartesian product $\times(\kappa_\sigma; \sigma < \lambda)$ of the κ_σ? (Thus families from ${}^\lambda\kappa$ correspond to the special case where $\kappa_\sigma = \kappa$ for all σ.) This new situation is particularly relevant if κ is a singular cardinal, $\lambda = \kappa'$ and the κ_σ form a sequence of cardinals below κ converging to κ.

There is the following lemma which will enable us (in certain circumstances) to construct large almost disjoint subfamilies of a cartesian product.

Lemma 1.2.8. *Let* $\langle \lambda_\nu; \nu < \eta \rangle$ *be any sequence of cardinals and let* $\langle A_\nu; \nu < \eta \rangle$ *be any sequence of sets with always* $|A_\nu| = \Pi(\lambda_\mu; \mu \leqslant \nu)$. *Then there is a subfamily* $\mathcal{F}$ *of* $\times(A_\nu; \nu < \eta)$ *with* $\delta(\mathcal{F}) \leqslant \eta$ *and* $|\mathcal{F}| = \Pi(\lambda_\nu; \nu < \eta)$.

Proof. For each ν, identify A_ν with $\times(\lambda_\mu; \mu \leqslant \nu)$. Put $\pi = \Pi(\lambda_\nu; \nu < \eta)$, and let $\langle f_\alpha; \alpha < \pi \rangle$ be an enumeration of $\times(\lambda_\nu; \nu < \eta)$. Define functions g_α in $\times(A_\nu;$

$\nu < \eta$) for α with $\alpha < \pi$ as follows: if $\nu < \eta$ then

$$g_\alpha(\nu) = f_\alpha \upharpoonright (\nu + 1) .$$

Put $\mathcal{F} = \{g_\alpha ; \alpha < \pi\}$ so $|\mathcal{F}| = \pi$. Further $\mathcal{F}$ is almost disjoint, for take distinct functions g_α, g_β from $\mathcal{F}$. If ν is the least ordinal such that $f_\alpha(\nu) \neq f_\beta(\nu)$ then $g_\alpha(\mu) \neq g_\beta(\mu)$ whenever $\mu \geqslant \nu$. Thus $|g_\alpha \cap g_\beta| \leqslant |\nu| < \eta$.

Corollary 1.2.9. *Suppose κ is singular with $\kappa' = \aleph_0$ and let $\langle \kappa_n ; n < \omega \rangle$ be any increasing sequence of cardinals all less than κ with $\Sigma(\kappa_n ; n < \omega) = \kappa$. Then there is an almost disjoint subfamily $\mathcal{F}$ of $\times(\kappa_n ; n < \omega)$ with $|\mathcal{F}| > \kappa$.*

Proof. Note that if $n < \omega$ then $\Pi(\kappa_m ; m \leqslant n) = \kappa_n$. By Lemma 1.2.8 there is then $\mathcal{F}$ with $\mathcal{F} \subseteq \times(\kappa_n ; n < \omega)$ such that $\delta(\mathcal{F}) \leqslant \aleph_0$ and $|\mathcal{F}| = \Pi(\kappa_n ; n < \omega)$. By König's theorem though,

$$\Pi(\kappa_n ; n < \omega) > \Sigma(\kappa_n ; n < \omega) = \kappa .$$

The result in Corollary 1.2.9 is special to singular cardinals κ with $\kappa' = \aleph_0$. When κ is singular with $\kappa' \geqslant \aleph_1$, for some increasing sequences $\langle \kappa_\sigma ; \sigma < \kappa' \rangle$ of cardinals below κ there is an almost disjoint subfamily $\mathcal{F}$ of $\times(\kappa_\sigma ; \sigma < \kappa')$ with $|\mathcal{F}| > \kappa$; for other sequences possibly $|\mathcal{F}| \leqslant \kappa$ for all such families $\mathcal{F}$. Examples are given in Theorems 1.2.15 and 1.2.13 below. However, first we need to examine some properties of closed unbounded subsets of the cardinal λ.

Definition 1.2.10. Let λ be an infinite cardinal. The subset C of λ is said to be *closed* if whenever $B \in [C]^{<\lambda'}$ then $\bigcup B \in C$. The subset C of λ is said to be *unbounded* (or *cofinal*) in λ if for all α with $\alpha < \lambda$ there is γ in C with $\gamma \geqslant \alpha$.

If λ is regular, it is easily seen that the intersection of any family of fewer than λ closed unbounded subsets of λ is again closed and unbounded in λ.

Definition 1.2.11. A subset S of λ is *stationary* in λ if S intersects every closed unbounded subset of λ.

It is clear that any stationary subset of a regular cardinal λ has power λ. We shall make use of the following result of Fodor [42].

Lemma 1.2.12. *Let S be a stationary subset of the uncountable regular cardinal λ and let $f : S \to \lambda$ be a regressive function (that is, $f(\alpha) < \alpha$ whenever $\alpha \neq 0$). Then there is a stationary subset T of S such that f is constant on T.*

Proof. Suppose the lemma is false, so for all β in S the set $f^{-1}(\beta)$ is not stationary in λ. Hence there is a closed unbounded subset C_β of λ such that $C_\beta \cap f^{-1}(\beta) = \emptyset$. Put

$$D = \{\alpha < \lambda; \alpha \in \cap \{C_\beta; \beta < \alpha\}\} .$$

I claim that D is a closed unbounded subset of λ. This claim will suffice to prove the lemma, for then S, being stationary, must meet D. Thus there is α in $D \cap S$. Then $\alpha \in C_\beta$ whenever $\beta < \alpha$, so $f(\alpha) \neq \beta$ whenever $\beta < \alpha$. This contradicts that f is regressive.

To see that D is closed, take any B in $[D]^{<\lambda}$. Choose any β with $\beta < \bigcup B$; we want to show that $\bigcup B \in C_\beta$ for then it would follow that $\bigcup B \in \cap \{C_\beta; \beta < \bigcup B\}$, that is, $\bigcup B \in D$. Since $\beta < \bigcup B$ then $\beta < \gamma$ for some γ in B. For any δ in $B - \gamma$ since $\beta < \delta$ and $\delta \in D$ it follows that $\delta \in C_\beta$. Thus $B - \gamma \subseteq C_\beta$ and since C_β is closed, $\bigcup(B - \gamma) \in C_\beta$. Since $\bigcup B = \bigcup(B - \gamma)$ thus $\bigcup B \in C_\beta$ as required.

To show that D is unbounded in λ, given α with $\alpha < \lambda$ we need to find β in D with $\beta \geqslant \alpha$. For n with $n < \omega$ define inductively ordinals β_n with $\beta_n < \lambda$ as follows: $\beta_0 = \alpha$; choose β_{n+1} such that $\beta_{n+1} \geqslant \beta_n$ and

$$\beta_{n+1} \in \cap \{C_\gamma; \gamma < \beta_n\} .$$

Such a choice of β_{n+1} is possible since $\cap \{C_\gamma; \gamma < \beta_n\}$, being an intersection of fewer that λ closed unbounded subsets of λ, is certainly unbounded. Put $\beta = \bigcup \{\beta_n; n < \omega\}$, so surely $\beta \geqslant \alpha$. Also $\beta \in D$. For take any γ with $\gamma < \beta$, and we shall show that $\beta \in C_\gamma$; from this it will follow that $\beta \in D$. Since $\gamma < \beta$ there is n such that $\gamma < \beta_n$ and hence $\gamma < \beta_m$ whenever $m \geqslant n$. The choice of β_m then ensures that $\beta_m \in C_\gamma$ whenever $m > n$. Then $\bigcup \{\beta_m; m > n\} \in C_\gamma$ since C_γ is closed, and $\beta = \bigcup \{\beta_m; m > n\}$ so in fact $\beta \in C_\gamma$. This completes the proof.

Given a subset C of the infinite carainal λ, one can form the closure $\overline{C}$ of C – the least closed subset D of λ with $C \subseteq D$. When C is infinite, $|\overline{C}| = |C|$. This may be seen by considering the map $f: \overline{C} - C \to C$ where for any α in $\overline{C} - C$, $f(\alpha) =$ least β in C with $\alpha \leqslant \beta$. Clearly f is one-to-one and hence $|\overline{C} - C| \leqslant |C|$. In particular, if λ is singular there is a closed sequence $\langle \lambda_\sigma; \sigma < \lambda' \rangle$ of cardinals below λ converging to λ. (That the sequence is closed means that the set $\{\lambda_\sigma; \sigma < \lambda'\}$ is closed in λ.)

We are now set to return to the investigation of families of functions.

Theorem 1.2.13. *Let κ be a singular cardinal with $\kappa' > \aleph_0$. Suppose $2^\lambda \leqslant \kappa$*

for all λ with $\lambda < \kappa$. Let $\langle \kappa_\sigma; \sigma < \kappa' \rangle$ be a closed strictly increasing sequence of cardinals below κ with $\kappa = \Sigma(\kappa_\sigma; \sigma < \kappa')$ and let $\mathcal{F}$ be an almost disjoint subfamily of $\times(\kappa_\sigma; \sigma < \kappa')$. Then $|\mathcal{F}| \leqslant \kappa$.

Proof. Since $\langle \kappa_\sigma; \sigma < \kappa' \rangle$ is closed and strictly increasing, whenever σ is a limit ordinal then $\kappa_\sigma = \bigcup \{\kappa_\tau; \tau < \sigma\}$. Take the family $\mathcal{F}$ with $\mathcal{F} \subseteq \times(\kappa_\sigma; \sigma < \kappa')$ and $\delta(\mathcal{F}) \leqslant \kappa'$. Fix f in $\mathcal{F}$. Since always $f(\sigma) < \kappa_\sigma$, when σ is a limit ordinal then $f(\sigma) < \kappa_\tau$ for some τ with $\tau < \sigma$. Thus if $L = \{\sigma < \kappa'; \sigma$ is a limit ordinal$\}$ then there is a regressive function $g : L \to \kappa'$ such that always $f(\sigma) < \kappa_{g(\sigma)}$. Now L is a stationary subset of κ' (for given any closed unbounded subset C of κ', chose inductively ordinals α_n in C with $\alpha_{n+1} > \alpha_n$; if $\alpha = \bigcup \{\alpha_n; n < \omega\}$ then $\alpha \in L \cap C$). Hence by Lemma 1.2.12 there is a stationary subset $S(f)$ of L such that g is constant on $S(f)$, say with value $\tau(f)$. Thus $f(\sigma) < \kappa_{\tau(f)}$ for all σ in $S(f)$.

For S in $[L]^{\kappa'}$ and τ with $\tau < \kappa'$, put

$$\mathcal{F}(S, \tau) = \{f \in \mathcal{F}; S(f) = S \text{ and } \tau(f) = \tau\} .$$

For any f in $\mathcal{F}$, since $S(f)$ is stationary in κ' then $|S(f)| = \kappa'$. If for some f_1, f_2 in $\mathcal{F}$ we have $S(f_1) = S(f_2) = S$ and f_1 and f_2 agree on S then f_1 and f_2 agree on a set of size κ' and so $f_1 = f_2$, by the property $\delta(\mathcal{F}) \leqslant \kappa'$. Hence

$$|\mathcal{F}(S, \tau)| \leqslant |{}^S\kappa_\tau| = \kappa_\tau^{\kappa'} \leqslant 2^{\kappa_\tau \cdot \kappa'} \leqslant \kappa .$$

Now the number of pairs (S, τ) is $(\kappa')^{\kappa'} \cdot \kappa' \leqslant 2^{\kappa' \cdot \kappa'} \cdot \kappa' \leqslant \kappa$. Hence

$$|\mathcal{F}| = |\bigcup \{\mathcal{F}(S, \tau); S \in [L]^{\kappa'} \text{ and } \tau < \kappa'\}| \leqslant \kappa \cdot \kappa = \kappa .$$

This completes the proof.

If, in the situation of Theorem 1.2.13, we consider almost disjoint subfamilies $\mathcal{F}$ of $\times(\kappa_\sigma^+; \sigma < \kappa')$, again we can find a bound for $|\mathcal{F}|$. We shall need to quote a couple of results which will not be proved until Chapters 2 and 3.

Theorem 1.2.14. *Let κ be a singular cardinal with $\kappa' > \aleph_0$. Suppose $2^\lambda \leqslant \kappa$ for all λ with $\lambda < \kappa$. Let $\langle \kappa_\sigma; \sigma < \kappa' \rangle$ be a closed strictly increasing sequence of cardinals below κ with $\kappa = \Sigma(\kappa_\sigma; \sigma < \kappa')$ and let $\mathcal{F}$ be an almost disjoint subfamily of $\times(\kappa_\sigma^+; \sigma < \kappa')$. Then $|\mathcal{F}| \leqslant \kappa^+$.*

Proof. Suppose for a contradiction that there is such a family $\mathcal{F}$ with $|\mathcal{F}| = \kappa^{++}$. Define a function H from $\mathcal{F}$ to $\mathfrak{P}\mathcal{F}$ as follows: for f in $\mathcal{F}$,

$$H(f) = \{g \in \mathcal{F}; \forall \sigma < \kappa'(g(\sigma) \leqslant f(\sigma))\} - \{f\} .$$

Then always $|H(f)| \leqslant \kappa$. For $H(f) \subseteq \times(f(\sigma)+1; \sigma<\kappa')$ where $|f(\sigma)+1| \leqslant \kappa_\sigma$, and $\delta(H(f)) \leqslant \delta(\mathcal{F}) = \kappa'$, so that Theorem 1.2.13 applies to the family $H(f)$, and shows $|H(f)| \leqslant \kappa$. Thus, in the terminology of Chapter 3, H is a set map of order κ^+ defined on $\mathcal{F}$. Since $|\mathcal{F}| = \kappa^{++}$, Theorem 3.1.1 shows that there is a subset $\mathcal{S}$ of $\mathcal{F}$ of power κ^{++} which is free for H. This means that $f \notin H(g)$ and $g \notin H(f)$ for all f, g in $\mathcal{S}$. In particular then, there is a sequence $\langle f_\mu;\ \mu<\kappa^{++}\rangle$ of elements of $\mathcal{F}$ such that $f_\mu \notin H(f_\nu)$ whenever $\mu<\nu<\kappa^{++}$. Thus there must be $\sigma_{\mu\nu}$ such that $f_\mu(\sigma_{\mu\nu}) > f_\nu(\sigma_{\mu\nu})$ whenever $\mu<\nu<\kappa^{++}$. Consider $[\kappa^{++}]^2$ partitioned into κ' classes according to the value of $\sigma_{\mu\nu}$. By virtue of Theorem 2.2.5 of the next chapter, the partition relation $(2^{\kappa'})^+ \to (\kappa'^+)^2_{\kappa'}$ holds; since $(2^{\kappa'})^+ < \kappa^{++}$ and $\kappa'^+ > \aleph_0$ also the relation $\kappa^{++} \to (\aleph_0)^2_{\kappa'}$ holds. This relation ensures that in the present situation there are ordinals μ_n (for n with $n<\omega$) such that $\mu_0<\mu_1<\ldots<\kappa^{++}$ and an ordinal σ with $\sigma<\kappa'$ such that $\sigma_{\mu_m\mu_n} = \sigma$ whenever $m \lessdot n<\omega$. This means that $f_{\mu_0}(\sigma) > f_{\mu_1}(\sigma) > f_{\mu_2}(\sigma) > \ldots$, which is impossible. This contradiction proves the theorem.

Let us impose the stronger condition that the GCH is true for all cardinals below κ, that is $2^\lambda = \lambda^+$ whenever $\lambda<\kappa$. Then we can use Lemma 1.2.8 to produce an almost disjoint subfamily of $\times(\kappa_\sigma^+; \sigma<\kappa')$ of power 2^κ.

Theorem 1.2.15. *Let κ be a singular cardinal with $\kappa' > \aleph_0$. Suppose $2^\lambda = \lambda^+$ for all λ with $\lambda<\kappa$. Let $\langle\kappa_\sigma; \sigma<\kappa'\rangle$ be any increasing sequence of cardinals below κ with $\kappa = \Sigma(\kappa_\sigma; \sigma<\kappa')$. Then there is a subfamily $\mathcal{F}$ of $\times(\kappa_\sigma^+; \sigma<\kappa')$ with $\delta(\mathcal{F}) \leqslant \kappa'$ and $|\mathcal{F}| = 2^\kappa$.*

Proof. For σ with $\sigma<\kappa'$, note that

$$\Pi(\kappa_\tau^+; \tau \leqslant \sigma) = \Pi(2^{\kappa_\tau}; \tau \leqslant \sigma) = 2^{\Sigma(\kappa_\tau; \tau \leqslant \sigma)} = 2^{\kappa_\sigma} = \kappa_\sigma^+ .$$

Thus Lemma 1.2.8 gives a family $\mathcal{F}$ of $\times(\kappa_\sigma^+; \sigma<\kappa')$ with $\delta(\mathcal{F}) \leqslant \kappa'$ and $|\mathcal{F}| = \Pi(\kappa_\sigma^+; \sigma<\kappa')$. However

$$\Pi(\kappa_\sigma^+; \sigma<\kappa') = \Pi(2^{\kappa_\sigma}; \sigma<\kappa') = 2^{\Sigma(\kappa_\sigma; \sigma<\kappa')} = 2^\kappa .$$

If we combine Theorems 1.2.14 and 1.2.15, then immediately we have the following.

Theorem 1.2.16. *Let κ be a singular cardinal with $\kappa' > \aleph_0$. Suppose $2^\lambda = \lambda^+$ for all λ with $\lambda<\kappa$. Then $2^\kappa = \kappa^+$.*

Theorem 1.2.16 is of particular interest in view of the principal result of Easton [12], which implies that no similar theorem is possible for any regular cardinal κ. Theorem 1.2.16 (in fact a more general result) was first proved recently by Silver [92], using powerful methods from mathematical logic. A combinatorial proof of Silver's theorem was then found by Baumgartner and Prikry [5], amongst others. Using similar methods, Galvin and Hajnal [45] gave general bounds for 2^κ (where $\kappa > \kappa' > \aleph_0$) in terms of the values of 2^{λ} for λ with $\lambda < \kappa$. The proofs of Theorems 1.2.14 and 1.2.15 come from Galvin and Hajnal [45]. Theorem 1.2.13 is from Erdös, Hajnal and Milner [36], where a necessary and sufficient condition is given on the sequence $\langle \kappa_\sigma ; \sigma < \kappa' \rangle$ for the existence of an almost disjoint family of more than κ functions from $\times(\kappa_\sigma ; \sigma < \kappa')$.

§3. Transversals of non-disjoint families

In the previous section, the discussion could be viewed as a consideration of the number of almost disjoint transversals of a family of disjoint sets. This section will be devoted to transversals of non-disjoint families. One of the first results will show that in this situation there is not even one transversal for certain families. Consequently, we shall extend the notation of a transversal as follows.

Definition 1.3.1. Given a family $\mathscr{A}$, an η-transversal of $\mathscr{A}$ is a set T such that $1 \leqslant |A \cap T| < \eta$ for each A in $\mathscr{A}$.

Thus a transversal in the previous sense is a 2-transversal by the meaning of this definition. When considering a family $\mathscr{A}$ of sets each of power κ, we shall be interested in the existence of an η-transversal for values of η only with $\eta \leqslant \kappa$, since a κ^+-transversal T would allow $T \cap A = A$ for some or all of the members A of $\mathscr{A}$. (In particular, $\bigcup \mathscr{A}$ itself would be a κ^+-transversal of $\mathscr{A}$.)

Given an arbitrary family $\mathscr{A}$ of sets all of power λ, with a specified degree of disjunction $\delta(\mathscr{A}) = \theta$, we shall seek the smallest η for which we can be sure that an η-transversal for $\mathscr{A}$ exists. One can then go on to ask, for this smallest η, what is the maximum size of an almost disjoint family of η-transversals of $\mathscr{A}$. The situation is rather similar to that found in the last section, and we shall not pursue the matter further.

The results of this section are to be found in the paper [19] of Erdös and Hajnal, expressed in rather different notation.

We start by giving examples of families with no transversals in the sense of Definition 1.3.1.

Theorem 1.3.2 (GCH).

(i) *Suppose* $\lambda < \kappa$. *Then there is a family* $\mathcal{A}$ *of* κ *sets each of power* κ *with* $\delta(\mathcal{A}) \leqslant \lambda$, *which has no* λ-*transversal.*

(ii) *There is a family* $\mathcal{A}$ *of* κ^+ *pairwise almost disjoint sets each of power* κ *which has no* κ-*transversal.*

Proof. For (i), take λ with $\lambda < \kappa$ and suppose for a contradiction that (i) is false, that is that every family $\mathcal{A}$ of κ sets each of power κ with $\delta(\mathcal{A}) \leqslant \lambda$ has a λ-transversal. Take any family $\mathcal{B} = \{B_\nu; \nu < \kappa\}$ of pairwise disjoint sets each of power κ. Construct a sequence T_μ where $\mu < \kappa^+$ by transfinite induction, so that T_μ is a λ-transversal of $\mathcal{B} \cup \{T_\nu; \nu < \mu\}$ with $T_\mu \subseteq \bigcup \mathcal{B}$. Consider $\mathcal{T} = \{T_\mu; \mu < \kappa^+\}$. Here $|\bigcup \mathcal{T}| = \kappa$, $|T_\mu| = \kappa$ for each μ, and $\delta(\mathcal{T}) \leqslant \lambda < \kappa$. Since $|\mathcal{T}| = \kappa^+$, this contradicts Corollary 1.1.7, and (i) is proved.

Likewise, if we assume (ii) to be false, this construction (with $\lambda = \kappa$) can be continued to yield a family $\mathcal{T} = \{T_\mu; \mu < \kappa^{++}\}$ with $|\bigcup \mathcal{T}| = \kappa < \kappa^+$, $|T_\mu| = \kappa$ for each μ, and $\delta(\mathcal{T}) \leqslant \kappa$. Since $(\kappa^+)' \neq \kappa'$, this contradicts Theorem 1.1.6.

When κ is regular, the value $|\mathcal{A}| = \kappa^+$ of (ii) in this last theorem cannot be decreased. Any family of fewer than κ^+ pairwise almost disjoint sets each of power κ has a κ-transversal. This follows from the next theorem.

Theorem 1.3.3. *Any family* $\mathcal{A}$ *of pairwise almost disjoint sets each of power* κ *with* $|\mathcal{A}| = \kappa'$ *has a* κ'-*transversal.*

Proof. This is almost trivial. Suppose $\mathcal{A} = \{A_\nu; \nu < \kappa'\}$ satisfies the conditions of the theorem. Then whenever $\mu < \kappa'$ one can choose x_μ from $A_\mu - \bigcup \{A_\nu; \nu < \mu\}$. If $T = \{x_\mu; \mu < \kappa'\}$ then for any ν it follows that $T \cap A_\nu \subseteq \{x_\mu; \mu \leqslant \nu\}$, and consequently T is a κ'-transversal for $\mathcal{A}$.

The result of Theorem 1.3.3 is the best possible, in the sense that the existence of a κ'-transversal cannot in general be improved to the existence of a λ-transversal for any λ with $\lambda < \kappa'$. When κ is regular, this follows already from Theorem 1.3.2, using GCH. However, GCH is not used in the construction below.

Theorem 1.3.4. *There is a family* $\mathcal{A}$ *of* κ' *pairwise almost disjoint sets each of power* κ *which has no* λ-*transversal for any* λ *with* $\lambda < \kappa'$.

Proof. Write κ as a disjoint union, $\kappa = \bigcup \{K_\sigma; \sigma < \kappa'\}$ where always $|K_\sigma| = \kappa$.

Choose ordinals α_σ where $\sigma < \kappa'$ such that $\alpha_0 < \alpha_1 < ... < \kappa$ and $\kappa = \sup\{\alpha_\sigma; \sigma < \kappa'\}$. Put $\mathscr{A} = \{\alpha_\sigma \cup K_{2\sigma}; \sigma < \kappa'\} \cup \{K_{2\sigma+1}; \sigma < \kappa'\}$. Then $|\mathscr{A}| = \kappa'$, $\delta(\mathscr{A}) = \kappa$ and $|A| = \kappa$ for each A in $\mathscr{A}$. Suppose T meets every member of $\mathscr{A}$; then $|T \cap \kappa| \geqslant \kappa'$. Consequently, for any λ with $\lambda < \kappa'$ there must be σ with $\sigma < \kappa'$ such that $|T \cap \alpha_\sigma| \geqslant \lambda$. Then $|T \cap (\alpha_\sigma \cup K_{2\sigma})| \geqslant \lambda$, and so T is not a λ-transversal of $\mathscr{A}$.

Together with the observation that any family $\mathscr{A}$ of fewer than κ' pairwise almost disjoint sets of power κ has a 2-transversal, Theorems 1.3.2(ii), 1.3.3, 1.3.4 provide a complete discussion of the situation pertaining to families of pairwise almost disjoint sets of power κ, in the case that κ is regular. When κ is singular, there are still some gaps to be filled. Specifically, what is the situation with a family of θ pairwise almost disjoint sets each of power κ, where $\kappa' < \theta \leqslant \kappa$? There is the trivial fact that one can always find a θ^+-transversal, and this is the best possible result, for there is an example of a family of θ pairwise almost disjoint sets each of power κ which has no θ-transversal. Such an example is given in Theorem 1.3.6 below. The construction is due to Erdös and Hajnal [19], and is quite complicated. We shall need the following lemma (see Hajnal [52, Theorem 9]).

Lemma 1.3.5 (GCH). *Let λ be a regular cardinal and let X be a set with $|X| = \lambda^+$. There is a pairwise almost disjoint family $\mathscr{X} = \{X_\nu; \nu < \lambda^+\}$ of subsets of X with always $|X_\nu| = \lambda$ such that for all subsets Y of X with $|Y| = \lambda^+$ there is X_ν in $\mathscr{X}$ with $X_\nu \subseteq Y$.*

Proof. Let $<$ be a well ordering of X. Let $\mathscr{Z}$ be the family of all subsets of X of order type λ^2 under $<$. (Here and for the rest of the proof, by λ^2 is meant the ordinal power.) Since $\mathscr{Z} \subseteq [X]^\lambda$, it follows from GCH that $|\mathscr{Z}| = \lambda^+$, and we may write $\mathscr{Z} = \{Z_\mu; \mu < \lambda^+\}$.

Inductively define subsets X_ν of Z_ν with $\text{tp}(X_\nu) = \lambda$, for ν with $\nu < \lambda^+$, as follows. Suppose that ν is given with $\nu < \lambda^+$ and that the X_μ where $\mu < \nu$ have already been defined. If $\nu < \lambda$ then $|Z_\nu - \bigcup\{X_\mu; \mu < \nu\}| = \lambda$; choose for X_ν any subset of $Z_\nu - \bigcup\{X_\mu; \mu < \nu\}$ with order type λ. Now suppose $\lambda \leqslant \nu < \lambda^+$. Let $\langle X_{\nu\alpha}; \alpha < \lambda\rangle$ give a well ordering of $\{X_\mu; \mu < \nu\}$. Note that since $\text{tp}(Z_\nu) = \lambda^2$ there are sets $Z_{\nu\alpha}$ such that $Z_\nu = \bigcup\{Z_{\nu\alpha}; \alpha < \lambda\}$ where $\text{tp}(Z_{\nu\alpha}) = \lambda$ and $Z_{\nu\beta} < Z_{\nu\alpha}$ whenever $\beta < \alpha$ (that is, $x < y$ if $x \in Z_{\nu\beta}$ and $y \in Z_{\nu\alpha}$). Since always $\text{tp}(X_{\nu\beta}) = \lambda$, there is at most one α for which $\text{tp}(X_{\nu\beta} \cap Z_{\nu\alpha}) = \lambda$. Define ordinals $\gamma(\alpha)$ with $\gamma(\alpha) < \lambda$ and elements $x_{\nu\alpha}$ with $x_{\nu\alpha} \in Z_{\nu\gamma(\alpha)}$ for α with $\alpha < \lambda$ by induction as follows. Choose $\gamma(\alpha)$ such that $\gamma(\alpha) > \gamma(\beta)$ and also $|X_{\nu\beta} \cap Z_{\nu\gamma(\alpha)}| < \lambda$ for all β with $\beta < \alpha$. This is possible, by the regularity of λ.

Now choose $x_{\nu\alpha}$ from $Z_{\nu\gamma(\alpha)} - \bigcup\{X_{\nu\beta}; \beta < \alpha\}$. The choice of $\gamma(\alpha)$ and the regularity of λ make this possible. Put $X_\nu = \{x_{\nu\alpha}; \alpha < \lambda\}$. Since $x_{\nu\beta} < x_{\nu\alpha}$ whenever $\beta < \alpha$ it follows that $\mathrm{tp}(X_\nu) = \lambda$, and surely $X_\nu \subseteq Z_\nu$. This completes the definition of the sets X_ν.

The X_ν are pairwise almost disjoint, for suppose say $\mu < \nu$. Then $X_\mu = X_{\nu\alpha}$ for some α, so $X_\mu \cap X_\nu \subseteq \{x_{\nu\beta}; \beta \leqslant \alpha\}$ and thus $|X_\mu \cap X_\nu| < \lambda$. Take any subset Y of X with $|Y| = \lambda^+$. Then there is a subset Z of Y with $\mathrm{tp}(Z) = \lambda^2$, and $Z = Z_\nu$ for some ν. Since $X_\nu \subseteq Z_\nu$ in fact $X_\nu \subseteq Y$. Thus if $\mathcal{X} = \{X_\nu; \nu < \lambda^+\}$ then $\mathcal{X}$ has the required properties, and the lemma is proved.

Theorem 1.3.6 (GCH). *For each θ where $\kappa' < \theta \leqslant \kappa$ there is a family of θ pairwise almost disjoint sets each of power κ which has no θ-transversal.*

Proof. Let κ and θ be cardinals with $\kappa' < \theta \leqslant \kappa$. Thus κ is singular, so there is a sequence $\langle \kappa_\sigma; \sigma < \kappa' \rangle$ of cardinals below κ with $\kappa = \Sigma(\kappa_\sigma; \sigma < \kappa')$.

If $\kappa' < \theta'$, the theorem is easily proved, without GCH. Start with a pairwise disjoint family $\{A_\sigma; \sigma < \kappa'\}$ where always $|A_\sigma| = \kappa$. For each σ with $\sigma < \kappa'$, choose a pairwise disjoint family $\{A_{\sigma\nu}; \nu < \theta\}$ of subsets of A_σ with always $|A_{\sigma\nu}| = \kappa_\sigma$. Put $B_\nu = \bigcup\{A_{\sigma\nu}; \sigma < \kappa'\}$, so $|B_\nu| = \Sigma(\kappa_\sigma; \sigma < \kappa') = \kappa$. Consider the family

$$\mathcal{A} = \{A_\sigma; \sigma < \kappa'\} \cup \{B_\nu; \nu < \theta\}.$$

The B_ν are pairwise disjoint and $A_\sigma \cap B_\nu = A_{\sigma\nu}$ so $|A_\sigma \cap B_\nu| = \kappa_\sigma < \kappa$. Thus $\mathcal{A}$ is a family of θ pairwise almost disjoint sets each of power κ. Let T be a transversal of $\mathcal{A}$. Since then always $T \cap B_\nu \neq \emptyset$ and $\bigcup\{B_\nu; \nu < \theta\} \subseteq \bigcup\{A_\sigma; \sigma < \kappa'\}$, surely $|T \cap \bigcup\{A_\sigma; \sigma < \kappa'\}| \geqslant \theta$. Since $\kappa' < \theta'$ there must be some σ with $|T \cap A_\sigma| \geqslant \theta$. Hence $\mathcal{A}$ has no θ-transversal.

Now consider the case $\theta' \leqslant \kappa'$. Start this time from a pairwise disjoint family $\mathcal{A}_1 = \{A_\sigma; \sigma < \kappa'^+\}$ with always $|A_\sigma| = \kappa$. Since κ' is regular, we can apply Lemma 1.3.5 to $\mathcal{A}_1$ to produce a family of families $\mathfrak{X} = \{\mathcal{X}_\nu; \nu < \kappa'^+\}$ where always $\mathcal{X}_\nu \in [\mathcal{A}_1]^{\kappa'}$ and $|\mathcal{X}_\mu \cap \mathcal{X}_\nu| < \kappa'$ whenever $\mu \neq \nu$, with the property that for any subfamily $\mathcal{Y}$ of $\mathcal{A}_1$ with $|\mathcal{Y}| = \kappa'^+$ there is $\mathcal{X}_\nu$ in $\mathfrak{X}$ with $\mathcal{X}_\nu \subseteq \mathcal{Y}$.

Write $\mathcal{X}_\nu = \{A_{\nu\tau}; \tau < \kappa'\}$. For each ν and τ, choose a pairwise disjoint family $\{A_{\nu\tau\gamma}; \gamma < \theta\}$ of subsets of $A_{\nu\tau}$ with always $|A_{\nu\tau\gamma}| = \kappa_\tau$. Put $B_{\nu\gamma} = \bigcup\{A_{\nu\tau\gamma}; \tau < \kappa'\}$, so $|B_{\nu\gamma}| = \Sigma(\kappa_\tau; \tau < \kappa') = \kappa$. Consider the family

$$\mathcal{B} = \{A_\sigma; \sigma < \kappa'^+\} \cup \{B_{\nu\gamma}; \nu < \kappa' \text{ and } \gamma < \theta\}.$$

We shall show that $\mathcal{B}$ is pairwise almost disjoint. Take distinct B_1 and B_2 from $\mathcal{B}$, and consider the various possibilities. If $B_1 = A_\sigma$ and $B_2 = A_\tau$ for some σ and τ, then $B_1 \cap B_2 = \emptyset$. If $B_1 = B_{\nu\gamma}$ and $B_2 = B_{\nu\delta}$ for some ν, γ and δ then again $B_1 \cap B_2 = \emptyset$. If say $B_1 = A_\sigma$ and $B_2 = B_{\nu\gamma}$ then $B_1 \cap B_2 = \emptyset$ unless $A_\sigma \in \mathcal{X}_\nu$, in which case $A_\sigma = A_{\nu\tau}$ for some τ so $A_\sigma \cap B_{\nu\gamma} = A_{\nu\tau\gamma}$ and hence $|B_1 \cap B_2| = |A_{\nu\tau\gamma}| = \kappa_\tau < \kappa$. The final possibility is $B_1 = B_{\nu\gamma}$ and $B_2 = B_{\mu\delta}$ where $\mu \neq \nu$. Since $\bigcup \mathcal{B} = \bigcup\{A_\sigma; \sigma < \kappa'^+\}$ it follows that $B_1 \cap B_2 = \bigcup\{A_\sigma \cap (B_{\nu\gamma} \cap B_{\mu\delta}); \sigma < \kappa'^+\}$. However, either $A_\sigma \cap B_{\nu\gamma} = \emptyset$ or $A_\sigma \cap B_{\mu\delta} = \emptyset$ unless $A_\sigma \in \mathcal{X}_\nu \cap \mathcal{X}_\mu$, so $B_1 \cap B_2 = \bigcup\{A_\sigma \cap B_{\nu\gamma} \cap B_{\mu\delta}; A_\sigma \in \mathcal{X}_\nu \cap \mathcal{X}_\mu\}$. Now always $|A_\sigma \cap B_{\nu\gamma}| < \kappa$, and $|\mathcal{X}_\nu \cap \mathcal{X}_\mu| < \kappa$ by the choice of $\mathfrak{X}$, so again $|B_1 \cap B_2| < \kappa$. Thus $\mathcal{B}$ is indeed pairwise almost disjoint. Hence further, $|\mathcal{B}| = \kappa'^+ \dotplus \kappa' \cdot \theta = \theta$.

Finally, we shall show that $\mathcal{B}$ has no θ-transversal. Since $\theta' \leqslant \kappa' < \theta$ we know θ is singular, so there is a sequence $\langle \theta_\zeta; \zeta < \theta' \rangle$ of cardinals below θ with $\theta = \Sigma(\theta_\zeta; \zeta < \theta')$. Suppose T is a set such that $|T \cap A_\sigma| < \theta$ for each σ with $\sigma < \kappa'^+$. So for each σ there is $\zeta(\sigma)$ with $\zeta(\sigma) < \theta'$ such that $|T \cap A_\sigma| < \theta_{\zeta(\sigma)}$. Since $\theta' < \kappa'^+$ and κ'^+ is regular there is a subset H of κ'^+ with $|H| = \kappa'^+$ such that $\zeta(\sigma)$ is constant for σ in H, say with value ζ. By the choice of the family $\mathfrak{X}$ there is $\mathcal{X}_\nu$ in $\mathfrak{X}$ with $\mathcal{X}_\nu \subseteq \{A_\sigma; \sigma \in H\}$. Thus $|T \cap A_\sigma| < \theta_\zeta$ for all A_σ in $\mathcal{X}_\nu$. Hence $|T \cap \bigcup \mathcal{X}_\nu| \leqslant \kappa' \cdot \theta_\zeta < \theta$. Now $\{B_{\nu\gamma}; \gamma < \theta\}$ consists of θ pairwise disjoint sets, each a subset of $\bigcup \mathcal{X}_\nu$. Hence there is some γ such that $T \cap B_{\nu\gamma} = \emptyset$. Thus $\mathcal{B}$ can have no θ-transversal. This completes the proof.

Let us return to the case of families of κ sets each of power κ but where the degree of disjunction is smaller than κ. We shall assume the Generalized Continuum Hypothesis for the rest of this discussion. Theorem 1.3.2 shows that for such a family $\mathcal{A}$ with $\delta(\mathcal{A}) = \lambda$, the best that can be hoped for in general is a λ^+-transversal. In fact this can usually be achieved. There is first a trivial lemma.

Lemma 1.3.7 (GCH). *Let ι and λ be cardinals such that there is no η with $\iota = \eta^+$ and $\lambda' = \eta'$. Suppose $\mathcal{B}$ is a family with $|\bigcup \mathcal{B}| < \iota$ and $\delta(\mathcal{B}) \leqslant \lambda$ and $|B| \geqslant \lambda$ for each B in $\mathcal{B}$. Then $|\mathcal{B}| < \iota$.*

Proof. We may suppose $\lambda < \iota$. Since $|\mathcal{B}| \leqslant |[\bigcup \mathcal{B}]^\lambda|$, certainly if $|\bigcup \mathcal{B}|^+ < \iota$ then $|\mathcal{B}| < \iota$. And if $|\bigcup \mathcal{B}|^+ = \iota$, then Theorem 1.1.6 applies to give the result.

The following theorem is slightly more general than our immediate need dictates, but the full strength will be used in the proof of Theorem 1.3.12. We shall adopt the temporary notation $\mathcal{A}[\kappa]$ for $\{A \in \mathcal{A}; |A| = \kappa\}$.

Theorem 1.3.8 (GCH). *Suppose κ and λ are such that there is no η with $\kappa = \eta^+$*

and $\lambda' = \eta'$ *(unless* $\kappa = \lambda^+$*). Let* $\mathscr{A}$ *be any family of* κ *sets with* $\delta(\mathscr{A}) \leqslant \lambda$ *and* $|A| \leqslant \kappa$ *for each* A *in* $\mathscr{A}$*. Then there is a* λ^+*-transversal* T *of* $A[\kappa]$ *with* $|T \cap A| \leqslant \lambda$ *for each* A *in* $\mathscr{A}$*.*

Corollary 1.3.9 (GCH). *Under the conditions of the Theorem, any family* $\mathscr{A}$ *of* κ *sets each of power* κ *with* $\delta(\mathscr{A}) \leqslant \lambda$ *has a* λ^+*-transversal.*

Proof (of Theorem 1.3.8). We need only consider λ with $\lambda < \kappa$. The case when $\kappa = \lambda^+$ may be shown by a construction similar to that in the proof of Theorem 1.3.3 (and in fact GCH is not used). So suppose henceforth that $\lambda^+ < \kappa$. Let $\mathscr{A}$ be well ordered, $\mathscr{A} = \{A_\nu; \nu < \kappa\}$. We still try to choose x_ν in A_ν, but do so only when this is possible without having x_ν in any A from $\mathscr{A}$ which already meets $\{x_\mu; \mu < \nu\}$ in λ points. This ensures that the set of all the x_ν meets any A in at most λ points.

So formally, use transfinite induction to define elements x_ν when $\nu < \kappa$ as follows. Write

$$\mathscr{B}_\nu = \{A \in \mathscr{A}; |A \cap \{x_\mu; \mu < \nu\}| \geqslant \lambda\}.$$

Choose x_0 from A_0. When $\nu > 0$, if $A_\nu - \bigcup\mathscr{B}_\nu = \emptyset$ put $x_\nu = x_0$ and otherwise choose x_ν from $A_\nu - \bigcup\mathscr{B}_\nu$.

By the property $\delta(\mathscr{A}) \leqslant \lambda$, any A in $\mathscr{B}_\nu$ is uniquely determined by its intersection with $\{x_\mu; \mu < \nu\}$. Since $|\{x_\mu; \mu < \nu\}| < \kappa$, Lemma 1.3.7 shows that $\{A \cap \{x_\mu; \mu < \nu\}; A \in \mathscr{B}_\nu\}$ has power less than κ; hence $|\mathscr{B}_\nu| < \kappa$. Again using that $\delta(\mathscr{A}) \leqslant \lambda < \kappa$, it follows that $|A_\nu \cap \bigcup\mathscr{B}_\nu| < \kappa$ provided $A_\nu \notin \mathscr{B}_\nu$. Thus if $|A_\nu| = \kappa$ and $|A_\nu \cap \{x_\mu; \mu < \nu\}| < \lambda$ then $A_\nu - \bigcup\mathscr{B}_\nu \neq \emptyset$ and so $x_\nu \in A_\nu$.

Put $T = \{x_\nu; \nu < \kappa\}$, then T meets every set in $\mathscr{A}[\kappa]$. And for any A in $\mathscr{A}$, if for some ρ it happens that $|A \cap \{x_\mu; \mu < \rho\}| = \lambda$ then for all ν with $\nu \geqslant \rho$ either $x_\nu = x_0$ or else $x_\nu \notin A$. Hence always $|T \cap A| \leqslant \lambda$. This completes the proof.

If λ is finite, the same proof holds, and in fact GCH is not needed. It is not known if Corollary 1.3.9 holds without restriction on κ and λ. The simplest case that is not settled is when $\kappa = \aleph_{\omega+1}$, $\lambda = \aleph_0$. Explicitly:

Question 1.3.10 (GCH). Is there a family $\mathscr{A}$ of $\aleph_{\omega+1}$ sets each of power $\aleph_{\omega+1}$ with $\delta(\mathscr{A}) = \aleph_0$ which has no $\aleph_1$-transversal?

One can always find a λ^{++}-transversal however.

Theorem 1.3.11 (GCH). *For all κ and λ, every family $\mathcal{A}$ of κ sets each of power κ with $\delta(\mathcal{A}) \leqslant \lambda$ has a λ^{++}-transversal.*

Proof. Modify the proof of Theorem 1.3.8 by putting

$$\mathcal{B}_\nu = \{A \in \mathcal{A}; |A \cap \{x_\mu; \mu < \nu\}| > \lambda\}.$$

Using now Corollary 1.1.7 one concludes that $|\mathcal{B}_\nu| < \kappa$; the rest of the proof carries over.

We are left finally with the problem of finding a transversal of a family of more than κ sets each of power κ, where $\delta(\mathcal{A}) = \lambda < \kappa$. Theorem 1.3.2 still shows that a λ^+-transversal is the best that can be hoped for in general (since any family extending a family without a λ-transversal has, of course, no λ-transversal). We can show that a λ^+-transversal is usually possible provided $|\mathcal{A}|$ is sufficiently small, and the problem is unsolved for larger values of $|\mathcal{A}|$. The proof will be by induction on $|\mathcal{A}|$, and so that the induction will work we need a more general statement, on the lines of that in Theorem 1.3.8.

Theorem 1.3.12 (GCH). *Suppose κ and λ are such that there is no η with $\kappa = \eta^+$ and $\lambda' = \eta'$ (unless $\kappa = \lambda^+$). Let κ_0 be the least cardinal greater than κ with $\kappa_0' = \lambda'$. Then for all ι with $\iota \leqslant \kappa_0$, if $\mathcal{A}$ is any family of ι sets with $\delta(\mathcal{A}) \leqslant \lambda$ and $|A| \leqslant \kappa$ for each A in $\mathcal{A}$, there is a λ^+-transversal T of $\mathcal{A}[\kappa]$ such that $|T \cap A| \leqslant \lambda$ for each A in $\mathcal{A}$.*

Corollary 1.3.13 (GCH). *Under the conditions of the Theorem, any family $\mathcal{A}$ of ι sets each of power κ with $\delta(\mathcal{A}) \leqslant \lambda$ has a λ^+-transversal.*

Proof of Theorem 1.3.12. Use induction on ι. By Theorem 1.3.8, the result is true when $\iota \leqslant \kappa$. So suppose ι is such that $\kappa < \iota \leqslant \kappa_0$ and that the theorem holds for all ι^* with $\iota^* < \iota$. Let $\mathcal{A} = \{A_\nu; \nu < \iota\}$ be a suitable family. Write $\mathcal{A}$ as a disjoint union, $\mathcal{A} = \bigcup\{\mathcal{A}_\nu; \nu < \iota'\}$ where always $|\mathcal{A}_\nu| < \iota$.

By the inductive assumption, there is a suitable set T_0 for $\mathcal{A}_0$, and we seek to extend T_0 to a suitable transversal of $\mathcal{A}$. However, it may happen that certain members of $\mathcal{A} - \mathcal{A}_0$ have an intersection of size greater than λ with $X_0 = \bigcup \mathcal{A}_0$; in particular they may meet T_0 in more than λ points and so T_0 could not be extended to $\mathcal{A}$. So first increase $\mathcal{A}_0$ by adding to it the family $\mathcal{C}_0$ of all sets A in $\mathcal{A}$ with $|A \cap X_0| \geqslant \lambda$, and look at the family $\mathcal{D}_0 = \{A \cap X_0; A \in \mathcal{A}_0 \cup \mathcal{C}_0\}$. It turns out that $|\mathcal{D}_0| < \iota$, and so there is a suitable transversal T_0 of $\mathcal{D}_0$, with $T_0 \subseteq X_0$. Now treat $\mathcal{A}_1$ similarly, putting $X_1 = \bigcup \mathcal{A}_1 - X_0$ and looking at $\mathcal{D}_1 = \{A \cap X_1; A \in \mathcal{A}_1 \cup \mathcal{C}_1\}$. We find T_1

with $T_1 \subseteq X_1$, and $T_0 \cup T_1$ extends T_0 and is suitable for $\mathcal{A}_0 \cup \mathcal{A}_1$. If we were to continue in this manner, possibly there would be a set A in some $\mathcal{A}_\nu$ such that $|A \cap X_\mu| \geqslant \lambda$ for more than λ values of μ with $\mu < \nu$, and so perhaps $|A \cap \bigcup\{T_\mu; \mu < \nu\}| > \lambda$, meaning that there would be no suitable extension of $(\bigcup\{T_\mu; \mu < \nu\})$ to all of $\mathcal{A}$. To prevent this happening, we should finish with such an A at the first stage ν where $|A \cap (\bigcup\{X_\mu; \mu < \nu\})| \geqslant \lambda$. So at any stage ν, we treat also the members of $\mathcal{B}_\nu$, where $\mathcal{B}_\nu$ includes all members of $\mathcal{A}$ with $|A \cap \bigcup\{X_\mu; \mu < \nu\}| \geqslant \lambda$, and form X_ν from $\mathcal{A}_\nu \cup \mathcal{B}_\nu$. The formal construction proceeds as follows.

Define sets X_ν where $\nu < \iota'$ with $X_\nu \in [\bigcup \mathcal{A}]^{<\iota}$ by induction as follows. Suppose for a particular ν with $\nu < \iota'$ that the X_μ where $\mu < \nu$ have already been defined. Put $X_\nu^* = \bigcup\{X_\mu; \mu < \nu\}$, so $|X_\nu^*| < \iota$. Write $\mathcal{B}_\nu =$ $\{A \in \mathcal{A}; |A \cap X_\nu^*| \geqslant \lambda\}$. Since $\delta(\mathcal{A}) \leqslant \lambda$, any A in $\mathcal{B}_\nu$ is uniquely determined by $A \cap X_\nu^*$. Since $|X_\nu^*| < \iota$, Lemma 1.3.7 shows that $|\{A \cap X_\nu^*; A \in \mathcal{B}_\nu\}| < \iota$. The definition of κ_0 and the fact that $\iota \leqslant \kappa_0$ ensure that the conditions of Lemma 1.3.7 are satisfied. Consequently $|\mathcal{B}_\nu| < \iota$. Put $X_\nu = (\bigcup \mathcal{A}_\nu \cup \bigcup \mathcal{B}_\nu)$ $- X_\nu^*$. Since $\kappa < \iota$ it follows that $|X_\nu| < \iota$. This completes the definition of X_ν.

Put $\mathcal{C}_\nu = \{A \in \mathcal{A}; |A \cap X_\nu| \geqslant \lambda\}$; just as we argued above for $\mathcal{B}_\nu$, it follows that $|\mathcal{C}_\nu| < \iota$. Write $\mathcal{D}_\nu = \{A \cap X_\nu; A \in \mathcal{A}_\nu \cup \mathcal{B}_\nu \cup \mathcal{C}_\nu\}$, so $|\mathcal{D}_\nu| < \iota$. By the inductive assumption, there is a set T_ν with $T_\nu \subseteq X_\nu$ such that T_ν is a λ^+-transversal of $\mathcal{D}_\nu[\kappa]$ and always $|T_\nu \cap D| \leqslant \lambda$ for each D in $\mathcal{D}_\nu$. Define $T = \bigcup\{T_\nu; \nu < \iota'\}$; I claim T is a suitable λ^+-transversal of $\mathcal{A}[\kappa]$.

Take any A in $\mathcal{A}$. We must show that $|T \cap A| \leqslant \lambda$ and that $T \cap A \neq \emptyset$ if $|A| = \kappa$. Clearly we may suppose $|A| > \lambda$. Let ξ be least such that $|A \cap X_\xi^*| \geqslant \lambda$. If $A \in \mathcal{A}_\nu$, certainly $A \subseteq X_\nu^* \cup X_\nu = X_{\nu+1}^*$, and so $\xi < \iota'$. The definition of X_ξ shows that $A \subseteq X_\xi^* \cup X_\xi$. There are two cases to consider, depending on whether or not ξ is a successor ordinal.

Suppose first that ξ is a successor, say $\xi = \nu + 1$. The choice of ξ shows $|A \cap X_\nu^*| < \lambda$. Since A is a disjoint union, $A = (A \cap X_\nu^*) \cup (A \cap X_\nu) \cup (A \cup X_\xi)$, certainly $A \cap T \subseteq (A \cap X_\nu^*) \cup (A \cap T_\nu) \cup (A \cap T_\xi)$. The choice of T_ν and T_ξ ensures that $|A \cap T_\nu|$, $|A \cap T_\xi| \leqslant \lambda$, and it follows that $|A \cap T| \leqslant \lambda$. Further, if $|A| = \kappa$ then since $\lambda < \kappa$, either $|A \cap X_\nu| = \kappa$ or $|A \cap X_\xi| = \kappa$; this ensures either $A \cap T_\nu \neq \emptyset$ or $A \cap T_\xi \neq \emptyset$, so $A \cap T \neq \emptyset$.

Now suppose ξ to be a limit ordinal. We have $|A \cap X_\nu^*| < \lambda$ whenever $\nu < \xi$. Since $A \cap X_\xi^* = \bigcup\{A \cap X_\nu^*; \nu < \xi\}$ and $A \cap X_\mu^* \subseteq A \cap X_\nu^*$ whenever $\mu < \nu$, it follows that here $|A \cap X_\xi^*| = \lambda$. Again A is a disjoint union, $A =$ $(A \cap X_\xi^*) \cup (A \cap X_\xi)$, so that $A \cap T \subseteq (A \cap X_\xi^*) \cap (A \cap T_\xi)$. Since $|A \cap T_\xi| \leqslant \lambda$, we have $|A \cap T| \leqslant \lambda$. Moreover if $|A| = \kappa$ then $|A \cap X_\xi| = \kappa$ and so $A \cap T_\xi \neq \emptyset$. This completes the proof.

If λ is finite, one can carry out the same construction to show the following by induction on $|\mathcal{A}|$ (and the GCH is not needed).

Theorem 1.3.14. *If λ is finite, every family $\mathcal{A}$ of sets each of power κ with $\delta(\mathcal{A}) \leqslant \lambda$ has an $\aleph_0$-transversal T (that is, $T \cap A$ is finite for each A in $\mathcal{A}$).*

The case when λ is finite has been analysed in more detail by Erdös and Hajnal in [19 pp. 109–113]. They show that any family $\mathcal{A}$ with $|\mathcal{A}| = \aleph_{\alpha+n}$ and $\delta(\mathcal{A}) \leqslant \lambda$ has a $[(\lambda - 1)(n + 1) + 2]$-transversal, and they construct an example to show that this is the best possible result.

It is not known if the result of Corollary 1.3.13 holds without restriction on the cardinals involved. It is easy to see that for $\kappa = \aleph_\alpha$, the number κ_0 is $\aleph_{\alpha+\lambda'}$. So the easiest cases left unanswered are the following. (See Problem 39 of [24].)

Question 1.3.15 (GCH). Is there a family $\mathcal{A}$ of $\aleph_{\omega+1}$ sets each of power $\aleph_1$ with $\delta(\mathcal{A}) = \aleph_0$ which has no $\aleph_1$-transversal? Is there a family $\mathcal{A}$ of $\aleph_{\omega_1+1}$ sets each of power $\aleph_{\omega+1}$ with $\delta(\mathcal{A}) = \aleph_1$ which has no $\aleph_2$-transversal?

Finally, we observe that Corollary 1.3.13 is always true, without restriction on ι, κ, λ if the demand for a λ^+-transversal be eased to require only a λ^{++}-transversal.

Theorem 1.3.16 (GCH). *For all ι, κ, λ any family $\mathcal{A}$ of ι sets each of power κ with $\delta(\mathcal{A}) \leqslant \lambda$ has a λ^{++}-transversal.*

Proof. Modify the proof of Theorem 1.3.12 by putting

$$\mathcal{B}_\nu = \{A \in \mathcal{A}; A \cap X_\nu^*| > \lambda\},$$

$$\mathcal{C}_\nu = \{A \in \mathcal{A}; |A \cap X_\nu| > \lambda\}.$$

Using Corollary 1.1.7, one sees $|\mathcal{B}_\nu|$, $|\mathcal{C}_\nu| < |\mathcal{A}|$ and the rest of the proof carries over.

CHAPTER 2

ORDINARY PARTITION RELATIONS

§1. The relations defined

The partition relations $\kappa \to (\eta_k; k < \gamma)^n$ and $\kappa \to (\eta)^n_\gamma$ defined below, which are the principal objects of study in this chapter, have now been discussed extensively in the literature. A special case of the symbols first appeared in Erdös and Rado [28], although several results that can be expressed in terms of these symbols had appeared before (see for example Sierpinski [85], Dushnik and Miller [11], Erdös [13], also Kurepa [64]). Detailed discussion is, amongst others, to be found in Erdös and Rado [29] and the encyclopedic Erdös, Hajnal and Rado [38], in both of which references to early results can be found.

The subject is concerned with generalizations of the following theorem due to Ramsey [80]: If the unordered pairs from an infinite set S are partitioned into finitely many classes, then there is an infinite subset H of S such that all the pairs from H fall in the one class of the partition.

In the following definition, κ is a cardinal, n a positive integer, γ an ordinal, and η_k (where $k < \gamma$) are cardinals.

Definition 2.1.1. The *ordinary partition symbol* $\kappa \to (\eta_k; k < \gamma)^n$ means: given any set S with $|S| = \kappa$, for all partitions $\mathbf{\Delta} = \{\Delta_k; k < \gamma\}$ of $[S]^n$ into γ parts, there exist k with $k < \gamma$ and a subset H of S with $|H| = \eta_k$ such that $[H]^n \subseteq \Delta_k$.

The special case in which $\eta_k = \eta$ for all k with $k < \gamma$ will be written $\kappa \to (\eta)^n_\gamma$. Symbols such as $\kappa \to (\eta_k; k < \gamma, \theta_l; l < \delta)^n$ have the obvious meaning.

The negation of the relation $\kappa \to (\eta_k; k < \gamma)^n$ is written $\kappa \not\to (\eta_k; k < \gamma)^n$.

Thus, in terms of this partition symbol, Ramsey's theorem quoted above can be expressed by: $\aleph_0 \to (\aleph_0)^2_m$ for any finite m.

Given a partition $[S]^n = \bigcup\{\Delta_k; k < \gamma\}$ of the n-element subsets of a set S, a subset H of S with $[H]^n \subseteq \Delta_k$ for some k is said to be *homogeneous* for that partition.

Before a more detailed discussion of the various partition relations, a few simple remarks are in order.

The partition symbol could equally well be defined in terms of disjoint partitions only; this is easily seen to be equivalent to the definition given.

The use of the ordinal γ to index the classes of the partition $\mathbf{\Delta}$ is not essential; any set with the same power as γ will serve equally well and the truth or the falsity of the relation will remain unchanged. In particular, the cardinals η_k in the relation $\kappa \to (\eta_k; k < \gamma)^n$ may be permuted without effecting the relation.

If κ has a partition property and λ is any cardinal at least as big as κ, then λ has that same property.

If κ enjoys a partition property with γ classes and δ is any ordinal smaller than γ, then the corresponding property with δ classes also holds for κ, since a partition with a small number of classes can be extended to a partition with a larger number of classes by adding superfluous empty parts.

If κ has a partition property in which the homogeneous sets have cardinality η_k, then for any cardinals θ_k with $\theta_k \leqslant \eta_k$ the corresponding property with η_k replaced by θ_k also holds for κ.

The truth or falsity of the relation $\kappa \to (\eta_k; k < \gamma)^n$ can be trivially determined unless the following conditions are in force:

$$n \geqslant 1,\ 2 \leqslant \gamma < \kappa,\ n \leqslant \eta_k \leqslant \kappa \text{ (for } k < \gamma) .$$

In fact we may suppose $n < \eta_k$, for if $\eta_k = n$ whenever $\delta \leqslant k < \gamma$ then the relation $\kappa \to (\eta_k; k < \gamma)^n$ is equivalent to the relation $\kappa \to (\eta_k; k < \delta)^n$. Given a set S with $|S| = \kappa$, any partition $[S]^n = \bigcup\{\Delta_k; k < \gamma\}$ trivially has a homogeneous set of size n if any Δ_k where $\delta \leqslant k < \gamma$ is non-empty; if these are all empty the question reduces to whether or not $\kappa \to (\eta_k; k < \delta)^n$.

There is the following substitution property.

Theorem 2.1.2. *Suppose* $\kappa \to (\eta_k; k < \gamma)^n$ *and* $\eta_0 \to (\theta_l; l < \delta)^n$. *Then* $\kappa \to (\theta_l; l < \delta, \eta_k; 1 \leqslant k < \gamma)^n$.

Proof. Given a partition

$$[S]^n = \bigcup\{\Delta_l; l < \delta\} \cup \bigcup\{\Gamma_k; 1 \leqslant k < \gamma\}$$

of a set S of power κ, put $\Gamma_0 = \bigcup\{\Delta_l; l < \delta\}$. By the relation $\kappa \to (\eta_k; k < \gamma)^n$, there is k with $k < \gamma$ and a subset H of S with $|H| = \eta_k$ and $[H]^n \subseteq \Gamma_k$. If $k > 0$, nothing more need be done. If $k = 0$, then $[H]^n \subseteq \bigcup\{\Delta_l; l < \delta\}$ and because of the relation $\eta_0 \to (\theta_l; l < \delta)^n$ there must be l with $l < \delta$ and a subset I of H with $|I| = \theta_l$ and $[I]^n \subseteq \Delta_l$. In any event, there is a suitable subset of S homogeneous for the original partition of $[S]^n$.

The restriction to finite exponent n is justified by the following theorem from [27], which shows that the weakest possible non trivial partition relation with infinite exponent, $\kappa \to (\aleph_0, \aleph_0)^{\aleph_0}$, is false for all κ.

Theorem 2.1.3. *Given any infinite set S, there is a partition $[S]^{\aleph_0} = \Delta_0 \cup \Delta_1$ of the denumerable subsets of S into two classes which has no infinite homogeneous set.*

Proof. Let $\prec$ be a relation which well orders $[S]^{\aleph_0}$. Define a partition $[S]^{\aleph_0} = \Delta_0 \cup \Delta_1$ as follows: for X in $[S]^{\aleph_0}$,

$$X \in \Delta_0 \Leftrightarrow \exists Y \in [X]^{\aleph_0} \, (Y \prec X) \, ,$$

$$X \in \Delta_1 \Leftrightarrow \forall Y \in [X]^{\aleph_0} \, (X \preccurlyeq Y) \, .$$

Take any set H in $[S]^{\aleph_0}$; we shall show H is not homogeneous for this partition. Let X be the least infinite subset of H; clearly $X \in \Delta_1$ and so $[H]^{\aleph_0} \not\subseteq \Delta_0$. On the other hand, write $H = \{h_i; i < \omega\}$ and for each finite k, put $H_k = \{h_0, h_2, ..., h_{2k}\} \cup \{h_{2i+1}; i < \omega\}$. Let H_{k_0} be the least of the H_k. Then $H_{k_0} \subseteq H_{k_0+1}$ and $H_{k_0} \prec H_{k_0+1}$, so $H_{k_0+1} \in \Delta_0$. Thus $[H]^{\aleph_0} \not\subseteq \Delta_1$, and so H is not homogeneous.

As a final remark in this section, note that the symbol $\kappa \to (\eta_k; k < \gamma)^n$ can also be defined when κ and the η_k are ordinals, or in fact arbitrary order types. Rather than speaking of the cardinalities of the sets involved, their order type is specified, for a suitable ordering. Some of the results for ordinal numbers will be discussed in Chapter 7. However, in this chapter we shall confine ourselves to the problems involving cardinal numbers.

§2. The Ramification Lemma

The method to be described in this section gives one of the most frequently applied techniques in establishing partition relations. The method was used in several early papers before it was expressed in the general form given below in the paper of Erdös, Hajnal and Rado [38]. The method has since found application outside the partition calculus, see for example the book of Juhasz [59] for several applications in general topology.

The technique hinges on the construction of a so-called ramification system. In order to describe what is meant by this, we look first at trees of sequences of ordinals.

Definition 2.2.1. A partially ordered set $\langle T, \leqslant\rangle$ is said to be a *tree* if for each x in T the set $\{y \in T; y < x\}$ of predecessors of x is well ordered (by $\leqslant$). The order type of $\{y \in T; y < x\}$ is called the *order* of x. The αth *level* of T is the set of all elements of T whose order is α. A maximal linearly ordered subset of T is called a *branch*; the members of T are called *nodes* in T.

We shall adopt the following notation. For each ordinal σ we shall denote by SEQ_σ the class of all sequences of length σ of ordinals, that is all sequences $\mathbf{v} = \langle \nu_\alpha; \alpha < \sigma\rangle$ where each ν_α is an ordinal. We shall write $\ell n(\mathbf{v})$ for the length of such a sequence $\mathbf{v}$. Bold face Greek letters $\boldsymbol{\mu}, \mathbf{v}$ will be used for sequences of ordinals. The *restriction* of $\mathbf{v}$ to τ, written $\mathbf{v} \upharpoonright \tau$ is defined as follows: if $\tau \geqslant \ell n(\mathbf{v})$ then $\mathbf{v} \upharpoonright \tau = \mathbf{v}$, whereas if $\tau < \ell n(\mathbf{v})$ then $\mathbf{v} \upharpoonright \tau = \boldsymbol{\mu}$ where $\ell n(\boldsymbol{\mu}) = \tau$ and $\boldsymbol{\mu}(\alpha) = \mathbf{v}(\alpha)$ for all α with $\alpha < \tau$.

Given ordinals ξ and ρ we construct a tree as follows: the nodes are all sequences $\mathbf{v}$ of length at most ρ with always $\mathbf{v}(\alpha) < \xi$, and the ordering is "is an initial segment of", that is $\boldsymbol{\mu} \leqslant \mathbf{v} \Leftrightarrow \exists\, \sigma(\boldsymbol{\mu} = \mathbf{v} \upharpoonright \sigma)$. It is easy to see that this gives a tree–referred to as the *full tree* on ξ of height ρ.

The trees that will concern us will not necessarily be full trees, but rather certain subtrees. In particular, they need not have the same number of branches passing through every node, nor even through every node at a given level.

The trees we shall use will be of the following form. Let an ordinal ρ be given, and for each sequence $\boldsymbol{\mu}$ with $\ell n(\boldsymbol{\mu}) < \rho$ let there be given an ordinal $n(\boldsymbol{\mu})$. Form the set $N = \{\mathbf{v}; \ell n(\mathbf{v}) \leqslant \rho$ and $\forall \alpha < \ell n(\mathbf{v})\, (\mathbf{v}(\alpha) < n(\mathbf{v} \upharpoonright \alpha))\}$, and order N as before. Then $\langle N, \leqslant\rangle$ is a tree, and it is on trees of this form that we shall build a ramification system. (The ordinals $n(\boldsymbol{\mu})$ indicate the amount of branching that occurs at the node $\boldsymbol{\mu}$ – one new branch for each ordinal less than $n(\boldsymbol{\mu})$.)

For example, suppose $n(\emptyset) = 2$ (here $\emptyset$ stands for the empty sequence), $n(\langle 0\rangle) = 1$, $n(\langle 1\rangle) = 3$, $n(\langle 0, 0\rangle) = 2$, $n(\langle 1, 0\rangle) = 1$, $n(\langle 1, 1\rangle) = 2$, $n(\langle 1, 2\rangle) = 1$. The lowest nodes of the resulting tree N appear as in Fig. 1 below.

Such a tree N of ordinal sequences is converted into a ramification system $\mathfrak{R}$ by associating with each node $\mathbf{v}$ of N two sets $S(\mathbf{v})$ and $F(\mathbf{v})$ by induction on the length of $\mathbf{v}$ as follows:

(i) If $\mathbf{v} \in N \cap SEQ_0$ (so $\mathbf{v} = \emptyset$) then $S(\mathbf{v})$ is some given set S.

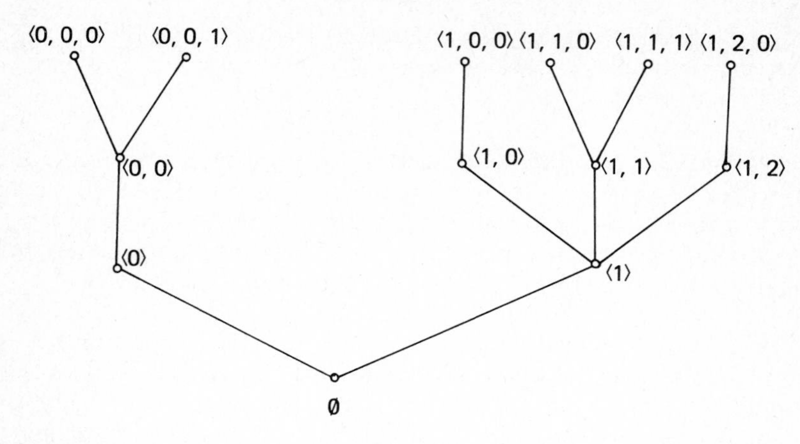

Fig. 1.

(ii) If $\mathbf{v} \in N \cap SEQ_\sigma$ and $S(\mathbf{v})$ is known, choose for $F(\mathbf{v})$ some subset of $S(\mathbf{v})$ and split $S(\mathbf{v}) - F(\mathbf{v})$ into (not necessarily disjoint) sets S_α where $\alpha < n(\mathbf{v})$. For $\boldsymbol{\mu}$ in $N \cap SEQ_{\sigma+1}$ with $\boldsymbol{\mu} \upharpoonright \sigma = \mathbf{v}$ (that is, for sequences in N which extend $\mathbf{v}$ by one place) put $S(\boldsymbol{\mu}) = S_{\boldsymbol{\mu}(\sigma)}$. Thus $S(\mathbf{v})$ is the disjoint union of two sets, $S(\mathbf{v}) = F(\mathbf{v}) \cup \bigcup\{S(\boldsymbol{\mu});\ \boldsymbol{\mu} \in N \cap SEQ_{\sigma+1}$ and $\boldsymbol{\mu} \upharpoonright \sigma = \mathbf{v}\}$.

(iii) If $\mathbf{v} \in N \cap SEQ_\sigma$ where σ is a limit ordinal, put $S(\mathbf{v}) = \bigcap\{S(\mathbf{v} \upharpoonright \alpha);\ \alpha < \sigma\}$

(Thus as one rises through the levels of the tree N, at any node $\mathbf{v}$ the set $S(\mathbf{v})$ arrives from below, the subset $F(\mathbf{v})$ is separated from $S(\mathbf{v})$ and kept at the node $\mathbf{v}$, and the set $S(\mathbf{v}) - F(\mathbf{v})$ is divided and pushed up to the nodes immediately above $\mathbf{v}$.)

The specification of a ramification system $\mathfrak{R}$ is thus seen to require:

(i) the ordinal ρ – the *height* of the ramification system;

(ii) the ordinals $n(\mathbf{v})$ for sequences $\mathbf{v}$ with $\ell n(\mathbf{v}) < \rho$, to construct the tree N. If all the $N(\mathbf{v})$ are equal, say to ξ, then the ramification system is said to have *order* ξ;

(iii) the sets $F(\mathbf{v})$ for $\mathbf{v}$ in N;

(iv) the sets $S(\mathbf{v})$ for $\mathbf{v}$ in N with length of $\mathbf{v}$ not a limit ordinal.

The ramification system is said to be *on* $S = S(\emptyset)$.

A ramification system $\mathfrak{R}$ and its supporting tree N can be constructed by induction as follows:

(a) The height ρ and the set $S = S(\emptyset)$ are given.

(b) For any ordinal σ with $\sigma < \rho$, one supposes $n(\boldsymbol{\mu})$ given for all sequences $\boldsymbol{\mu}$ with $\ell n(\boldsymbol{\mu}) < \sigma$ so $N \cap SEQ_\sigma$ is known, and for each $\mathbf{v}$ in $N \cap SEQ_\sigma$ one supposes $S(\mathbf{v})$ already given, and defines $n(\mathbf{v})$, $F(\mathbf{v})$ and $S(\boldsymbol{\mu})$ for all sequences $\boldsymbol{\mu}$ with $\boldsymbol{\mu} \in SEQ_{\sigma+1}$, $\boldsymbol{\mu} \restriction \sigma = \mathbf{v}$ and $\boldsymbol{\mu}(\sigma) < n(\mathbf{v})$. (The value of $n(\mathbf{v})$ for $\mathbf{v}$ in $SEQ_\sigma - N$ is immaterial.)

To save repetition, in all references to a ramification system $\mathfrak{R}$ the notation $F(\mathbf{v}), S(\mathbf{v}), n(\mathbf{v}), N, \rho$ is to be understood to be as in the description just given.

The effect of a ramification system $\mathfrak{R}$ of height ρ on a set S is to divide S into the sets $F(\mathbf{v})$ for those $\mathbf{v}$ in N with $\ell n(\mathbf{v}) < \rho$ together with the $S(\mathbf{v})$ for those $\mathbf{v}$ in N of length ρ and moreover the $F(\mathbf{v})$ along any given branch are pairwise disjoint. This is the content of the following lemma.

Lemma 2.2.2. *Let $\mathfrak{R}$ be a ramification system on S of height ρ. Then*
(i) $S = \bigcup\{F(\mathbf{v}); \mathbf{v} \in N$ *and* $\ell n(\mathbf{v}) < \rho\} \cup \bigcup\{S(\mathbf{v}); \mathbf{v} \in N$ *and* $\ell n(\mathbf{v}) = \rho\}$.
(ii) *If* $\mathbf{v} \in N$ *and* $\sigma, \tau \leqslant \ell n(\mathbf{v})$ *with* $\sigma \neq \tau$, *then* $F(\mathbf{v} \restriction \sigma) \cap F(\mathbf{v} \restriction \tau) = \emptyset$.

Proof. For (i), take x in S and suppose $x \notin F(\mathbf{v})$ for all $\mathbf{v}$ in N with $\ell n(\mathbf{v}) < \rho$. Use induction to define ordinals ν_α where $\alpha < \rho$ such that $\langle \nu_\alpha; \alpha \leqslant \tau\rangle \in N$ and $x \in S(\nu_\alpha; \alpha \leqslant \tau)$ for all τ with $\tau < \rho$. Suppose σ is given with $\sigma < \rho$ and the ν_α where $\alpha < \sigma$ have already been suitably defined. Then it follows that

$$\langle \nu_\alpha; \alpha < \sigma\rangle \in N \text{ and } x \in S(\nu_\alpha; \alpha < \sigma) . \tag{1}$$

If σ is a successor ordinal this is trivial. If σ is a limit ordinal, for any β with $\beta < \sigma$ since $\langle \nu_\alpha; \alpha \leqslant \beta\rangle \in N$ we know $\nu_\beta < n(\nu_\alpha; \alpha < \beta)$, and so $\langle \nu_\beta; \beta < \sigma\rangle \in N$. Also

$$S(\nu_\alpha; \alpha < \sigma) = \bigcap\{S(\nu_\alpha; \alpha < \beta); \beta < \sigma\} = \bigcap\{S(\nu_\alpha; \alpha \leqslant \beta); \beta < \sigma\}$$

and so $x \in S(\nu_\alpha; \alpha < \sigma)$. Thus (1) holds.

Since by assumption $x \notin F(\nu_\alpha; \alpha < \sigma)$, from the definition of a ramification system there must be $\boldsymbol{\mu}$ in $N \cap SEQ_{\sigma+1}$ with $\boldsymbol{\mu} \restriction \sigma = \langle \nu_\alpha; \alpha < \sigma\rangle$ for which $x \in S(\boldsymbol{\mu})$. Choose such a sequence $\boldsymbol{\mu}$, and define ν_σ by $\nu_\sigma = \boldsymbol{\mu}(\sigma)$. This completes the inductive definition.

It follows as above that $\langle \nu_\alpha; \alpha < \rho\rangle \in N$ and that $x \in S(\nu_\alpha; \alpha < \rho)$; and hence (i) holds.

To prove (ii), suppose $\tau < \sigma \leqslant \ell n(\mathbf{v})$ for some $\mathbf{v}$ in N. Then $\mathbf{v} \restriction \tau \in N$ and

$$F(\mathbf{v} \restriction \tau) \cap \bigcup\{S(\boldsymbol{\mu}); \boldsymbol{\mu} \in N \cap SEQ_{\tau+1} \text{ and } \boldsymbol{\mu} \restriction \tau = \mathbf{v} \restriction \tau\} = \emptyset ,$$

so in particular $F(\mathbf{v} \restriction \tau) \cap S(\mathbf{v} \restriction \tau + 1) = \emptyset$. However,

$$F(\mathbf{v} \restriction \sigma) \subseteq S(\mathbf{v} \restriction \sigma) \subseteq S(\mathbf{v} \restriction \tau + 1) ,$$

so $F(\mathbf{v} \restriction \tau) \cap F(\mathbf{v} \restriction \sigma) = \emptyset$. This completes the proof.

In the applications that we shall make of ramification systems, we shall be interested in having a branch through to the top of the tree such that some elements of S get pushed right up the branch, that is, they are never put in any of the $F(\boldsymbol{\mu})$ at the nodes along the way. In a system of height ρ, such a situation is reached by having a sequence $\mathbf{v}$ in $N \cap SEQ_\rho$ with $S(\mathbf{v})$ non-empty; the initial segments $\mathbf{v}\lceil\alpha$ for α with $\alpha \leqslant \rho$ give the nodes along the branch. The following theorem gives conditions under which there is such a branch.

Theorem 2.2.3 (The Ramification Lemma). *Let $\mathfrak{R}$ be a ramification system of height ρ on a set S with $|S| \geqslant \kappa$, for some infinite cardinal κ. Suppose either*

(i) $\rho < \kappa'$; $|F(\mathbf{v})| < \kappa$ *if* $\mathbf{v} \in N$; *there are cardinals* λ_σ *for* σ *with* $\sigma < \rho$ *such that* $\lambda_\sigma^{|\sigma|} < \kappa'$ *and* $|n(\mathbf{v}\lceil\tau)| \leqslant \lambda_\sigma$ *whenever* $\mathbf{v} \in N \cap SEQ_\sigma$ *and* $\tau < \sigma$, *or*

(ii) $\kappa = \lambda^+$; $\rho \leqslant \lambda'$; $|F(\mathbf{v})| \leqslant \lambda$ *and* $|n(\mathbf{v})| \leqslant \lambda$ *whenever* $\mathbf{v} \in N \cap SEQ_\sigma$ *for any* σ *with* $\sigma < \rho$; $\theta < \lambda \to 2^\theta \leqslant \lambda$, *or*

(iii) κ *is strongly inaccessible*; $\rho < \kappa$; $|F(\mathbf{v})| < \kappa$ *and* $|n(\mathbf{v})| < \kappa$ *whenever* $\mathbf{v} \in N \cap SEQ_\sigma$ *for any* σ *with* $\sigma < \rho$. $\Rightarrow$

Then there is $\mathbf{v}$ *in* $N \cap SEQ_\rho$ *for which* $S(\mathbf{v})$ *is non-empty.*

Proof. Suppose the conditions in (i) hold. For any σ with $\sigma < \rho$, we have

$$|N \cap SEQ_\sigma| \leqslant (\sup\{n(\mathbf{v}\lceil\tau); \tau < \sigma \text{ and } \mathbf{v} \in N \cap SEQ_\sigma\})^{|\sigma|}$$

so $|N \cap SEQ_\sigma| \leqslant \lambda_\sigma^{|\sigma|} < \kappa'$. Thus since $\rho < \kappa'$ and κ' is regular,

$$|\{\mathbf{v} \in N; \ell n(\mathbf{v}) < \rho\}| < \kappa' .$$

Consequently $|\bigcup\{F(\mathbf{v}); \mathbf{v} \in N \text{ and } \ell n(\mathbf{v}) < \rho\}| < \kappa$, and so from Lemma 2.2.2 it follows that

$$|\bigcup\{S(\mathbf{v}); \mathbf{v} \in N \cap SEQ_\rho\}| = \kappa .$$

So there must be $\mathbf{v}$ in $N \cap SEQ_\rho$ with $S(\mathbf{v}) \neq \emptyset$.

Now suppose the conditions in (ii) hold. This reduces to (i), with $\lambda_\sigma = \lambda$ for all σ with $\sigma < \rho$. Note by the last condition on λ, that if $\sigma < \rho$ we have $|\sigma| < \lambda'$, so that $\lambda^{|\sigma|} = \lambda < \kappa = \kappa'$.

Finally, suppose (iii) in force. Use induction on σ to show that $|N \cap SEQ_\sigma| < \kappa$ whenever $\sigma < \rho$. If $\sigma = 0$ then $|N \cap SEQ_\sigma| = 1$. If $\sigma = \tau + 1$, since

$$\mathbf{v} \in N \cap SEQ_\sigma \Leftrightarrow \mathbf{v}\lceil\tau \in N \cap SEQ_\tau \text{ and } \mathbf{v}(\tau) < n(\mathbf{v}\lceil\tau) ,$$

we have

$$|N \cap SEQ_\sigma| \leqslant \Sigma(|n(\boldsymbol{\mu})|; \boldsymbol{\mu} \in N \cap SEQ_\tau) ,$$

so $|N \cap SEQ_\sigma| < \kappa$ by the inductive assumption and the regularity of κ. And if σ is a limit ordinal, since

$$\mathbf{v} \in N \cap SEQ_\sigma \Leftrightarrow \forall \tau < \sigma(\mathbf{v} \upharpoonright \tau \in N \cap SEQ_\tau)$$

we have

$$|N \cap SEQ_\sigma| \leqslant \Pi(|n(\boldsymbol{\mu})|;\ \exists\, \tau < \sigma(\boldsymbol{\mu} \in N \cap SEQ_\tau))\ ,$$

so that $|N \cap SEQ_\sigma| < \kappa$ by the inductive hypothesis and the inaccessibility of κ. This establishes that $|N \cap SEQ_\sigma| < \kappa$ whenever $\sigma < \rho$.

Thus $|\{\mathbf{v} \in N;\ \ell n(\mathbf{v}) < \rho\}| < \kappa$, and so $|\bigcup\{F(\mathbf{v});\ \mathbf{v} \in N \text{ and } \ell n(\mathbf{v}) < \rho\}| < \kappa$. As in part (i), it now follows that there must be $\mathbf{v}$ in $N \cap SEQ_\rho$ for which $S(\mathbf{v}) \neq \emptyset$.

We shall make immediate use of the Ramification Lemma to establish a couple of partition relations, and illustrate the method used.

Theorem 2.2.4 (GCH). *If* $\gamma < \kappa'$ *then* $\kappa^+ \to (\kappa)^2_\gamma$.

Proof. Take any set S with $|S| = \kappa^+$, and suppose a partition $[S]^2 = \bigcup\{\Delta_k;\ k < \gamma\}$ is given. We have to find a homogeneous set H with $|H| = \kappa$. Define a ramification system $\mathfrak{R}$ on S of height κ and order γ as follows. Take σ with $\sigma < \kappa$ and suppose $S(\mathbf{v})$ given for all $\mathbf{v}$ in $N \cap SEQ_\sigma$. Take $\boldsymbol{\mu}$ from $N \cap SEQ_{\sigma+1}$ such that $\boldsymbol{\mu} \upharpoonright \sigma = \mathbf{v}$. If $S(\mathbf{v}) = \emptyset$, put $F(\mathbf{v}) = \emptyset$ and $S(\boldsymbol{\mu}) = \emptyset$. If $S(\mathbf{v}) \neq \emptyset$, choose $x(\mathbf{v})$ in $S(\mathbf{v})$, put $F(\mathbf{v}) = \{x(\mathbf{v})\}$ and

$$S(\boldsymbol{\mu}) = \{y \in S(\mathbf{v}) - F(\mathbf{v});\ \{x(\mathbf{v}), y\} \in \Delta_{\boldsymbol{\mu}(\sigma)}\}\ .$$

This defines $\mathfrak{R}$.

Then $|S| = \kappa^+$; $\rho = \kappa < \kappa^+ = (\kappa^+)'$; $|F(\mathbf{v})| \leqslant 1 < \kappa^+$ and $|n(\mathbf{v})| = |\gamma|$ for all $\mathbf{v} \in N$; and if $\sigma < \kappa$ then $|\gamma|^{|\sigma|} < \kappa^+$ by GCH. Thus Theorem 2.2.3 applies, and there is $\mathbf{v}$ in $N \cap SEQ_\kappa$ with $S(\mathbf{v}) \neq \emptyset$. Choose such a $\mathbf{v}$. Since $S(\mathbf{v}) \neq \emptyset$ we must have $S(\mathbf{v} \upharpoonright \sigma) \neq \emptyset$ for each σ with $\sigma < \kappa$, and so $F(\mathbf{v} \upharpoonright \sigma) \neq \emptyset$; thus $F(\mathbf{v} \upharpoonright \sigma) = \{x(\mathbf{v} \upharpoonright \sigma)\}$. For brevity, write $x_\sigma = x(\mathbf{v} \upharpoonright \sigma)$ for σ with $\sigma < \kappa$, and note $x_\sigma \neq x_\tau$ whenever $\sigma \neq \tau$ by Lemma 2.2.2(ii).

When $\sigma < \tau < \kappa$, we have

$$\{x_\tau\} = F(\mathbf{v} \upharpoonright \tau) \subseteq S(\mathbf{v} \upharpoonright \tau) \subseteq S(\mathbf{v} \upharpoonright \sigma + 1)\ ,$$

and so $\{x_\sigma, x_\tau\} \in \Delta_{\mathbf{v}(\sigma)}$ by the definition of $S(\mathbf{v} \upharpoonright \sigma + 1)$. Since always $\mathbf{v}(\sigma) < \gamma < \kappa'$, there must be K in $[\kappa]^\kappa$ and ν with $\nu < \gamma$ such that $\mathbf{v}(\sigma) = \nu$ for all σ in K. Put $H = \{x_\sigma;\ \sigma \in K\}$, so $|H| = \kappa$ and $[H]^2 \subseteq \Delta_\nu$. Thus H is a homogeneous set of power κ as required.

If the Generalized Continuum Hypothesis is not assumed, the same method will yield the following.

Theorem 2.2.5. *For all infinite cardinals* κ, $(2^\kappa)^+ \to (\kappa^+)^2_\kappa$.

Proof. Given a partition $[S]^2 = \bigcup\{\Delta_k; k < \kappa\}$ of a set S with $|S| = (2^\kappa)^+$, define a ramification system $\mathfrak{R}$ on S of height κ^+ and order κ in a manner similar to that in the proof of Theorem 2.2.4. This time $|S| = (2^\kappa)^+$; $\rho = \kappa^+ < (2^\kappa)^+$; $|n(\mathbf{v})| = \kappa$ for all $\mathbf{v}$ in N; and if $\sigma < \rho = \kappa^+$ then $\kappa^{|\sigma|} \leqslant (2^\kappa)^\kappa = 2^\kappa < (2^\kappa)^+$, so again the Ramification Lemma applies to $\mathfrak{R}$, and the result follows as before.

We turn now to another application of the Ramification Lemma, which has some useful corollaries.

Theorem 2.2.6. *Suppose* $\lambda_1 \to (\eta_k; k < \gamma)^2$ *and* $\lambda_2 \to (\theta_l; l < \delta)^1$. *Let* κ *be a regular cardinal with* $\lambda_1 \leqslant \kappa$ *and* $\lambda_2 < \kappa$ *and suppose* $(\lambda \cdot |\delta|)^\iota < \kappa$ *whenever* $\lambda < \lambda_1$ *and* $\iota < \lambda_2$. *Then* $\kappa \to (\eta_k; k < \gamma, \theta_l; l < \delta)^2$.

Proof. Take any set S with $|S| = \kappa$ and suppose that $[S]^2$ is partitioned,

$$[S]^2 = \bigcup\{\Delta_k; k < \gamma\} \cup \bigcup\{\Gamma_l; l < \delta\}.$$

It may be assumed for all subsets H of S with $|H| \geqslant \lambda_1$ that $[H]^2 \not\subseteq \bigcup\{\Delta_k; k < \gamma\}$, since otherwise the result follows from the property $\lambda_1 \to (\eta_k; k < \gamma)^2$. It suffices to find l and H with $l < \delta$, $H \subseteq S$, $|H| = \theta_l$ and $[H]^2 \subseteq \Gamma_l$.

Define a ramification system $\mathfrak{R}$ on S of height λ_2 as follows. Take σ with $\sigma < \lambda_2$ and $\mathbf{v}$ in $N \cap SEQ_\sigma$ and suppose $S(\mathbf{v})$ already defined. If $S(\mathbf{v}) = \emptyset$, put $F(\mathbf{v}) = \emptyset$ and $n(\mathbf{v}) = 0$. Otherwise, choose as $F(\mathbf{v})$ a subset of $S(\mathbf{v})$ maximal with the property $[F(\mathbf{v})]^2 \subseteq \bigcup\{\Delta_k; k < \gamma\}$, so $|F(\mathbf{v})| < \lambda_1$. Then if $y \in S'(\mathbf{v}) - F(\mathbf{v})$, by the maximality of $F(\mathbf{v})$ there is x in $F(\mathbf{v})$ such that $\{x, y\} \in \Gamma_l$ for some l with $l < \delta$. Thus there is a decomposition

$$S(\mathbf{v}) - F(\mathbf{v}) = \bigcup\{S(\boldsymbol{\mu}); \boldsymbol{\mu} \in SEQ_{\sigma+1},\ \boldsymbol{\mu} \upharpoonright \sigma = \mathbf{v} \text{ and } \boldsymbol{\mu}(\sigma) < n(\mathbf{v})\},$$

where for $\boldsymbol{\mu}$ in $SEQ_{\sigma+1}$ with $\boldsymbol{\mu} \upharpoonright \sigma = \mathbf{v}$, for some $x(\boldsymbol{\mu})$ in $F(\mathbf{v})$ and $l(\boldsymbol{\mu})$ with $l(\boldsymbol{\mu}) < \delta$,

$$S(\boldsymbol{\mu}) = \{y \in S(\mathbf{v}) - F(\mathbf{v}); \{x(\boldsymbol{\mu}), y\} \in \Gamma_{l(\boldsymbol{\mu})}\}.$$

This defines $F(\mathbf{v})$, $n(\mathbf{v})$ and $S(\boldsymbol{\mu})$ for all $\boldsymbol{\mu}$ in $N \cap SEQ_{\sigma+1}$ with $\boldsymbol{\mu} \upharpoonright \sigma = \mathbf{v}$, and so completes the definition of $\mathfrak{R}$.

In this description, $|n(\mathbf{v})| \leqslant |F(\mathbf{v})| \cdot |\delta|$. This means that the conditions of

Theorem 2.2.3 are satisfied by $\mathfrak{R}$, and so there is $\mathbf{v}$ in $N \cap SEQ_{\lambda_2}$ for which $S(\mathbf{v}) \neq \emptyset$. So $F(\mathbf{v} \upharpoonright \sigma) \neq \emptyset$ for each σ with $\sigma < \lambda_2$. Put $x_\sigma = x(\mathbf{v} \upharpoonright \sigma + 1)$, so $x_\sigma \neq x_\tau$ when $\sigma \neq \tau$ (since $x_\sigma \in F(\mathbf{v} \upharpoonright \sigma)$ and $x_\tau \in F(\mathbf{v} \upharpoonright \tau)$).

When $\sigma < \tau < \lambda_2$, we have

$$x_\tau \in F(\mathbf{v} \upharpoonright \tau) \subseteq S(\mathbf{v} \upharpoonright \tau) \subseteq S(\mathbf{v} \upharpoonright \sigma + 1)\,,$$

and so $\{x_\sigma, x_\tau\} \in \Gamma_{l(\mathbf{v} \upharpoonright \sigma+1)}$, by the definition of $S(\mathbf{v} \upharpoonright \sigma + 1)$. However, $\lambda_2 = \bigcup\{\Lambda_l ; l < \delta\}$ where $\Lambda_l = \{\sigma < \lambda_2 ; l(\mathbf{v} \upharpoonright \sigma + 1) = l\}$. By the relation $\lambda_2 \to (\theta_l ; l < \delta)^1$, there are l_0 with $l_0 < \delta$ and a subset H_0 of λ_2 with $|H_0| = \theta_{l_0}$ and $l(\mathbf{v} \upharpoonright \sigma + 1) = l_0$ for all σ in H_0. This means that if $\sigma < \tau < \kappa$ and $\sigma \in H_0$ then $\{x_0, x_\tau\} \in \Gamma_{l_0}$. Thus if $H = \{x_\sigma ; \sigma \in H_0\}$, then $[H]^2 \subseteq \Gamma_{l_0}$. Since $|H| = \theta_{l_0}$, this proves the theorem.

Corollary 2.2.7 (GCH). *If $\delta < \kappa'$ then $\kappa^+ \to (\kappa^+, (\kappa')_\delta)^2$.*

Proof. We shall use Theorem 2.2.6 with $\kappa = \kappa^+$, with $\lambda_1 = \kappa^+$ (noting that trivially $\kappa^+ \to (\kappa^+)^2_1$) and with $\lambda_2 = \kappa'$ (noting that $\kappa' \to (\kappa')^1_\delta$). By GCH, if $\lambda < \kappa^+$ and $\iota < \kappa'$ then $(\lambda \cdot |\delta|)^\iota \leqslant \kappa^\iota = \kappa < \kappa^+$, so the conditions of Theorem 2.2.6 are satisfied, and $\kappa^+ \to (\kappa^+, (\kappa')_\delta)^2$ follows.

Corollary 2.2.8. *Suppose κ is strongly inaccessible, $\gamma < \kappa$ and $\eta_k < \kappa$ for $k < \gamma$. Then $\kappa \to (\kappa, (\eta_k ; k < \gamma))^2$.*

Proof. Since κ is regular there is a cardinal θ with $\theta < \kappa, \gamma \leqslant \theta$ and $\eta_k \leqslant \theta$ for all k. From the inaccessibility of κ, we have $\theta^+ < \kappa$. Since θ^+ is regular, $\theta^+ \to (\theta)^1_\gamma$ so in particular $\theta^+ \to (\eta_k ; k < \gamma)^1$. Apply Theorem 2.2.6 with $\lambda_1 = \kappa$ (noting that trivially $\kappa \to (\kappa)^2_1$) and $\lambda_2 = \theta^+$. Because κ is strongly inaccessible, certainly $(\lambda \cdot |\gamma|)^\iota < \kappa$ whenever $\lambda < \lambda_1$ and $\iota < \lambda_2$, so we conclude that

$$\kappa \to (\kappa, (\eta_k ; k < \gamma))^2\,.$$

We conclude this section with a proof of Ramsey's theorem [80].

Theorem 2.2.9. *For all finite m, $\aleph_0 \to (\aleph_0)^2_m$.*

Proof. Take any set S with $|S| = \aleph_0$, and suppose a partition $[S]^2 = \bigcup\{\Delta_k ; k < m\}$ is given. Define a ramification system $\mathfrak{R}$ on S of height ω as follows. Take σ with $\sigma < \omega$ and suppose $S(\mathbf{v})$ has already been defined for each $\mathbf{v}$ in $N \cap SEQ_\sigma$. If $S(\mathbf{v}) = \emptyset$, put $F(\mathbf{v}) = \emptyset$ and $n(\mathbf{v}) = 0$. Otherwise, put $n(\mathbf{v}) = m$, choose $x(\mathbf{v})$ in $S(\mathbf{v})$ and put $F(\mathbf{v}) = \{x(\mathbf{v})\}$. For $\boldsymbol{\mu}$ in $N \cap SEQ_{\sigma+1}$ with

$\mathbf{\mu} \lceil \sigma = \mathbf{v}$, define $S(\mathbf{v}) = \{x \in S(\mathbf{v}) - F(\mathbf{v}); \{x, x(\mathbf{v})\} \in \Delta_{\mathbf{\mu}(\sigma)}\}$. This defines the ramification system $\mathfrak{R}$.

Choose numbers ν_i for i with $i < \omega$ by induction so that always $S(\langle \nu_j; j < i\rangle)$ is infinite, as follows. Supposing ν_j already defined for all j with $j < i$ so that $S(\langle \nu_j; j < i\rangle)$ is infinite (this is trivial if $i = 0$), since $S(\langle \nu_j; j < i\rangle)$ is split into only finitely many classes $S(\langle \nu_0, ..., \nu_{i-1}, \nu\rangle)$ one of these must be infinite. Choose ν_i so that $S(\langle \nu_0, ..., \nu_{i-1}, \nu_i\rangle)$ is one of these. Then for the sequence $\mathbf{v} = \langle \nu_i; i < \omega\rangle$, the element $x(\mathbf{v} \lceil \sigma)$ is defined for all σ with $\sigma < \omega$. Write $x_\sigma = x(\mathbf{v} \lceil \sigma)$. Note that whenever $\sigma < \tau < \omega$ then $x_\tau \in S(\mathbf{v} \lceil \tau) \subseteq S(\mathbf{v} \lceil \sigma + 1)$, and so $\{x_\tau, x_\sigma\} \in \Delta_{\mathbf{v}(\sigma)}$ by the definition of $S(\mathbf{v} \lceil \sigma + 1)$. There must be ν with $\nu < m$ and an infinite subset K of ω such that $\mathbf{v}(\sigma) = \nu$ for all σ in K. Put $H = \{x_\sigma; \sigma \in K\}$; then $|H| = \aleph_0$ and $[H]^2 \subseteq \Delta_\nu$. So H is an infinite homogeneous subset of S, as required.

The original theorem of Ramsey asserts more than Theorem 2.2.9, namely that the relation $\aleph_0 \to (\aleph_0)^n_m$ holds for all finite m and n. This may be proved by induction on n, combining the proof just given for $n = 2$ with the method of the Stepping-up Lemma from the next section.

§3. The stepping-up lemma

This section is devoted to a theorem which enables the results obtained in the last section for $n = 2$ to be "stepped-up" to results in the case $n \geqslant 3$.

Theorem 2.3.1. (The Stepping-up Lemma). *Let* κ *and* λ *be infinite cardinals such that* $\lambda < \kappa'$. *Let* $n \geqslant 1$ *and suppose* $\lambda \to (\eta_k; k < \gamma)^n$ *where* $|\gamma|^{|\sigma|} < \kappa'$ *whenever* $\sigma < \lambda$. *Then* $\kappa \to (\eta_k \dotplus 1; k < \gamma)^{n+1}$.

Proof. Take any set S with $|S| = \kappa$ and suppose given a disjoint partition $\mathbf{\Delta} = \{\Delta_k; k < \gamma\}$ of $[S]^{n+1}$. Define a ramification system $\mathfrak{R}$ on S of height λ as follows. Take σ with $\sigma < \lambda$, and suppose $S(\mathbf{v})$ has already been defined for each $\mathbf{v}$ in $N \cap SEQ_\sigma$. If $S(\mathbf{v}) = \emptyset$, put $F(\mathbf{v}) = \emptyset$ and $n(\mathbf{v}) = 0$. Otherwise, choose $x(\mathbf{v})$ in $S(\mathbf{v})$ and put $F(\mathbf{v}) = \{x(\mathbf{v})\}$. Write $G(\mathbf{v}) = \bigcup\{F(\mathbf{v} \lceil \tau); \tau \leqslant \sigma\}$, so $|G(\mathbf{v})| \leqslant |\sigma + 1|$. Define a partition $\mathbf{\Gamma}(\mathbf{v})$ of $S(\mathbf{v}) - F(\mathbf{v})$ by asserting

$$y \equiv z (\mathrm{mod}\ \mathbf{\Gamma}(\mathbf{v})) \Leftrightarrow a \cup \{y\} \equiv a \cup \{z\} \ (\mathrm{mod}\ \mathbf{\Delta}) \text{ for all } a \text{ in } [G(\mathbf{v})]^n .$$

Put $n(\mathbf{v}) = |\mathbf{\Gamma}(\mathbf{v})|$, so $n(\mathbf{v}) \leqslant |\gamma|^{|\sigma+1|^n}$, and let $S(\mathbf{\mu})$ for $\mathbf{\mu}$ in $N \cap SEQ_{\sigma+1}$ with

$\mu\restriction\sigma = \mathbf{v}$ range over the classes of $\boldsymbol{\Gamma}(\mathbf{v})$. This completes the definition of $\mathfrak{R}$.

If $\sigma < \lambda$, put $\lambda_\sigma = |\gamma|^{|\sigma|^n}$; then $n(\mathbf{v}\restriction\tau) \leqslant \lambda_\sigma$ whenever $\tau < \sigma$. Since λ is infinite, the assumption $|\gamma|^{|\sigma|} < \kappa'$ whenever $\sigma < \lambda$ ensures that $\lambda_\sigma^{|\sigma|} < \kappa'$ for $\sigma < \lambda$. Thus the conditions of the Ramification Lemma are fulfilled by $\mathfrak{R}$. Hence there is $\mathbf{v}$ in $N \cap SEQ_\lambda$ for which $S(\mathbf{v}) \neq \emptyset$. Choose such a sequence $\mathbf{v}$; then always $S(\mathbf{v}\restriction\sigma) \neq \emptyset$ so $F(\mathbf{v}\restriction\sigma) = \{x(\mathbf{v}\restriction\sigma)\}$. Choose x_λ from $S(\mathbf{v})$, and when $\sigma < \lambda$ put $x_\sigma = x(\mathbf{v}\restriction\sigma)$. Then $x_\sigma \neq x_\tau$ when $\tau < \sigma \leqslant \lambda$.

If $\tau < \sigma < \lambda$ then $x_\sigma \in F(\mathbf{v}\restriction\sigma) \subseteq S(\mathbf{v}\restriction\sigma) \subseteq (\mathbf{v}\restriction\tau + 1)$, and $x_\lambda \in S(\mathbf{v}) \subseteq S(\mathbf{v}\restriction\tau + 1)$ so both $x_\sigma, x_\lambda \in S(\mathbf{v}\restriction\tau + 1)$. Thus both x_σ and x_λ are in the same class of $\boldsymbol{\Gamma}(\mathbf{v}\restriction\tau)$, and from the definition of $\boldsymbol{\Gamma}(\mathbf{v}\restriction\tau)$, if $\tau_1 < \tau_2 < ... < \tau_n \leqslant \tau$,

$$\{x_{\tau_1}, ..., x_{\tau_n}, x_\sigma\} \equiv \{x_{\tau_1}, ..., x_{\tau_n}, x_\lambda\} \pmod{\boldsymbol{\Delta}}.$$

Put $X = \{x_\tau; \tau < \lambda\}$; then there is a partition $[X]^n = \bigcup\{\Delta'_k; k < \lambda\}$ where

$$\Delta'_k = \{a \in [X]^n; a \cup \{x_\lambda\} \in \Delta_k\}.$$

By the relation $\lambda \to (\eta_k; k < \gamma)^n$, there are k with $k < \gamma$ and a subset H of X with $|H| = \eta_k$ such that $[H]^n \subseteq \Delta'_k$. But then $[H \cup \{x_\lambda\}]^{n+1} \subseteq \Delta_k$, and $|H \cup \{x_\lambda\}| = \eta_k \dot{+} 1$, so the theorem is proved.

The results obtained in the last section may now be extended as follows.

Theorem 2.3.2 (GCH). *Suppose* $n \geqslant 1$ *and* $\gamma < \kappa'$. *Then*

(i) $\kappa^{(n+)} \to (\kappa)^{n+1}_\gamma$,

(ii) $\kappa^{(n+)} \to (\kappa^+, (\kappa')_\gamma)^{n+1}$.

Proof. Of course, when κ is regular then (i) follows from (ii). Use induction on n. The cases $n = 1$ are Theorem 2.2.4 and Corollary 2.2.7. To go from n to $n + 1$, use Theorem 2.3.1, noting that $\kappa^{((n+1)+)}$ is regular and that when $\sigma < \kappa^{(n+)}$ by GCH,

$$|\gamma|^{|\sigma|} \leqslant |\gamma|^+ \cdot |\sigma|^+ \leqslant \kappa^{(n+)} < \kappa^{((n+1)+)}$$

Without assuming the Generalized Continuum Hypothesis, Theorem 2.2.5 steps up as follows.

Theorem 2.3.3. *If* $n \geqslant 1$ *then* $(\beth_n(\kappa))^+ \to (\kappa^+)^{n+1}_\kappa$.

Proof. Again by induction on n. The case $n = 1$ is Theorem 2.2.5. To go from n to $n + 1$, use the Stepping-up Lemma, noting that $(\beth_{n+1}(\kappa))^+$ is regular and that when $\sigma < (\beth_n(\kappa))^+$,

$$\kappa^{|\sigma|} \leqslant \kappa^{\beth_n(\kappa)} \leqslant 2^{\kappa \cdot \beth_n(\kappa)} = 2^{\beth_n(\kappa)} = \beth_{n+1}(\kappa) < (\beth_{n+1}(\kappa))^+.$$

4. Results for singular cardinals

We turn now to partition relations which hold for singular cardinals κ. A special technique, that of canonical partitions, is available which reduces the question to a consideration of partition properties of the regular cardinal κ'.

We shall assume that Generalized Continuum Hypothesis throughout. Without this assumption, the methods developed in this section work if we assume that, as well as being singular, the cardinal κ in question is a strong limit cardinal, that is, $2^\lambda < \kappa$ whenever $\lambda < \kappa$.

Definition 2.4.1. Let $\mathbf{\Delta}$ be a disjoint partition of $[S]^n$ and let $\mathcal{C}$ be a pairwise disjoint family of subsets of S. The partition $\mathbf{\Delta}$ is said to be *canonical* in $\mathcal{C}$ if for all a, b in $[\bigcup \mathcal{C}]^n$, if $|a \cap C| = |b \cap C|$ for all C in $\mathcal{C}$, then $a \equiv b \pmod{\mathbf{\Delta}}$.

The following lemma gives a method of constructing canonical partitions. See Erdös, Hajnal and Rado [38, §9].

Lemma 2.4.2 (GCH). *Let κ be singular and let $\langle \kappa_\sigma; \sigma < \kappa' \rangle$ be an increasing sequence of cardinals below κ with $\kappa = \Sigma(\kappa_\sigma; \sigma < \kappa')$. Let S be any set with $|S| = \kappa$ and take any pairwise disjoint partition $\mathbf{\Delta}$ of $[S]^n$ into fewer than κ classes. Then there is a pairwise disjoint family $\mathcal{C} = \{C_\sigma; \sigma < \kappa'\}$ of subsets of S with always $|C_\sigma| = \kappa_\sigma$ such that $\mathbf{\Delta}$ is canonical in $\mathcal{C}$.*

Proof. We shall give the proof for $n = 3$; an entirely similar proof can be given for general n. So let the disjoint partition $\mathbf{\Delta} = \{\Delta_k; k < \gamma\}$ of $[S]^3$ be given, where $\gamma < \kappa'$. To produce the required family $\{C_\sigma; \sigma < \kappa'\}$ we first find a family $\{A_\sigma; \sigma < \kappa'\}$ of subsets of S such that for each i with $1 \leqslant i \leqslant 3$,

$$\text{if } a, b \in [A_\sigma]^i \text{ and } c \in [\bigcup\{A_\tau; \tau < \sigma\}]^{3-i},$$

$$\text{then } a \cup c \equiv b \cup c \pmod{\mathbf{\Delta}}. \tag{1}$$

This family is refined to a family $\{B_\sigma; \sigma < \kappa'\}$ such that for some chosen sets $c_{\tau j}$ with $|c_{\tau j}| = j$, for all i, j with $1 \leqslant i, j \leqslant 2$ and $i + j \leqslant 3$, and for all τ with $\tau < \kappa'$

$$\text{if } a, b \in [B_\sigma]^i \text{ and } d \in [\bigcup\{B_\tau; \tau < \sigma\}]^{3-i-j},$$

$$\text{then } a \cup c_{\tau j} \cup d \equiv b \cup c_{\tau j} \cup d \pmod{\mathbf{\Delta}}. \tag{2}$$

Finally $\{B_\sigma; \sigma < \kappa'\}$ is refined to $\{C_\sigma; \sigma < \kappa'\}$ such that for some chosen sets d_ψ with $|d_\psi| = 1$, for all τ, ψ where $\tau, \psi < \kappa'$

$$\text{if } a, b \in [C_\sigma]^1,$$

$$\text{then } a \cup c_{\tau 1} \cup d_\psi \equiv b \cup c_{\tau 1} \cup d_\psi \pmod{\mathbf{\Delta}}. \tag{3}$$

Then $\mathbf{\Delta}$ is shown to be canonical in $\{C_\sigma; \sigma < \kappa'\}$.

We start by taking an arbitrary well ordering $\prec$ of S and a pairwise disjoint decomposition $S = \bigcup\{S_\sigma; \sigma < \kappa'\}$ with always $|S_\sigma| = \kappa_\sigma$. Use transfinite induction to define ordinals $\xi(\sigma)$ with $\xi(\sigma) < \kappa'$ and sets A_σ in $[S_{\xi(\sigma)}]^{\kappa\sigma}$ for σ with $\sigma < \kappa'$ as follows. Put $A(\sigma) = \bigcup\{A_\tau; \tau < \sigma\}$. For each a in $[S - A(\sigma)]^3$, say $a = \{x_1, x_2, x_3\}$ where $x_1 \prec x_2 \prec x_3$, define a function $f_a : \bigcup\{[A(\sigma)]^i; i<3\} \to \gamma$ by, for c in $[A(\sigma)]^i$,

$$f_a(c) = k \Leftrightarrow \{x_1, ..., x_{3-i}\} \cup c \in \Delta_k .$$

Now $|A(\sigma)| = \Sigma(\kappa_\tau; \tau < \sigma) \leqslant \kappa_\sigma$ so the number η of possible functions is $\gamma^{\kappa\sigma}$, so $\eta < \kappa$ by GCH. Choose $\xi(\sigma)$ such that $\xi(\sigma) > \xi(\tau)$ for all τ with $\tau < \sigma$, and also $\kappa_{\xi(\sigma)} \geqslant \eta^{+++}, \kappa_\sigma^{+++}$. It follows from Theorem 2.3.2 that $\lambda^{+++} \to (\lambda^+)^3_\lambda$, so certainly $\kappa_{\xi(\sigma)} \to (\kappa_\sigma)^3_\eta$. Hence there is A_σ in $[S_{\xi(\sigma)}]^{\kappa\sigma}$ such that f_a is the same for all a in $[A_\sigma]^3$. This defines A_σ so that (1) holds.

Choose, for $j = 1,2$, some set $c_{\tau j}$ from $[A(\tau)]^j$ for each τ with $\tau < \kappa'$. Inductively define ordinals $\pi(\sigma)$ with $\pi(\sigma) < \kappa'$ and sets B_σ in $[A_{\pi(\sigma)}]^{\kappa\sigma}$ where $\sigma < \kappa'$ as follows. With $B(\sigma) = \bigcup\{B_\tau; \tau < \sigma\}$, for each a in $[S - B(\sigma)]^2$, say $a = \{x_1, x_2\}$ where $x_1 \prec x_2$, define a function g_a mapping from $\{\langle c_{\tau j}, d\rangle;$ $d \subseteq B(\sigma), \tau < \kappa', j = 1,2$ and $|c_{\tau j} \cup d| \leqslant 2\}$ into γ by, if $|c_{\tau j} \cup d| = i$,

$$g_a(c_{\tau j}, d) = k \Leftrightarrow \{x_1, x_i\} \cup c_{\tau j} \cup d \in \Delta_k .$$

Again, the number of possible such functions is less than κ, so $\pi(\sigma)$ can be chosen so large that $\pi(\sigma) > \pi(\tau)$ whenever $\tau < \sigma$ and so that there is B_σ in $[A_{\pi(\sigma)}]^{\kappa\sigma}$ such that all a in $[B_\sigma]^2$ have the same g_a. This gives B_σ such that (2) holds.

Now choose d_τ in $[B(\tau)]^1$ for each τ, and inductively define ordinals $\rho(\sigma)$ and sets C_σ in $[B_{\rho(\sigma)}]^{\kappa\sigma}$ as follows. Put $C(\sigma) = \bigcup\{C_\tau; \tau < \sigma\}$. For each x in $S - C(\sigma)$ define a function h_x where $h_x : \{\langle c_{\tau 1}, d_\psi\rangle; \tau, \psi < \kappa'\} \to \gamma$ by

$$h_x(c_{\tau 1}, d_\psi) = k \Leftrightarrow \{x\} \cup c_{\tau 1} \cup d_\psi \in \Delta_k .$$

As before, $\rho(\sigma)$ can be chosen so large that $\rho(\sigma) > \rho(\tau)$ whenever $\tau < \sigma$ and so that there is C_σ in $[B_{\rho(\sigma)}]^{\kappa\sigma}$ such that h_x is the same for all x in C_σ. This defines C_σ such that (3) holds.

It remains to show that $\mathbf{\Delta}$ is canonical in $\mathcal{C} = \{C_\sigma; \sigma < \kappa'\}$. Take a, b in $[\bigcup \mathcal{C}]^3$ with $|a \cap C_\sigma| = |b \cap C_\sigma|$ for all σ. We must show $a \equiv b (\mathrm{mod}\ \mathbf{\Delta})$. There are at most three ordinals ζ, σ, τ with $\zeta < \sigma < \tau$ such that a meets $C_\zeta, C_\sigma, C_\tau$. Write $a = a_1 \cup a_2 \cup a_3$ where $a_1 = a \cap C_\zeta$, $a_2 = a \cap C_\sigma$, $a_3 = a \cap C_\tau$ (allowing $a_1 = \emptyset$ or $a_1 = a_2 = \emptyset$ when $a \subseteq C_\sigma \cup C_\tau$ or $a \subseteq C_\tau$). Similarly write $b = b_1 \cup b_2 \cup b_3$.

If $a = a_3$ then $a, b \in [C_\tau]^3$ so by (1), $a \equiv b$ (mod $\mathbf{\Delta}$). Otherwise, write

$c_3 = c_{\pi(\rho(\tau))j}$ where $|a_3| = j$. Since $a_3 \in C_\tau \subseteq B_{\rho(\tau)} \subseteq A_{\pi(\rho(\tau))}$ both $a_3, c_3 \in A_{\pi(\rho(\tau))}$ with $|a_3| = |c_3|$. And $a_1 \cup a_2 \subseteq A(\pi(\rho(\tau)))$ since both π and ρ are increasing functions so by (1)

$$a_1 \cup a_2 \cup a_3 \equiv a_1 \cup a_2 \cup c_3 \pmod{\boldsymbol{\Delta}} .$$

Similarly,

$$b_1 \cup b_2 \cup b_3 \equiv b_1 \cup b_2 \cup c_3 \pmod{\boldsymbol{\Delta}} .$$

If $a_1 = \emptyset$, it follows from (2) that $a_2 \cup c_3 \equiv b_2 \cup c_3 \pmod{\boldsymbol{\Delta}}$, and so $a \equiv b \pmod{\boldsymbol{\Delta}}$. The only remaining possibility is that $|a_1| = |a_2| = |a_3| = 1$. In this case, put $c_2 = d_{\rho(\sigma)}$. Here $a_2 \in C_\sigma \subseteq B_{\rho(\sigma)}$, $c_2 \in B_{\rho(\sigma)}$ and $a_1 \subseteq B(\rho(\sigma))$, so by (2)

$$a_1 \cup a_2 \cup c_3 \equiv a_1 \cup c_2 \cup c_3 \pmod{\boldsymbol{\Delta}} .$$

Similarly,

$$b_1 \cup b_2 \cup c_3 \equiv b_1 \cup c_2 \cup c_3 \pmod{\boldsymbol{\Delta}} .$$

Finally, because of (3),

$$a_1 \cup c_2 \cup c_3 \equiv b_1 \cup c_2 \cup c_3 \pmod{\boldsymbol{\Delta}} ,$$

so again $a \equiv b \pmod{\boldsymbol{\Delta}}$. Thus is all cases, $a \equiv b \pmod{\boldsymbol{\Delta}}$, and the lemma is proved.

The following remarks concerning the proof of Lemma 2.4.2 will be relevant to later applications. The pairwise disjoint decomposition $S = \bigcup\{S_\sigma; \sigma < \kappa'\}$ with $|S_\sigma| = \kappa_\sigma$ of S could have been specified in advance. In the course of the proof an increasing function $f : \kappa' \to \kappa'$ was constructed (where $f(\sigma) = \xi\pi\rho(\sigma)$) such that $C_\sigma \in [S_{f(\sigma)}]^{\kappa_\sigma}$. In particular, if with respect to the well ordering $\prec$ of S, we had specified the S_σ so that $S_\sigma \prec S_\tau$ whenever $\sigma < \tau$ then it follows that $C_\sigma \prec C_\tau$ whenever $\sigma < \tau$.

Theorem 2.4.3 (GCH). *Let* κ *be singular. Then*

$$\kappa \to (\kappa, \eta_k; k < \gamma)^2 \text{ if and only if } \kappa' \to (\kappa', \eta_k; k < \gamma)^2 .$$

Proof. For the easier direction, suppose $\kappa \to (\kappa, \eta_k; k < \gamma)^2$, and take any partition

$$[\kappa']^2 = \Delta \cup \bigcup\{\Delta_k; k < \gamma\} .$$

Take pairwise disjoint sets S_σ for σ with $\sigma < \kappa'$ such that $|S_\sigma| < \kappa$ and if $S = \bigcup\{S_\sigma; \sigma < \kappa'\}$ then $|S| = \kappa$. For k with $k < \gamma$ define

$$\Gamma_k = \{\{x, y\}; \exists\ \{\sigma, \tau\} \in [\kappa']^2 (x \in S_\sigma, y \in S_\tau \text{ and } \{x, y\} \in \Delta_k)\}\ ,$$

and put $\Gamma = [S]^2 - \bigcup\{\Gamma_k; k < \gamma\}$. Then

$$[S]^2 = \Gamma \cup \bigcup\{\Gamma_k; k < \gamma\}\ .$$

Since $|S| = \kappa$, by the assumption $\kappa \to (\kappa, \eta_k; k < \gamma)^2$, either there is H in $[S]^\kappa$ with $[H]^2 \subseteq \Gamma$, or else there are k with $k < \gamma$ and H in $[S]^{\eta_k}$ with $[H]^2 \subseteq \Gamma_k$. In either event, put $I = \{\sigma < \kappa'; H \cap S_\sigma \neq \emptyset\}$. In the first case, $|I| = \kappa'$ and $[I]^2 \subseteq \Delta$. In the second, for each σ we must have $|H \cap S_\sigma| \leqslant 1$ and so $|I| = \eta_k$, and $[I]^2 \subseteq \Delta_k$. Thus $\kappa' \to (\kappa', \eta_k; k < \gamma)^2$.

For the other direction, we shall use Lemma 2.4.2. Suppose $\kappa' \to (\kappa', \eta_k; k < \gamma)^2$, and for a set S with $|S| = \kappa$ let any partition

$$[S]^2 = \Delta \cup \bigcup\{\Delta_k; k < \gamma\}$$

be given. Suppose for each k with $k < \gamma$ that there is no subset H of S with $|H| = \eta_k$ and $[H]^2 \subseteq \Delta_k$. We must show that then there is H in $[S]^\kappa$ with $[H]^2 \subseteq \Delta$. Take a family $\mathcal{C} = \{C_\sigma; \sigma < \kappa'\}$ such that the given partition is canonical in $\mathcal{C}$. We may suppose that always $|C_\sigma| > \kappa'$.

Since $\eta_k \leqslant \kappa' < |C_\sigma|$, for each σ we have $[C_\sigma]^2 \not\subseteq \Delta_k$ for each k. Thus from the definition of canonicity, always $[C_\sigma]^2 \subseteq \Delta$. Choose C with $C \subseteq \bigcup\{C_\sigma; \sigma < \kappa'\}$ such that $|C \cap C_\sigma| = 1$ for each σ, so $|C| = \kappa'$. The partition of $[S]^2$ restricts to give a partition of $[C]^2$, and when $k < \gamma$ by assumption there is no subset H of C with $|H| = \eta_k$ for which $[H]^2 \subseteq \Delta_k$. By the relation $\kappa' \to (\kappa', \eta_k; k < \gamma)^2$, there must then be H_0 in $[C]^{\kappa'}$ with $[H_0]^2 \subseteq \Delta$. Put $H = \bigcup\{C_\sigma; C_\sigma \cap H_0 \neq \emptyset\}$, so $|H| = \kappa$. Then if $x \in H \cap C_\sigma$ and $u \in H \cap C_\tau$, since $[H_0]^2 \subseteq \Delta$, by canonicity $\{x, u\} \in \Delta$. Thus it follows that $[H]^2 \subseteq \Delta$, and the proof is complete.

Corollary 2.4.4 (GCH). *Suppose κ is singular with $\kappa' = \lambda^+$ for some λ. Then $\kappa \to (\kappa, (\lambda')_\delta)^2$ for any δ with $\delta < \lambda'$.*

Proof. From Corollary 2.2.7 we have $\lambda^+ \to (\lambda^+, (\lambda'))^2_\delta$, and the result follows from Theorem 2.4.3.

Corollary 2.4.5 (GCH). *Suppose κ is singular with $\kappa' = \lambda^+$ for some λ. If $n \geqslant 0$ and $\delta < \lambda'$, then*

$$\kappa^{(n+)} \to (\kappa, (\lambda')_\delta)^{n+2}\ .$$

Proof. By induction on n, starting from Corollary 2.4.4 and using Theorem 2.3.1.

§5. The case $n = 2$

This section starts with some special partitions of the pairs from an infinite set which serve to demonstrate the failure of certain partition relations with $n = 2$. Using these and the positive results from earlier sections, an almost complete discussion is then given of the truth of the relation $\kappa \to (\eta_k; k < \gamma)^2$.

Theorem 2.5.1. *Suppose $\gamma \geqslant 2$ and $\kappa_k < \eta_k$ for k with $k < \gamma$. If $\kappa \leqslant \Pi(\kappa_k; k < \gamma)$ then $\kappa \nrightarrow (\eta_k; k < \gamma)^2$.*

Proof. Take sets S_k with $|S_k| = \kappa_k$ for $k < \gamma$. Let S be a subset of the cartesian product $\times \{S_k; k < \gamma\}$ with $|S| = \kappa$. Define a partition $\mathbf{\Delta} = \{\Delta_k; k < \gamma\}$ of $[S]^2$ as follows: for f, g in S,

$$\{f, g\} \in \Delta_k \Leftrightarrow f(k) \neq g(k)\,.$$

If H is a subset of S homogeneous for $\mathbf{\Delta}$, say with $[H]^2 \subseteq \Delta_k$, then $|H| \leqslant |\Delta_k| \leqslant \kappa_k < \eta_k$.

Corollary 2.5.2. *If κ is any cardinal, then $2^\kappa \nrightarrow (3)^2_\kappa$.*

Corollary 2.5.3 (GCH). *Let κ be singular and suppose there are cardinals η_k where $k < \gamma$ with $2 < \eta_k < \kappa$ and $\kappa < \Pi(\eta_k; k < \gamma)$. Then $\kappa^+ \nrightarrow (\eta_k; k < \gamma)^2$.*

Proof. By virtue of Corollary 2.5.2, we need only consider the case when $\gamma < \kappa$. Since κ is singular, there is an increasing sequence $\langle \kappa_\sigma; \sigma < \kappa' \rangle$ of cardinals κ_σ all less than κ with $\kappa = \Sigma(\kappa_\sigma; \sigma < \kappa')$. We can choose inductively distinct ordinals $k(\sigma)$ where $\sigma < \kappa'$ with $k(\sigma) < \gamma$ such that $\eta_{k(\sigma)} > \kappa_\sigma$, for otherwise there would be τ with $\tau < \kappa'$ such that $\eta_k \leqslant \kappa_\tau$ whenever $k \notin \{k(\sigma); \sigma < \tau\}$. But since $\gamma < \kappa$ and $\tau < \kappa'$, by GCH this would lead to the contradiction

$$\kappa < \Pi(\eta_k; k < \gamma) \leqslant \kappa_\tau^{|\gamma|} \cdot \Pi(\eta_{k(\sigma)}; \sigma < \tau) \leqslant \kappa_\tau^{|\gamma|} \cdot \kappa^{|\tau|} \leqslant \kappa\,.$$

However, GCH implies that $\Pi(\kappa_\sigma; \sigma < \kappa') = \kappa^+$, so by Theorem 2.5.1 we have $\kappa^+ \nrightarrow (\kappa_\sigma; \sigma < \kappa')$, and consequently $\kappa^+ \nrightarrow (\eta_k; k < \gamma)$.

Corollary 2.5.4. *If κ is singular then $\kappa^+ \nrightarrow (\kappa)^2_{\kappa'}$.*

The basic method of proof of the next theorem is due to Sierpinski, who used it in what appears to have been the first proof of the relation $2^{\aleph_0} \nrightarrow (\aleph_1, \aleph_1)^2$. The more general setting appears in Erdös and Rado [28].

Theorem 2.5.5. *Let κ and λ be infinite cardinals. Suppose $\Sigma(\theta^{|\alpha|}; \alpha<\lambda)<\kappa$. Then $\theta^\lambda \nrightarrow (\kappa, \lambda^+)^2$.*

Proof. Let $\prec$ be the lexicographic ordering on the functions in ${}^\lambda\theta$; that is if f, g are distinct members of ${}^\lambda\theta$ and if α is the least ordinal where $f(\alpha)\neq g(\alpha)$, then

$$f\prec g \Leftrightarrow f(\alpha)<g(\alpha)\,.$$

Notice the following facts about well ordered and conversely well ordered subsets (under $\prec$) of ${}^\lambda\theta$:

If $X\subseteq {}^\lambda\theta$ and X is well ordered by $\prec$, then $|X|<\kappa$; (1)

If $X\subseteq {}^\lambda\theta$ and X is well ordered by $\succ$, then $|X|\leqslant\lambda$. (2)

To prove (1), suppose $X = \{f_\mu; \mu<\xi\}$ where $f_\mu\prec f_\nu$ whenever $\mu<\nu<\xi$, and suppose without loss of generality that ξ is a limit ordinal. For each μ with $\mu<\xi$ define f_μ^* in ${}^\lambda\theta$ as follows: if α is least such that $f_\mu(\alpha)\neq f_{\mu+1}(\alpha)$, then (if $\beta<\lambda$)

$$f_\mu^*(\beta) = \begin{cases} f_{\mu+1}(\beta) & \text{if } \beta\leqslant\alpha\,, \\ 0 & \text{if } \beta>\alpha\,. \end{cases}$$

Thus $f_\mu\prec f_\mu^*\preccurlyeq f_{\mu+1}$, and in fact if $\mu<\nu<\xi$ then $f_\mu^*\preccurlyeq f_{\mu+1}\preccurlyeq f_\nu\prec f_\nu^*$. Hence the map which sends f to f^* is one-to-one from X into Y, where $Y = \{f\in{}^\lambda\theta$; for some $\alpha<\lambda$, $f(\beta) = 0$ whenever $\beta\geqslant\alpha\}$. Thus $|X|\leqslant|Y|$, and $|Y| = \Sigma(\theta^{|\alpha|}; \alpha<\lambda)$ so by assumption $|Y|<\kappa$. This proves (1).

To establish (2), suppose for a contradiction that there is a subset X of ${}^\lambda\theta$ where $X = \{f_\mu; \mu<\lambda^+\}$ with $f_\mu\succ f_\nu$ whenever $\mu<\nu<\lambda^+$. We shall show that the functions in X with large μ have increasingly greater initial sections in common. More precisely, for each α with $\alpha<\lambda$ there is $\mu(\alpha)$ with $\mu(\alpha)<\lambda^+$ such that

if $\mu\geqslant\mu(\alpha)$ and $\gamma\leqslant\alpha$, then $f_\mu(\gamma) = f_{\mu(\alpha)}(\gamma)$. (3)

We shall define $\mu(\alpha)$ by transfinite induction. So let α be given (with $\alpha<\lambda$) and suppose that $\mu(\beta)$ has already been suitably defined for all β with $\beta<\alpha$. Since $\Sigma(|\mu(\beta)|; \beta<\alpha)\leqslant\lambda\cdot|\alpha| = \lambda$, there is ν with $\nu<\lambda^+$ such that $\mu(\beta)<\nu$ whenever $\beta<\alpha$. Thus if $\mu\geqslant\nu$ and $\gamma<\alpha$, then $f_\mu(\gamma) = f_\nu(\gamma)$, and if $\nu\leqslant\mu_1<\mu_2$, then $f_{\mu_1}(\alpha)\geqslant f_{\mu_2}(\alpha)$.

There must be $\mu(\alpha)$ such that $\nu\leqslant\mu(\alpha)$ and $f_\mu(\alpha) = f_{\mu(\alpha)}(\alpha)$ whenever $\mu\geqslant\mu(\alpha)$. For if not, inductively one could find ordinals $\nu(n)$ with $\nu = \nu(0)<$

$\nu(1) < \nu(2) < \ldots$ where $f_{\nu(0)}(\alpha) > f_{\nu(1)}(\alpha) > f_{\nu(2)}(\alpha) > \ldots$, giving an infinite descending chain of ordinals (all less than θ), which is impossible. This defines $\mu(\alpha)$ so that (3) holds, and completes the definition. Now

$$\Sigma(|\mu(\alpha)|; \alpha < \lambda) \leqslant \lambda^2 = \lambda ,$$

and so there is ν with $\nu < \lambda^+$ for which $\mu(\alpha) \leqslant \nu$ for all α with $\alpha < \lambda$. But then $f_\nu(\alpha) = f_{\nu+1}(\alpha)$ for all α; that is $f_\nu = f_{\nu+1}$, which is ridiculous. Thus (2) is proved.

We are now ready to prove the theorem. Take any well ordering $\leqslant$ of ${}^\lambda\theta$, and use it to define the following partition $\{\Delta_0, \Delta_1\}$ of $[{}^\lambda\theta]^2$:

$$\{f, g\} \in \Delta_0 \Leftrightarrow f \leqslant g \text{ and } f \prec g ,$$

$$\{f, g\} \in \Delta_1 \Leftrightarrow f \leqslant g \text{ and } f \succ g .$$

If X is a subset of ${}^\lambda\theta$ with $[X]^2 \subseteq \Delta_0$, then X is well ordered by $\prec$ and so $|X| < \kappa$ by (1). And if X is a subset of ${}^\lambda\theta$ with $[X]^2 \subseteq \Delta_1$, then X is conversely well ordered by $\prec$ so $|X| \leqslant \lambda$ by (2).

Corollary 2.5.6. *For all κ, if λ is the least cardinal for which $\kappa < 2^\lambda$, then $2^\lambda \nrightarrow (\kappa^+, \lambda^+)^2$.*

Proof. Apply Theorem 2.5.5 with $\theta = 2$ and κ replaced by κ^+. By choice of λ we have $\lambda \leqslant \kappa$ and $2^{|\alpha|} \leqslant \kappa$ whenever $\alpha < \lambda$. Thus $\Sigma(2^{|\alpha|}; \alpha < \lambda) \leqslant \kappa \cdot \kappa = \kappa < \kappa^+$, and the result follows.

Corollary 2.5.7 (GCH). *If κ is any cardinal, then $\kappa^+ \nrightarrow (\kappa^+, (\kappa')^+)^2$.*

Proof. From the GCH, $\Sigma(\kappa^{|\alpha|}; \alpha < \kappa') = \kappa \cdot \kappa' = \kappa < \kappa^+$ so Theorem 2.5.5 applies to show $\kappa^{\kappa'} \nrightarrow (\kappa^+, \kappa')^+)^2$.

Theorem 2.5.8. *If κ is uncountable and not strongly inaccessible then $\kappa \nrightarrow (\kappa)^2_2$.*

Proof. If κ is singular, then there is an increasing sequence $\langle \kappa_\sigma; \sigma < \kappa' \rangle$ of cardinals with always $\kappa_\sigma < \kappa$ which is cofinal in κ. Define a partition $\mathbf{\Delta} = \{\Delta_0, \Delta_1\}$ of $[\kappa]^2$ as follows: if $\alpha < \beta < \kappa$ then

$$\{\alpha, \beta\} \in \Delta_0 \Leftrightarrow \exists\, \sigma < \kappa' \, (\alpha \leqslant \kappa_\sigma < \beta) ,$$

$$\{\alpha, \beta\} \in \Delta_1 \text{ otherwise.}$$

If H is homogeneous for $\mathbf{\Delta}$, then $|H| < \kappa$. For if $[H]^2 \subseteq \Delta_0$ then H can have at most one element in $\{\alpha < \kappa; \kappa_\sigma \leqslant \alpha < \kappa_{\sigma+1}\}$ for each σ so that $|H| \leqslant \kappa' < \kappa$, and if $[H]^2 \subseteq \Delta_1$ then H must be completely contained in some $\{\alpha < \kappa; \kappa_\sigma \leqslant \alpha < \kappa_{\sigma+1}\}$ so $|H| \leqslant \kappa_{\sigma+1} < \kappa$.

If κ is regular but not strongly inaccessible then there is λ with $\lambda < \kappa \leqslant 2^\lambda$. If we take the least such λ then $\Sigma(2^{|\alpha|}; \alpha < \lambda) < \kappa$, so by Theorem 2.5.5 we have $2^\lambda \nrightarrow (\kappa, \lambda^+)^2$. In particular then, $\kappa \nrightarrow (\kappa, \kappa)^2$.

We conclude this list of negative relations with the following theorem.

Theorem 2.5.9 (GCH). *If κ is singular then $\kappa^+ \nrightarrow (\kappa^+, (3)_{\kappa'})^2$.*

Proof. By the GCH, $|[\kappa^+]^\kappa| = \kappa^+$ so we may put $[\kappa^+]^\kappa = \{A_\alpha; \alpha < \kappa^+\}$. For each α, put $\mathcal{A}_\alpha = \{A_\beta; \beta < \alpha \text{ and } A_\beta \subseteq \alpha\}$, so $|\mathcal{A}_\alpha| \leqslant \kappa$ and $[\kappa^+]^\kappa = \bigcup\{\mathcal{A}_\alpha; \alpha < \kappa^+\}$. Choose an increasing sequence of cardinals $\langle \kappa_\sigma; \sigma < \kappa' \rangle$ all less than κ such that $\kappa = \Sigma(\kappa_\sigma; \sigma < \kappa')$. Choose subfamilies $\mathcal{A}_{\alpha\sigma}$ of $\mathcal{A}_\alpha$ with $|\mathcal{A}_{\alpha\sigma}| \leqslant \kappa_\sigma$ such that $\mathcal{A}_\alpha = \bigcup\{\mathcal{A}_{\alpha\sigma}; \sigma < \kappa'\}$, for each α.

Let σ be fixed with $\sigma < \kappa'$. Use induction on α where $\alpha < \kappa^+$ to define subsets $f_\sigma(\alpha)$ of α with $|f_\sigma(\alpha)| < \kappa_{\sigma+1}$ as follows. Put $f_\sigma(\alpha) = \emptyset$ if $\alpha < \kappa$. Suppose $\kappa \leqslant \alpha < \kappa^+$ and $f_\sigma(\beta)$ has been defined for all β with $\beta < \alpha$. Since $\beta \notin \beta$ certainly $\beta \notin f_\sigma(\beta)$. Thus, in the notation of Chapter 3, § 1, we can consider $f_\sigma : \alpha \to \mathfrak{P}\alpha$ as a set mapping on α, and f_σ has order at most $\kappa_{\sigma+1}$. We appeal to Theorem 3.1.2. Since $|\mathcal{A}_{\alpha\sigma}| \leqslant \kappa_\sigma < \kappa$, there is a subset $F_{\alpha\sigma}$ of α which is free for f_σ and for which $|F_{\alpha\sigma} \cap A| = \kappa$ for each A in $\mathcal{A}_{\alpha\sigma}$. To say that $F_{\alpha\sigma}$ is *free* for f_σ means that $F_{\alpha\sigma} \cap \bigcup f_\sigma[F_{\alpha\sigma}] = \emptyset$. Choose $f_\sigma(\alpha)$ from $[F_{\alpha\sigma}]^{\leqslant \kappa_\sigma}$ so that $A \cap f_\sigma(\alpha) \neq \emptyset$ for each A in $\mathcal{A}_{\alpha\sigma}$. Then $f_\sigma(\alpha) \subseteq \alpha$ and $|f_\sigma(\alpha)| \leqslant \kappa_\sigma < \kappa_{\sigma+1}$. This completes the definition of $f_\sigma(\alpha)$.

We are now ready to define a partition

$$[\kappa^+]^2 = \Delta \cup \bigcup\{\Delta_\sigma; \sigma < \kappa'\}$$

which will prove the theorem. For σ with $\sigma < \kappa'$, if $\alpha, \beta \in \kappa^+$ with $\beta < \alpha$ then define

$$\{\alpha, \beta\} \in \Delta_\sigma \Leftrightarrow \beta \in f_\sigma(\alpha),$$

and put $\Delta = [\kappa^+]^2 - \bigcup\{\Delta_\sigma; \sigma < \kappa'\}$. Take any set H from $[\kappa^+]^{\kappa^+}$. Take A with $A \in [H]^\kappa$ so $A = A_\beta$ for some β. Since $|H| = \kappa^+$, we can choose α with $\alpha \in H$ such that $\beta < \alpha$ and $A \subseteq \alpha$. Then $A \in \mathcal{A}_{\alpha\sigma}$ for some σ with $\sigma < \kappa'$. The definition of $f_\sigma(\alpha)$ ensures that there is γ with $\gamma \in A \cap f_\sigma(\alpha)$. Then $\gamma \in H$ and $\{\alpha, \gamma\} \in \Delta_\sigma$; hence $[H]^2 \nsubseteq \Delta$. And if for any σ there were α, β, γ in κ^+, say with $\gamma < \beta < \alpha$, such that $[\{\alpha, \beta, \gamma\}]^2 \subseteq \Delta_\sigma$ then $\beta, \gamma \in f_\sigma(\alpha) \subseteq F_{\alpha\sigma}$ and $\gamma \in f_\sigma(\beta)$ so that $\gamma \in F_{\alpha\sigma} \cap \bigcup f_\sigma[F_{\alpha\sigma}]$ contrary to the fact that $F_{\alpha\sigma}$ is free for f_σ. This shows that the above partition has the correct properties to prove the theorem.

We are now ready to discuss the symbol $\kappa \to (\eta_k; k<\gamma)^2$ (where $2 \leqslant \gamma < \kappa$ and $2 < \eta_k \leqslant \kappa$). We shall assume the Generalized Continuum Hypothesis throughout. We shall restrict the discussion to the case where at most one of the η_k has value κ. For it follows from Theorem 2.5.8 that if even $\kappa \to (\kappa)_2^2$ then κ is inaccessible (or $\kappa = \aleph_0$), and so, excluding $\kappa = \aleph_0$, it is consistent with the usual axioms for set theory that there are no such cardinals. Cardinals with the property $\kappa \to (\kappa)_2^2$ are known to be weakly compact, and consequently are larger even than the first inaccessible cardinal (the last a result due to Hanf [57]).

Case A. $\eta_0 = \kappa$ and $\eta_k < \kappa$ for k with $1 \leqslant k < \gamma$.

If κ is inaccessible then $\kappa \to (\kappa, \eta_k; 1 \leqslant k < \gamma)^2$ provided $\gamma < \kappa$, by Corollary 2.2.8. This is the best possible result, because of the trivial relation $\kappa \not\to (\kappa, (3)_\kappa)^2$.

If κ is a successor cardinal, say $\kappa = \lambda^+$, then $\kappa \to (\kappa, (\lambda')_\gamma)^2$ for any γ with $\gamma < \lambda'$, by Corollary 2.2.7. This is the best possible, in view of $\lambda^+ \not\to (\lambda^+, (\lambda')^+)^2$ from Corollary 2.5.7 and $\lambda^+ \not\to (\lambda^+, (3)_{\lambda'})^2$. The last relation holds by Corollary 2.5.2 if λ is regular, and Theorem 2.5.9 otherwise.

If κ is singular, suppose first that κ' is a successor cardinal, say $\kappa' = \lambda^+$. Then by Corollary 2.4.4 we have $\kappa \to (\kappa, (\lambda')_\gamma)^2$ provided $\gamma < \lambda'$, and this is the best possible positive result. This follows since $\kappa \not\to (\kappa, (\lambda')^+)^2$ by virtue of Theorem 2.4.1 applied to Corollary 2.5.7, and $\kappa \not\to (\kappa, (3)_{\lambda'})^2$ by applying Theorem 2.4.3 to the relation $\lambda^+ \not\to (\lambda, (3)_{\lambda'})^2$ established above. On the other hand, should κ' be a non-successor cardinal then κ' is inaccessible or countable. By Theorem 2.4.3, the relation $\kappa \to (\kappa, \eta_k; 1 \leqslant k < \gamma)^2$ is equivalent to $\kappa' \to (\kappa', \eta_k; 1 \leqslant k < \gamma)^2$. This second relation is false if any of the η_k exceed κ', involves the relation $\kappa' \to (\kappa')_2^2$ if any η_k is κ', and has been discussed above otherwise.

Case B. $\eta_k < \kappa$ whenever $k < \gamma$.

This case is completely settled by the following theorem.

Theorem 2.5.10 (GCH). *Suppose* $\eta_k < \kappa$ *whenever* $k < \gamma$. *Then* $\kappa \to (\eta_k; k<\gamma)^2$ *if and only if* $\Pi(\eta_k; k<\gamma) < \kappa$.

Proof. Suppose $\Pi(\eta_k; k<\gamma) < \kappa$. Then certainly $\gamma < \kappa$ and $\eta_k < \kappa$ for all k.

If κ is inaccessible, by Corollary 2.2.8 we have $\kappa \to (\kappa, \eta_k; k<\gamma)^2$ so indeed $\kappa \to (\eta_k; k<\gamma)^2$.

If κ is singular, in fact the η_k must be bounded away from κ. For suppose some set $\{\eta_k; k \in K\}$ is cofinal in κ, where we may suppose $|K| = \kappa'$. From König's Lemma follows the contradiction

$$\kappa \leqslant \Sigma(\eta_k; k \in K) < \Pi(\eta_k; k \in K) \leqslant \Pi(\eta_k; k < \gamma) < \kappa .$$

Thus there is θ with $\theta < \kappa$ such that $\eta_k \leqslant \theta$ whenever $k < \gamma$, and we may suppose in addition that $\gamma \leqslant \theta$. From Theorem 2.2.4 it follows that $\theta^{++} \to (\theta^+)^2_\theta$, and hence $\kappa \to (\eta_k; k < \gamma)^2$ since $\theta^{++} < \kappa$.

Thus we may suppose κ is a successor cardinal, say $\kappa = \lambda^+$. Since $\Pi(\eta_k; k < \gamma) \leqslant \lambda$, it follows that $\gamma < \lambda$ and always $\eta_k \leqslant \lambda$. If λ is regular, by Theorem 2.2.4 we have that $\lambda^+ \to (\lambda)^2_\gamma$ provided $\gamma < \lambda$, and so it follows that $\kappa \to (\eta_k; k < \gamma)^2$. If λ is singular, since $\Pi(\eta_k; k < \gamma) \leqslant \lambda$ certainly $|\{k; \eta_k = \lambda\}| < \lambda'$. Moreover, there can be no set $\{\eta_k; k \in K\}$ of those η_k with $\eta_k < \lambda$ which is cofinal in λ, since otherwise (assuming $|K| = \kappa'$) we have the contradiction

$$\lambda \leqslant \Sigma(\eta_k; k \in K) < \Pi(\eta_k; k \in K) \leqslant \Pi(\eta_k; k < \gamma) \leqslant \lambda .$$

Thus there is θ with $\theta < \lambda$ such that $\eta_k \leqslant \theta$ whenever $\eta_k < \lambda$, and we may suppose also $\lambda' \leqslant \theta$ and $\gamma \leqslant \theta$. So to show $\kappa \to (\eta_k; k < \gamma)$ it would be enough to establish that $\lambda^+ \to ((\theta)_\gamma, (\lambda)_\delta)^2$, where $\gamma < \lambda$, $\delta < \lambda'$. However, from Theorem 2.2.4 it follows that

$$\lambda^+ \to (\lambda, (\lambda)_\delta)^2 \text{ and } \theta^{++} \to (\theta^+)^2_\theta, \text{ so } \lambda \to (\theta)^2_\gamma ,$$

whence $\lambda^+ \to ((\theta)_\gamma, (\lambda)_\delta)^2$ by Theorem 2.1.2. This completes the proof of the sufficiency of the condition $\Pi(\eta_k; k < \gamma) < \kappa$.

Suppose now $\Pi(\eta_k; k < \gamma) \geqslant \kappa$ (where $\eta_k < \kappa$), and we show that $\kappa \nrightarrow (\eta_k; k < \gamma)^2$.

If κ is inaccessible, we must have $|\gamma| \geqslant \kappa$ and the result follows from the trivial relation $\kappa \nrightarrow (3)^2_\kappa$.

If κ is singular, suppose first that some set $\{\eta_k; k \in K\}$ is cofinal in κ, where we may take $|K| = \kappa'$. Then

$$\kappa = \Sigma(\eta_k; k \in K) < \Pi(\eta_k; k \in K) \leqslant \Pi(\eta_k; k \in \gamma) .$$

By Corollary 2.5.3 then $\kappa^+ \nrightarrow (\eta_k; k < \gamma)^2$, so all the more $\kappa \nrightarrow (\eta_k; k < \gamma)^2$. If no such set of the η_k is cofinal in κ, there is κ_0 with $\kappa_0 < \kappa$ such that always $\eta_k \leqslant \kappa_0$. Then $\Pi(\eta_k; k < \gamma) \leqslant \kappa_0^{|\gamma|}$, so $|\gamma| \geqslant \kappa$ and $\kappa \nrightarrow (\eta_k; k < \gamma)^2$.

Finally, suppose κ is a successor cardinal, say $\kappa = \lambda^+$. If λ is regular, $\Pi(\eta_k; k < \gamma) \leqslant \lambda^{|\gamma|}$ so $|\gamma| \geqslant \lambda$. The result follows from the relation $\lambda^+ \nrightarrow (3)^2_\lambda$, given by Corollary 2.5.2. So we may further suppose that λ is singular. If $|\{k < \gamma; \eta_k = \lambda\}| \geqslant \lambda'$, since by Corollary 2.5.4 we have $\lambda^+ \nrightarrow (\lambda)^2_{\lambda'}$, it follows that $\kappa \nrightarrow (\eta_k; k < \gamma)^2$. Thus we may assume $|\{k < \gamma; \eta_k = \lambda\}| = \theta$, where $\theta < \lambda'$. As before, if there is some set $\{\eta_k; k \in K\}$ with $\eta_k < \lambda$ for $k \in K$ which is cofinal in λ then $\lambda < \Pi(\eta_k; k < \gamma)$, so $\kappa \nrightarrow (\eta_k; k < \gamma)^2$ by Corollary 2.5.3. And if there is no such set of the η_k cofinal in λ, there must be λ_0 with $\lambda_0 < \lambda$ and $\eta_k \leqslant \lambda_0$ whenever $\eta_k \neq \lambda$. Since $\Pi(\eta_k; k < \gamma) \leqslant \lambda^\theta \cdot \lambda_0^{|\gamma|}$ and $\lambda^\theta = \lambda$ it follows that $|\gamma| \geqslant \lambda$. However, $\lambda^+ \nrightarrow (3)^2_\lambda$ from Corollary 2.5.2, so here also $\kappa \nrightarrow (\eta_k; k < \gamma)^2$. This completes the proof of Theorem 2.5.10.

§6. Results when $n \geq 3$

The situation with regard to partitions of pairs now being settled, we can turn our attention to partitions of the n-element subsets for n with $n \geq 3$. Here a few unsolved problems remain. The Stepping-up Lemma (Theorem 2.3.1) seems to be essentially the only method of proving positive relations in this case.

If GCH is assumed, the Stepping-up Lemma ensures that if $\kappa \to (\eta_k; k < \gamma)^n$ then also $2^\kappa \to (\eta_k \dotplus 1; k < \gamma)^{n+1}$. One seeks to prove the converse to this, namely that if $\kappa \nrightarrow (\eta_k; k < \gamma)^n$ then $2^\kappa \nrightarrow (\eta_k \dotplus 1; k < \gamma)^{n+1}$. This has not been done in general; various restrictions have to be imposed on the η_k. We shall not aim to give a full discussion even of the known results. Apart from the first few theorems, for simplicity we shall limit the discussion to the case $n = 3$. Many of the results for $n = 3$ can be extended without undue difficulty to higher values of n; on the other hand further new methods are needed as well. The reader is referred to Erdös, Hajnal and Rado [38, §§ 13, 14] for details.

We shall start by considering relations of the form $\kappa \to (\kappa, \eta_k; k < \gamma)^n$ where $n \geq 3$. In Theorem 2.6.2 below we shall show that even the apparently weakest non-trivial relation of this form, namely $\kappa \to (\kappa, 4)^3$, implies that $\kappa \to (\kappa)^2_2$ and so by Theorem 2.5.8 κ must be strongly inaccessible. Thus for uncountable accessible cardinals κ, no non-trivial relation of the above form is true, as opposed to the positive results which are possible when $n = 2$. (See Case A of the discussion in the preceeding section.)

Lemma 2.6.1. *If κ is singular then $\kappa \nrightarrow (\kappa, 4)^3$.*

Proof. Take any set S of power κ, and write S as a disjoint union, $S = \bigcup\{S_\sigma; \sigma < \kappa'\}$ where always $|S_\sigma| < \kappa$. Define a partition $\{\Delta_0, \Delta_1\}$ of $[S]^3$ as follows: for a in $[S]^3$,

$$a \in \Delta_1 \Leftrightarrow (\exists\ \sigma, \tau < \kappa')\,(\sigma < \tau \text{ and } |a \cap S_\sigma| = 2 \text{ and } |a \cap S_\tau| = 1)\,,$$

$$a \in \Delta_0 \quad \text{otherwise.}$$

If $H \in [S]^4$ it is easy to see that $[H]^3 \not\subseteq \Delta_1$. And if H is a subset of S with $[H]^3 \subseteq \Delta_0$, whenever $H \cap S_\tau \neq \emptyset$ then $|H \cap S_\sigma| \leq 1$ for all σ with $\sigma < \tau$. It follows that $|H| < \kappa'$ or $|H| < \kappa' \dotplus |S_\tau|$ for some τ; in any event $|H| < \kappa$.

Theorem 2.6.2. *Suppose for some n with $n \geq 3$ that $\kappa \to (\kappa, n+1)^n$. Then $\kappa \to (\kappa)^2_2$.*

Proof. If $\kappa \to (\kappa, n+1)^n$ then clearly $\kappa \to (\kappa, 4)^3$ and so it is enough to show that if $\kappa \to (\kappa, 4)^3$ then $\kappa \to (\kappa)^2_2$. So suppose $\kappa \to (\kappa, 4)^3$. Take any disjoint partition $[\kappa]^2 = \Delta_0 \cup \Delta_1$ and suppose that there is no set I in $[\kappa]^\kappa$ such that $[I]^2 \subseteq \Delta_1$. We must show that there is I in $[\kappa]^\kappa$ with $[I]^2 \subseteq \Delta_0$.

Define a disjoint partition $[\kappa]^3 = \Gamma_0 \cup \Gamma_1$ as follows: for α, β, γ from κ with $\alpha < \beta < \gamma$,

$$\{\alpha, \beta, \gamma\} \in \Gamma_0 \Leftrightarrow \{\alpha, \beta\} \in \Delta_0 \text{ or } \{\beta, \gamma\} \in \Delta_1 ,$$

$$\{\alpha, \beta, \gamma\} \in \Gamma_1 \Leftrightarrow \{\alpha, \beta\} \in \Delta_1 \text{ and } \{\beta, \gamma\} \in \Delta_0 .$$

Suppose $H = \{\alpha, \beta, \gamma, \delta\}$ is a subset of κ with $[H]^3 \subseteq \Gamma_1$. If we assume $\alpha < \beta < \gamma < \delta$, since both $\{\alpha, \beta, \gamma\} \in \Gamma_1$ and $\{\beta, \gamma, \delta\} \in \Gamma_1$ we find both $\{\beta, \gamma\} \in \Delta_0$ and $\{\beta, \gamma\} \in \Delta_1$, which is impossible. So there is no such set H.

From the relation $\kappa \to (\kappa, 4)^3$ it follows that there must be then a set H from $[\kappa]^\kappa$ with $[H]^3 \subseteq \Gamma_0$. For α from H, put

$$H_1(\alpha) = \{\beta \in H; \alpha < \beta \text{ and } \{\alpha, \beta\} \in \Delta_1\} .$$

Take β, γ from $H_1(\alpha)$. If $\{\beta, \gamma\} \in \Delta_0$ then $\{\alpha, \beta, \gamma\} \in \Gamma_1$ contradicting that $[H]^3 \subseteq \Gamma_0$, and so $\{\beta, \gamma\} \in \Delta_1$. Thus $[H_1(\alpha)]^2 \subseteq \Delta_1$ and hence $|H_1(\alpha)| < \kappa$. Inductively choose elements α_ν for ν with $\nu < \kappa$ so that

$$\alpha_\nu \in H - (\{\alpha_\mu; \mu < \nu\} \cup \bigcup\{H_1(\alpha_\mu); \mu < \nu\}) .$$

Since $|H| = \kappa$ and κ is regular (by Lemma 2.6.1) such a choice is always possible. And if $I = \{\alpha_\nu; \nu < \kappa\}$ then $I \in [\kappa]^\kappa$ and $[I]^2 \subseteq \Delta_0$, as required.

Theorem 2.6.2 is a result due to Hajnal [54, Theorem 2].

We shall conclude this part of the discussion by stating the following theorem.

Theorem 2.6.3. *If $n \geqslant 3$ and $\gamma < \kappa$ then $\kappa \to (\kappa)^n_\gamma$ if and only if $\kappa \to (\kappa)^2_2$.*

This theorem was first stated in Rowbottom [81]. For a detailed proof see, for example, Kleinberg [61]. From Theorems 2.6.2 and 2.6.3 it follows that all non-trivial relations of the form $\kappa \to (\kappa, \eta_k; k < \gamma)^n$ where $n \geqslant 3$ are equivalent to the relation $\kappa \to (\kappa)^2_2$.

We now turn to relations of the form $\kappa \to (\eta_k; k < \gamma)^n$ where $\eta_k < \kappa$ for all k. There is a simple result first.

Theorem 2.6.4. *Suppose $\kappa \nrightarrow (\eta_k; k < \gamma)^n$ where all the η_k are finite. Then $\kappa^+ \nrightarrow (\eta_k + 1; k < \gamma)^{n+1}$.*

Proof. Assume $\kappa \nrightarrow (\eta_k; k < \gamma)^n$. Let S be any set of power κ^+, and take a well ordering $<$ of S of order type κ^+. For each x in S, put $L(x) = \{y \in S; y < x\}$, so always $|L(x)| \leqslant \kappa$. Thus for each x there is a partition $\mathbf{\Delta}(x) = \{\Delta_k(x); k < \gamma\}$ of $[L(x)]^n$ which for each k has no set H with $H \in [L(x)]^{\eta_k}$ such that $[H]^n \subseteq \Delta_k(x)$. Define a partition $\mathbf{\Delta} = \{\Delta_k; k < \gamma\}$ of $[S]^{n+1}$ as follows: if $x_1, ..., x_{n+1} \in S$ with $x_1 < ... < x_{n+1}$, then

$$\{x_1, ..., x_{n+1}\} \in \Delta_k \Leftrightarrow \{x_1, ..., x_n\} \in \Delta_k(x_{n+1}) .$$

For any k, take a finite subset H of S and suppose $[H]^{n+1} \subseteq \Delta_k$. Let x be the largest element in H. Then $[H - \{x\}]^n \subseteq \Delta_k(x)$, so by the choice of $\mathbf{\Delta}(x)$ it follows that $|H - \{x\}| < \eta_k$. Thus $|H| < \eta_k + 1$, and the theorem is proved.

Starting from Corollary 2.5.2, namely that $2^\kappa \nrightarrow (3)^2_\kappa$, repeated applications of Theorem 2.6.4 yield that $\kappa^{(n+)} \nrightarrow (n + 2)^{n+1}_\kappa$ (when $n \geqslant 1$). This shows that (assuming the GCH) the result $\kappa^{(n+)} \to (\kappa)^{n+1}_\gamma$ whenever $\gamma < \kappa'$ from Theorem 2.3.2 is the best possible in γ (at least for κ regular). With a different method (see Corollary 2.6.8 for the case $n = 2$) the related negative relation $\beth_n(\kappa) \nrightarrow (n + 2)^{n+1}_\kappa$ can be established. This is the same as the relation above if the GCH is assumed, and is otherwise stronger. It should be compared with the relation $(\beth_n(\kappa))^+ \to (\kappa^+)^{n+1}_\kappa$ of Theorem 2.3.3.

For the rest of this section, we shall be considering partitions of ${}^\kappa 2$. As before, let $\prec$ denote the lexicographic ordering of ${}^\kappa 2$. For distinct elements f, g in ${}^\kappa 2$, let $\delta(f, g)$ be the least α where $f(\alpha) \neq g(\alpha)$. Then if $f \prec g \prec h$, certainly $\delta(f, g) \neq \delta(g, h)$. Choose and fix a well ordering $\leqslant$ of ${}^\kappa 2$. The notation $\{f, g, h\}_\leqslant$ will be used to indicate that in the set $\{f, g, h\}$ the relation $f \leqslant g \leqslant h$ holds.

We introduce the following notation:

$$K_{00} = \{\{f, g, h\}_\leqslant \subseteq {}^\kappa 2\,; f \prec g \text{ and } g \prec h\} ,$$

$$K_{01} = \{\{f, g, h\}_\leqslant \subseteq {}^\kappa 2; f \prec g \text{ and } h \prec g\} ,$$

$$K_{10} = \{\{f, g, h\}_\leqslant \subseteq {}^\kappa 2; g \prec f \text{ and } g \prec h\} ,$$

$$K_{11} = \{\{f, g, h\}_\leqslant \subseteq {}^\kappa 2; g \prec f \text{ and } h \prec g\} ,$$

$$K = K_{00} \cup K_{11} ,$$

$$P_0 = \{\{f, g, h\}_\leqslant \subseteq K; \delta(f, g) < \delta(g, h)\} ,$$

$$P_1 = \{\{f, g, h\}_\leqslant \subseteq K; \delta(f, g) > \delta(g, h)\} .$$

There are a number of simple observations to be made.

Firstly, if $[X]^3 \subseteq K$ then either $[X]^3 \subseteq K_{00}$ or $[X]^3 \subseteq K_{11}$, so that X is either well ordered or conversely well ordered by the lexicographic ordering $\prec$. For we may suppose $|X| \geqslant 4$, and take distinct sets a, b from $[X]^3$. Let $a \cup b = \{f_0, f_1, ..., f_p\}_{\leqslant}$, so $p \geqslant 3$. Since $[X]^3 \subseteq K$, always $\{f_q, f_{q+1}, f_{q+2}\} \in K$, for $q = 0, 1, ..., p-2$. Thus either $f_q \prec f_{q+1} \prec f_{q+2}$ or $f_q \succ f_{q+1} \succ f_{q+2}$, and it follows that either $f_0 \prec f_1 \prec ... \prec f_p$ or $f_0 \succ f_1 \succ ... \succ f_p$. Thus both a, b are in K_{00} or both a, b are in K_{11}, and the result holds.

Suppose X is an infinite subset of ${}^{\kappa}2$. say $|X| = \lambda$, with either $[X]^3 \cap K_{01} = \emptyset$ or $[X]^3 \cap K_{10} = \emptyset$. Then there is X' in $[X]^{\lambda}$ with $[X']^3 \subseteq K$. For we may suppose $X = \{f_\alpha; \alpha < \lambda\}$ where $f_\alpha \leqslant f_\beta$ whenever $\alpha < \beta$. Assume in fact $[X]^3 \cap K_{01} = \emptyset$. Suppose there are α, β with $a < \beta < \lambda$ for which $f_\alpha \prec f_\beta$. Take any γ, δ with $\beta < \gamma < \delta < \lambda$. Then $f_\beta \prec f_\gamma$ since otherwise $\{f_\alpha, f_\beta, f_\gamma\} \in$ $\in [X]^3 \cap K_{01}$, and so $f_\gamma \prec f_\delta$ since otherwise $\{f_\beta, f_\gamma, f_\delta\} \in [X]^3 \cap K_{01}$. Thus on the set $X' = \{f_\gamma; \beta \leqslant \gamma < \lambda\}$ the orders $\leqslant$ and $\prec$ agree, so $[X']^3$ $\subseteq K_{00} \subseteq K$. We are left with the case that always $f_\alpha \succ f_\beta$ whenever $\alpha < \beta < \lambda$. But then the orders $\leqslant$ and $\succ$ agree on X, so $[X]^3 \subseteq K_{11} \subseteq K$. The argument is similar if $[x]^3 \cap K_{10} = \emptyset$.

The last remark is not quite as easy to prove, so we state it as a lemma.

Lemma 2.6.5. *Suppose $|X| \geqslant \lambda$ where λ is regular, and $[X]^3 \subseteq K$. Then there is X' in $[X]^{\lambda}$ with $[X']^3 \subseteq P_0$.*

Proof. Again we may suppose $X = \{f_\alpha; \alpha < \lambda\}$ where $f_\alpha \leqslant f_\beta$ whenever $\alpha < \beta < \lambda$. Use transfinite induction to define ordinals $\sigma(\alpha)$ where $\alpha < \kappa$ with always $\sigma(\alpha) < \lambda$ as follows:

(i) $\sigma(0) = 0$.

(ii) Let $\xi(\alpha)$ be the least value of $\delta(f_{\sigma(\alpha)}, f_\beta)$ for β with $\beta > \sigma(\alpha)$, then $\sigma(\alpha+1)$ is to be the least β with $\beta > \sigma(\alpha)$ for which $\delta(f_{\sigma(\alpha)}, f_\beta) = \xi(\alpha)$.

(iii) If α is a limit ordinal, $\sigma(\alpha) = \bigcup\{\sigma(\gamma); \gamma < \alpha\}$; the regularity of λ ensures that $\sigma(\alpha) < \lambda$.

We shall prove the following:

$$\text{if } \gamma \geqslant \sigma(\alpha+1), \text{ then } \delta(f_{\sigma(\alpha)}, f_\gamma) = \xi(\alpha)\,, \tag{1}$$

$$\text{if } \alpha < \beta, \text{ then } \xi(\alpha) < \xi(\beta)\,. \tag{2}$$

From (1) and (2) it follows that

$$\text{if } \alpha < \beta < \gamma < \lambda, \text{ then } \delta(f_{\sigma(\alpha)}, f_{\sigma(\beta)}) < \delta(f_{\sigma(\beta)}, f_{\sigma(\gamma)})\,, \tag{3}$$

for if $\alpha < \beta < \gamma$ then $\sigma(\beta) \geqslant \sigma(\alpha+1)$ and $\sigma(\gamma) \geqslant \sigma(\beta+1)$, so from (1) it follows that $\delta(f_{\sigma(\alpha)}, f_{\sigma(\beta)}) = \xi(\alpha)$ and $\delta(f_{\sigma(\beta)}, f_{\sigma(\gamma)}) = \xi(\beta)$. Thus (3) follows from

(2). Hence if

$$X' = \{f_{\sigma(\alpha)}; \alpha < \lambda\},$$

then $X' \in [X]^\lambda$ and $[X']^3 \subseteq P_0$, and this would establish the lemma.

To prove (1), take γ with $\gamma > \sigma(\alpha+1)$ (the case $\gamma = \sigma(\alpha+1)$ being trivial). Certainly $\gamma > \sigma(\alpha)$ so the definition of $\xi(\alpha)$ ensures that $\xi(\alpha) \leqslant \delta(f_{\sigma(\alpha)}, f_\gamma)$. If in fact $\xi(\alpha) < \delta(f_{\sigma(\alpha)}, f_\gamma)$, then $f_{\sigma(\alpha)}(\xi) = f_\gamma(\xi)$ certainly when $\xi \leqslant \xi(\alpha)$. Since $f_{\sigma(\alpha)}(\xi) = f_{\sigma(\alpha+1)}(\xi)$ when $\xi < \xi(\alpha)$ and $f_{\sigma(\alpha)}(\xi(\alpha)) \neq f_{\sigma(\alpha+1)}(\xi(\alpha))$, we have $f_{\sigma(\alpha+1)}(\xi) = f_\gamma(\xi)$ when $\xi < \xi(\alpha)$ and $f_{\sigma(\alpha+1)}(\xi(\alpha)) \neq f_\gamma(\xi(\alpha))$. Thus $\delta(f_{\sigma(\alpha+1)}, f_\gamma) = \xi(\alpha)$. So in fact

$$\delta(f_{\sigma(\alpha)}, f_{\sigma(\alpha+1)}) = \delta(f_{\sigma(\alpha+1)}, f_\gamma)\,.$$

This is impossible when $f_{\sigma(\alpha)} \prec f_{\sigma(\alpha+1)} \prec f_\gamma$ or when $f_{\sigma(\alpha)} \succ f_{\sigma(\alpha+1)} \succ f_\gamma$ (using the first remark above) and so cannot happen since $[X]^3 \subseteq K_{00}$ or $[X]^3 \subseteq K_{11}$. We conclude that $\xi(\alpha) \nless \delta(f_{\sigma(\alpha)}, f_\gamma)$, and (1) is proved.

To establish (2), it is enough to observe that $\xi(\alpha) < \xi(\alpha+1)$ and that if α is a limit ordinal then $\xi(\gamma) < \xi(\alpha)$ whenever $\gamma < \alpha$. For the first, from (1) we have $\delta(f_{\sigma(\alpha)}, f_\gamma) = \delta(f_{\sigma(\alpha)}, f_{\sigma(\alpha+1)}) = \xi(\alpha)$ whenever $\gamma > \sigma(\alpha+1)$. Hence $f_\gamma(\xi) = f_{\sigma(\alpha)}(\xi) = f_{\sigma(\alpha+1)}(\xi)$ when $\xi < \xi(\alpha)$, and $f_\gamma(\xi(\alpha)) \neq f_{\sigma(\alpha)}(\xi(\alpha))$, $f_{\sigma(\alpha+1)}(\xi(\alpha)) \neq f_{\sigma(\alpha)}(\xi(\alpha))$ so $f_\gamma(\xi(\alpha)) = f_{\sigma(\alpha+1)}(\xi(\alpha))$ (because $f(\xi(\alpha))$ takes only the values 0 or 1). Thus the definition of $\xi(\alpha+1)$ ensures that $\xi(\alpha) < \xi(\alpha+1)$. And if $\gamma < \alpha$ where α is a limit ordinal, whenever $\beta > \sigma(\alpha)$ both $\beta, \sigma(\alpha) > \sigma(\gamma+2)$ so from (1) we have $\delta(f_{\sigma(\gamma+1)}, f_\beta) = \xi(\gamma+1) = \delta(f_{\sigma(\gamma+1)}, f_{\sigma(\alpha)})$. Thus $f_\beta(\xi) = f_{\sigma(\gamma+1)}(\xi) = f_{\sigma(\alpha)}(\xi)$ whenever $\xi < \xi(\gamma+1)$. Hence $\xi(\alpha) \geqslant \xi(\gamma+1) > \xi(\gamma)$, so that (2) holds. This completes the proof.

We are now ready to state and prove three theorems which give part-converses to the Stepping-up Lemma of §3.

Theorem 2.6.6. *Suppose η_0, η_1 are infinite with η_0 regular. If $\kappa \nrightarrow (\eta_k; k < \gamma)^2$ then $2^\kappa \nrightarrow (\eta_k \dotplus 1; k < \gamma)^3$.*

Proof. From the assumption $\kappa \nrightarrow (\eta_k; k < \gamma)^2$, there is a partition $\mathbf{\Delta} = \{\Delta_k; k < \gamma\}$ of $[\kappa]^2$ such that for each k with $k < \gamma$, there is no H in $[\kappa]^{\eta_k}$ with $[H]^2 \subseteq \Delta_k$. We shall use this partition to construct a partition of $[{}^\kappa 2]^3$. For each k, put

$$\Delta_k^* = \{\{f, g, h\}_\leqslant \in P_0;\ \{\delta(f, g), \delta(g, h)\} \in \Delta_k\}\,.$$

Define a partition $\{\Gamma_k; k < \gamma\}$ of $[^\kappa 2]^3$ as follows:

$$\Gamma_k = \Delta_k^* \text{ if } 2 \leq k < \gamma ,$$

$$\Gamma_1 = K_{01} \cup \Delta_1^* ,$$

$$\Gamma_0 = [^\kappa 2]^3 - \bigcup\{\Gamma_k; 1 \leq k < \gamma\} .$$

Suppose there was X in $[^\kappa 2]^{\eta_1}$ with $[X]^3 \subseteq \Gamma_1$. Then $[X]^3 \cap K_{10} \neq \emptyset$, so there would be X' in $[X]^{\eta_1}$ with $[X']^3 \subseteq K$. Since $K \cap K_{01} = \emptyset$ and $P_0 \subseteq K$ we would have $[X']^3 \subseteq \Delta_1^*$. Likewise, suppose there was X in $[^\kappa 2]^{\eta_0}$ with $[X]^3 \subseteq \Gamma_0$. Then $[X]^3 \cap K_{01} = \emptyset$, so there would be X' in $[X]^{\eta_0}$ with $[X']^3 \subseteq K$. By Lemma 2.6.5 there would be further X'' in $[X']^{\eta_0}$ with $[X'']^3 \subseteq P_0$. But then we would have $[X'']^3 \subseteq \Delta_0^*$, since $[X'']^3 \subseteq \Gamma_0 \cap P_0 \subseteq \Delta_0^*$.

So to prove the theorem, it is enough to show:

$$\text{for every } k \text{ with } k < \gamma, \text{ if } [X]^3 \subseteq \Delta_k^* \text{ then } |X| < \eta_k \dotplus 1 . \qquad (1)$$

Suppose on the contrary that for some k there is X with $X \in [^\kappa 2]^{\eta_k}$ and $[X]^3 \subseteq \Delta_k^*$. We may suppose $X = \{f_\alpha; \alpha < \eta_k \dotplus 1\}$ where $f_\alpha \ll f_\beta$ whenever $\alpha < \beta$. Since $[X]^3 \subseteq P_0 \subseteq K$, we know X is either well ordered or conversely well ordered by $\prec$, and $\delta(f_\alpha, f_\beta) < \delta(f_\beta, f_\gamma)$ whenever $\alpha < \beta < \gamma$. From this it is easy to see that if $\alpha < \beta < \gamma$ then $\delta(f_\alpha, f_\beta) = \delta(f_\alpha, f_\gamma)$. Consequently, if $\alpha < \beta$ then $\{\delta(f_\alpha, f_{\alpha+1}), \delta(f_\beta, f_{\beta+1})\} = \{\delta(f_\alpha, f_\beta), \delta(f_\beta, f_{\beta+1})\}$, and so $\{\delta(f_\alpha, f_{\alpha+1}), \delta(f_\beta, f_{\beta+1})\} \in \Delta_k$, since $\{f_\alpha, f_\beta, f_{\beta+1}\} \in \Delta_k^*$. Put $H = \{\delta(f_\alpha, f_{\alpha+1}); \alpha < \eta_k\}$, so $[H]^2 \subseteq \Delta_k$. However $H \in [\kappa]^{\eta_k}$, and this contradicts the choice of the partition $\mathbf{\Delta}$. This proves (1), and completes the proof of Theorem 2.6.6.

Theorem 2.6.7. *Suppose η_0 is infinite and regular. If $\kappa \nrightarrow (\eta_k; k < \gamma)^2$ then $2^\kappa \nrightarrow (4, \eta_k \dotplus 1; k < \gamma)^3$.*

Proof. Take a partition $\mathbf{\Delta} = \{\Delta_k; k < \gamma\}$ of $[\kappa]^2$ as in the proof of Theorem 2.6.6, and define Δ_k^* as above. Consider the partition $\{\Gamma\} \cup \{\Gamma_k; k < \gamma\}$ of $[^\kappa 2]^3$ where:

$$\Gamma = K_{01} ,$$

$$\Gamma_k = \Delta_k^* \text{ if } 1 \leq k < \gamma ,$$

$$\Gamma_0 = [^\kappa 2]^3 - (\Gamma \cup \bigcup\{\Gamma_k; 1 \leq k < \gamma\}) .$$

If $[X]^3 \subseteq \Gamma$ then $|X| < 4$, since $\{f_0, f_1, f_2\}_\ll \in K_{01}$ and $\{f_1, f_2, f_3\}_\ll \in K_{01}$ would require both $f_2 \prec f_1$ and $f_1 \prec f_2$. If there is X in $[^\kappa 2]^{\eta_0}$ with $[X]^3 \subseteq \Gamma_0$

then $[X]^3 \cap K_{01} = \emptyset$, so as in the preceeding proof there would be X'' in $[X]^{\eta_0}$ with $[X'']^3 \subseteq \Delta_0^*$. Just as above, for no k can there be X in $[{}^\kappa 2]^{\eta_k}$ with $[X]^3 \subseteq \Delta_k^*$. Thus the partition $\{\Gamma\} \cup \{\Gamma_k; k < \gamma\}$ serves to prove the theorem.

Theorem 2.6.8. *Suppose* $\kappa \nrightarrow (\eta_k; k < \gamma)^2$. *Then* $2^\kappa \nrightarrow (4, 4, \eta_k \dot{+} 1; k < \gamma, \eta_k \dot{+} 1; k < \gamma)^3$.

Proof. We again start from a partition $\Delta = \{\Delta_k; k < \gamma\}$ of $[\kappa]^2$ as in the proof of Theorem 2.6.5, and define Δ_k^* as before. Put also

$$\Delta_k^{**} = \{\{f, g, h\}_\preccurlyeq \in P_1;\ \{\delta(f, g), \delta(g, h)\} \in \Delta_k\}\ .$$

This time, consider the partition $\{\Gamma_0, \Gamma_1\} \cup \{\Gamma_{ik}; k < \gamma \text{ and } i = 0,1\}$ of $[{}^\kappa 2]^3$, where

$$\Gamma_0 = K_{01},\ \Gamma_1 = K_{10}\ ,$$

$$\Gamma_{0k} = \Delta_k^* \text{ for } k \text{ with } k < \gamma\ ,$$

$$\Gamma_{1k} = \Delta_k^{**} \text{ for } k \text{ with } k < \gamma\ .$$

This is indeed a partition of $[{}^\kappa 2]^3$, since $[{}^\kappa 2]^3 = K_{01} \cup K_{10} \cup K$ and $K = P_0 \cup P_1$.

If $[X]^3 \subseteq \Gamma_0$ or $[X]^3 \subseteq \Gamma_1$, we know $|X| < 4$. As before, for no k is there X with $X \in [{}^\kappa 2]^{\eta_k}$ and $[X]^3 \subseteq \Delta_k^*$. An entirely similar argument shows that for no k is there X in $[{}^\kappa 2]^{\eta_k}$ with $[X]^3 \subseteq \Delta_k^{**}$. So for no k is there X in $[{}^\kappa 2]^{\eta_k}$ with $[X]^3 \subseteq \Gamma_{0k}$ or $[X]^3 \subseteq \Gamma_{1k}$. This proves the theorem.

Corollary 2.6.9. $\beth_2(\kappa) \nrightarrow (4)_\kappa^3$.

Proof. By Corollary 2.5.2, $\beth_1(\kappa) \nrightarrow (3)_\kappa^2$ so by Theorem 2.6.8 we have $2^{\beth_1(\kappa)} \nrightarrow (4,4,(4)_\kappa, (4)_\kappa)^3$, that is, $\beth_2(\kappa) \nrightarrow (4)_\kappa^3$.

As the final theorem in the discussion of the case $n = 3$, we mention without proof the following theorem from [38].

Theorem 2.6.10 (GCH). *Let* κ *be singular and let* λ *be the immediate predecessor of* κ' *if this exists; otherwise let* $\lambda = \kappa'$. *Then* $\kappa^+ \nrightarrow (\kappa, (4)_\lambda)^3$.

By using the Stepping-up Lemma from §3 to obtain positive results and the negative relations established in this section, a discussion of the case $n = 3$ can be given, similar to that for $n = 2$ of the last section. The only problems to which a solution is not available are of a technical nature. Apart from questions involving inaccessible cardinals, the simplest unsolved problem is the following. (This is Problem 1 in the list [24].)

Problem 2.6.11 (GCH). Is $\aleph_{\omega_{\omega+1}+1} \nrightarrow (\aleph_{\omega_{\omega+1}}, (4)_{\aleph_0})^3$ true?

§7. Square bracket relations

The square bracket relation is in some sense the opposite to the partition relation of the earlier sections. This relation was first introduced and discussed by Erdös, Hajnal and Rado [38, §18].

Definition 2.7.1. The relation $\kappa \to [\eta_k; k<\gamma]^n$ holds if given any set S of power κ, for all disjoint partitions $\boldsymbol{\Delta} = \{\Delta_k; k<\gamma\}$ of $[S]^n$ into γ parts, there exist k with $k<\gamma$ and a subset H of S with $|H| = \eta_k$ such that $[H]^n \cap \Delta_k = \emptyset$.

The special case in which $\eta_k = \eta$ for all k is written $\kappa \to [\eta]^n_\gamma$, and in general the notation conventions of §1 will be adopted for the square bracket symbol as well.

Clearly, the relations $\kappa \to (\eta_0, \eta_1)^n$ and $\kappa \to [\eta_0, \eta_1]^n$ are equivalent. And if $\kappa \to (\eta_0, \eta_1)^n$ then $\kappa \to [\eta_k; k<\gamma]^n$ (for $\gamma \geqslant 2$).

The simple remarks made in §1 about the relation $\kappa \to (\eta_k; k<\gamma)^n$ have analogues for the relation $\kappa \to [\eta_k; k<\gamma]^n$. We remark only that given cardinals η_k where $k<\gamma$, if the relation $\kappa \to [\eta_k; k<\delta]^n$ is true for some δ with $\delta \leqslant \gamma$, then also the relation $\kappa \to [\eta_k; k<\gamma]^n$ holds. The relation $\kappa \to [\eta_k; k<\gamma]^n$ is trivially true when $\gamma > \kappa$, so we always suppose $\gamma \leqslant \kappa$.

We shall concentrate on the case $n = 2$, and assume the GCH throughout our discussion. From Theorem 2.5.10 we know that $\kappa \to [\kappa_0, \kappa_1]^2$ if $\kappa_0, \kappa_1 < \kappa$. Thus all relations $\kappa \to [\eta_k; k<\gamma]^2$ are true if more than one of the η_k are less than κ.

Consider first the case when κ is a successor cardinal, say $\kappa = \lambda^+$. From Corollary 2.2.7 we have $\kappa \to (\kappa, \lambda')^2$, and so $\kappa \to [\lambda', \kappa]^2$. We know from Corollary 2.5.7 that $\kappa \nrightarrow (\kappa, (\lambda')^+)^2$, and hence $\kappa \nrightarrow [(\lambda')^+, \kappa]^2$. This leaves unsettled only relations of the form $\kappa \to [(\lambda')^+, (\kappa)_\gamma]^2$ where $2 \leqslant \gamma \leqslant \kappa$. In fact even the weakest of these, namely $\kappa \to [(\lambda')^+, (\kappa)_\kappa]^2$, is false – see Theorem 2.7.4 below. This we shall prove now.

We start with a trivial lemma.

Lemma 2.7.2. *Let $\mathcal{A} = \{A_\alpha; \alpha<\kappa\}$ be a family of at most κ sets (possibly* *listed with repetitions) each of power κ. Then there are pairwise disjoint sets B_α with $B_\alpha \in [A_\alpha]^\kappa$.*

Proof. Take any one-to-one and onto map $p: \kappa \times \kappa \to \kappa$. Choose elements $x_{\alpha\beta}$ for α, β with $\alpha,\beta < \kappa$ by induction on $p(\alpha, \beta)$ so that $x_{\alpha\beta} \in A_\alpha - \{x_{\gamma\delta}; p(\gamma, \delta) < p(\alpha, \beta)\}$. Put $B_\alpha = \{x_{\alpha\beta}; \beta < \kappa\}$.

Lemma 2.7.3 (GCH). *Let S be any set with $|S| = \kappa^+$. Then there is a disjoint partition $[S]^2 = \bigcup\{\Gamma_k; k < \kappa^+\}$ with the property that whenever $A \in [S]^\kappa$, $B \in [S]^{\kappa^+}$ and $k < \kappa^+$ then there are a, b with $a \in A$, $b \in B$ for which $\{a, b\} \in \Gamma_k$.*

Proof. Note that such a partition provides an example to show that $\kappa^+ \nrightarrow [\kappa^+]^2_{\kappa^+}$.

Identify S with κ^+. Let $\langle A_\alpha; \alpha < \kappa^+\rangle$ be a well ordering of $[\kappa^+]^\kappa$. For each τ with $\tau < \kappa^+$, put

$$\mathcal{A}_\tau = \{A_\alpha; \alpha < \tau \text{ and } A_\alpha \subseteq \tau\},$$

so $|\mathcal{A}_\tau| \leqslant \kappa$. Suppose for each τ with $\tau < \kappa^+$ that we could find a function $f_\tau: \tau \to \tau$ such that

$$\text{if } A \in \mathcal{A}_\tau,\ k < \tau \text{ then } f_\tau(\sigma) = k \text{ for some } \sigma \text{ in } A\ . \tag{1}$$

We could then construct a suitable partition $\{\Gamma_k; k < \kappa^+\}$ as follows: given σ, τ in κ^+ with $\sigma < \tau$,

$$\{\sigma, \tau\} \in \Gamma_k \Leftrightarrow f_\tau(\sigma) = k\ .$$

This partition has the required property, for take A from $[\kappa^+]^\kappa$, B from $[\kappa^+]^{\kappa^+}$ and k with $k < \kappa^+$. Let τ_0 be the least ordinal for which $k < \tau_0$ and $A \in \mathcal{A}_{\tau_0}$. Since $|B| = \kappa^+$, there is τ in B with $\tau_0 \leqslant \tau$. By (1), there is σ in A such that $f_\tau(\sigma) = k$, that is, $\{\sigma, \tau\} \in \Gamma_k$, as required.

Thus to prove the lemma, it suffices to find functions $f_\tau: \tau \to \tau$ with the property (1). Clearly we may suppose $\mathcal{A}_\tau \neq \emptyset$, for otherwise any function $f_\tau: \tau \to \tau$ satisfies (1). Thus $\tau \geqslant \kappa$, so $|\tau| = \kappa$ and we may write $\mathcal{A}_\tau = \{A_{\tau\alpha}; \alpha < \kappa\}$. Applying Lemma 2.7.2, there are pairwise disjoint sets $B_{\tau\alpha}$ where $\alpha < \kappa$ with $B_{\tau\alpha} \in [A_{\tau\alpha}]^\kappa$. Take one-to-one and onto maps $q_{\tau\alpha}: B_{\tau\alpha} \to \tau$, and define f_τ by the following: if for some (necessarily unique) α we have $\sigma \in B_{\tau\alpha}$ then $f_\tau(\sigma) = q_{\tau\alpha}(\sigma)$, otherwise the value of $f_\tau(\sigma)$ is arbitrary. To see that (1) holds, suppose $A \in \mathcal{A}_\tau$ and $k < \tau$. Thus $A = A_{\tau\alpha}$ for some α. There is σ in $B_{\tau\alpha}$ with $q_{\tau\alpha}(\sigma) = k$, that is, $f_\tau(\sigma) = \alpha$. Since $B_{\tau\alpha} \subseteq A_{\tau\alpha}$, this proves (1) and completes the proof of the lemma.

Theorem 2.7.4 (GCH). *For all* κ, $\kappa^+ \not\to [(\kappa')^+, (\kappa^+)_{\kappa^+}]^2$.

Proof. If κ is regular, the result follows from Lemma 2.7.3. Se we need consider only the case that κ is singular. We shall use the partition from Lemma 2.7.3 to refine the partition used in Corollary 2.5.7 to show that $\kappa^+ \not\to (\kappa^+, (\kappa')^+)^2$.

For Corollary 2.5.7 we defined the disjoint partition $[{}^{\kappa'}\kappa]^2 = \Delta_0 \cup \Delta_1$ with

$$\{f, g\} \in \Delta_0 \Leftrightarrow f \leqslant g \text{ and } f \prec g\,,$$

$$\{f, g\} \in \Delta_0 \Leftrightarrow f \leqslant g \text{ and } f \succ g\,,$$

where $\prec$ is the lexicographic ordering and $\leqslant$ an arbitrary well ordering of the functions in ${}^{\kappa'}\kappa$. We now choose $\leqslant$ to have order type κ^+. From §5 we know that if H is a subset of ${}^{\kappa'}\kappa$ with $|H| = (\kappa')^+$ then $[H]^2 \cap \Delta_0 \neq \emptyset$.

Let $\{\Gamma_k; k < \kappa^+\}$ be a partition of $[{}^{\kappa'}\kappa]^2$ as provided for by Lemma 2.7.3. Put $\Delta_{1k} = \Delta_1 \cap \Gamma_k$. Then $\{\Delta_{1k}; k < \kappa^+\}$ is a disjoint partition of Δ_1, which we shall show has the property

$$\text{if } H \in [{}^{\kappa'}\kappa]^{\kappa^+} \text{ and } k < \kappa^+ \text{ then } [H]^2 \cap \Delta_{1k} \neq \emptyset. \tag{1}$$

Thus the partition $[{}^{\kappa'}\kappa]^2 = \Delta_0 \cup \bigcup\{\Delta_{1k}; k < \kappa^+\}$ is a partition which serves to prove the theorem.

Take any subset H of ${}^{\kappa'}\kappa$ with $|H| = \kappa^+$. For f in H, put $L_H(f) = \{g \in H; g \lessdot f\}$ and $R_H(f) = \{g \in H; f \lessdot g\}$. We first observe

$$\text{there is } f \text{ in } H \text{ with } |L_H(f)| = |R_H(f)| = \kappa^+ . \tag{2}$$

For suppose (2) is false, so for every f in H either $|L_H(f)| \leqslant \kappa$ or $|R_H(f)| \leqslant \kappa$. There must be a subset I of H with $|I| = \kappa^+$ such that either $|L_H(f)| \leqslant \kappa$ for all f in I or else $|R_H(f)| \leqslant \kappa$ for all f in I. Suppose in fact the first alternative holds – the argument is similar in the other case. We can choose inductively elements f_α from I, for α with $\alpha < \kappa^+$, so that $f_\alpha \in I - \bigcup\{L_H(f_\beta); \beta < \alpha\}$. Then $f_\beta \prec f_\alpha$ whenever $\beta < \alpha$ so that $\{f_\alpha; \alpha < \kappa^+\}$ is well ordered by $\prec$, contradicting that any subset of ${}^{\kappa'}\kappa$ well ordered by $\prec$ (or by $\succ$) has power less than κ^+ (as follows from (1) and (2) in the proof of Theorem 2.5.5). This establishes (2).

We are now ready to prove (1). Given H in $[{}^{\kappa'}\kappa]^{\kappa^+}$, choose f in H with $|L_H(f)| = |R_H(f)| = \kappa^+$. Take any A in $[R_H(f)]^\kappa$. Since $|A| = \kappa$, $|L_H(f)| = \kappa^+$ and $\leqslant$ has order type κ^+, A is not cofinal in $L_H(f)$ under the ordering $\leqslant$, and so there is B in $[L_H(f)]^{\kappa^+}$ such that $A \lessdot B$. Let k with $k < \kappa^+$ be given. By Theorem 2.7.3, there are a in A and b in B such that $\{a, b\} \in \Gamma_k$. And for such a pair $\{a, b\}$ also $b \prec f \prec a$ so that $\{a, b\} \in \Delta_1$. Thus $\{a, b\} \in \Delta_1 \cap \Gamma_k = \Delta_{1k}$. This proves (1), and completes the proof of the theorem.

We shall now consider the square bracket relation for singular cardinals κ. Let us suppose further that κ' is a successor cardinal, say $\kappa' = \lambda^+$. From §5 we know $\kappa \to (\lambda', \kappa)^2$ and $\kappa \nrightarrow ((\lambda')^+, \kappa)^2$ so the two relations $\kappa \to [\lambda', \kappa]^2$ and $\kappa \nrightarrow [(\lambda')^+, \kappa]$ are in force. This means we must discuss relations of the form $\kappa \to [(\lambda')^+, (\kappa)_\gamma]^2$ where $2 \leqslant \gamma \leqslant \kappa$. We shall find from Theorems 2.7.5 and 2.7.6 below that $\kappa \nrightarrow [(\lambda')^+, (\kappa)_\gamma]^2$ when $\gamma \leqslant \kappa'$ and $\kappa \to [\kappa]^2_\gamma$ when $\gamma > \kappa'$, which settles all problems in this case.

Theorem 2.7.5 (GCH). *If κ is singular and $n \geqslant 2$ then $\kappa \to [\kappa]^n_\gamma$ for all γ with $\gamma > \kappa'$.*

Proof. Take any set S with $|S| = \kappa$, and let a disjoint partition $\mathbf{\Delta} = \{\Delta_k; k < \gamma\}$ of $[S]^n$ be given, where $\gamma > \kappa'$. By Lemma 2.4.2, there is a family $\mathcal{C} = \{C_\sigma; \sigma < \kappa'\}$ of pairwise disjoint subsets of S with $|\bigcup \mathcal{C}| = \kappa$ such that $\mathbf{\Delta}$ is canonical in $\mathcal{C}$. Thus for each sequence $\langle n_\sigma; \sigma < \kappa' \rangle$ where $0 \leqslant n_\sigma \leqslant n$ and $\Sigma(n_\sigma; \sigma < \kappa') = n$ there is an ordinal $k(n_\sigma; \sigma < \kappa')$ in γ such that if $a \in [\bigcup \mathcal{C}]^n$ and $|a \cap C_\sigma| = n_\sigma$ for each σ, then $a \in \Delta_{k(n_\sigma; \sigma < \kappa')}$. Thus $[\bigcup \mathcal{C}]^n \subseteq \bigcup \{\Delta_{k(n_\sigma; \sigma < \kappa')}; 0 \leqslant n_\sigma \leqslant n \text{ and } \Sigma(n_\sigma; \sigma < \kappa') = n\}$. The number of such sequences $\langle n_\sigma; \sigma < \kappa' \rangle$ is at most $(\kappa')^n = \kappa'$. Because $\gamma > \kappa'$, there must be some class from $\mathbf{\Delta}$ having empty intersection with $[\bigcup \mathcal{C}]^n$. Since $|\bigcup \mathcal{C}| = \kappa$, this proves the theorem.

Theorem 2.7.6 (GCH). *Let κ be singular with $\kappa' = \lambda^+$. Then $\kappa \nrightarrow [(\lambda')^+, (\kappa)_{\kappa'}]^2$.*

Proof. Take any set S with $|S| = \kappa$ and write S as a disjoint union, $S = \bigcup \{S_\sigma; \sigma < \lambda^+\}$, where always $|S_\sigma| < \kappa$.

By Theorem 2.7.4, we know $\lambda^+ \nrightarrow [(\lambda')^+, (\lambda^+)_{\lambda^+}]^2$, and so there is a disjoint partition

$$[\lambda^+]^2 = \Gamma \cup \{\Gamma_k; k < \lambda^+\}$$

such that if $I \in [\lambda^+]^{(\lambda')^+}$ then $[I]^2 \cap \Gamma \neq \emptyset$, and if $I \in [\lambda^+]^{\lambda^+}$ then $[I]^2 \cap \Gamma_k \neq \emptyset$ for each k with $k < \lambda^+$. We use this partition of $[\lambda^+]^2$ to define a partition of $[S]^2$. When $k < \lambda^+$, put

$$\Delta_k = \{\{x, y\} \subseteq S; \text{ for some } \{\sigma, \tau\} \text{ in } [\lambda^+]^2, \; x \in S_\sigma, y \in S_\tau \text{ and } \{\sigma, \tau\} \in \Gamma_k\},$$

$$\Delta = [S]^2 - \bigcup \{\Delta_k; k < \lambda^+\}.$$

Then $\{\Delta\} \cup \{\Delta_k; k < \lambda^+\}$ gives a disjoint partition of $[S]^2$ into $\lambda^+ = \kappa'$ classes. This partition provides an example to prove the theorem.

Suppose H is a subset of S with $[H]^2 \subseteq \bigcup\{\Delta_k; k < \lambda^+\}$. Then $|H \cap S_\sigma| \leqslant 1$ for each σ, and $[\{\sigma; H \cap S_\sigma \neq \emptyset\}]^2 \cap \Gamma = \emptyset$. Thus $|H| = |\{\sigma; H \cap S_\sigma \neq \emptyset\}| \leqslant \lambda'$. Thus if $H \subseteq S$ with $|H| = (\lambda')^+$ then $[H]^2 \cap \Delta \neq \emptyset$. And if $H \subseteq S$ with $[H]^2 \cap \Delta_k = \emptyset$ (for any k), then $[\{\sigma; H \cap S_\sigma \neq \emptyset\}]^2 \cap \Gamma_k = \emptyset$, so $|\{\sigma; H \cap S_\sigma \neq \emptyset\}| < \lambda^+ = \kappa'$, and so $|H| < \kappa$.

There remains the case of singular κ with κ' not a successor cardinal. We shall prove a result for $\kappa' = \aleph_0$, but shall not attempt a discussion otherwise, that is when κ' is inaccessible. When $\kappa' = \aleph_0$, there is the following theorem, much stronger than Theorem 2.7.5.

Theorem 2.7.7 (GCH). *Suppose κ is singular with $\kappa' = \aleph_0$. Then $\kappa \to [\kappa]^n_\gamma$ whenever $n \geqslant 2$ and $\gamma > 2^{n-1}$.*

Proof. Let S be any set with $|S| = \kappa$, and take a disjoint partition $\mathbf{\Delta} = \{\Delta_k; k < \gamma\}$ of $[S]^n$, where $\gamma = 2^{n-1} + 1$. From Lemma 2.4.2, there is a pairwise disjoint family $\mathcal{C} = \{C_\sigma; \sigma < \omega\}$ with $\bigcup \mathcal{C} \in [S]^\kappa$ and $|C_\sigma| < |C_\tau|$ when $\sigma < \tau$, such that $\mathbf{\Delta}$ is canonical in $\mathcal{C}$. Thus

$$\text{if } a, b \in [\textstyle\bigcup \mathcal{C}]^n \text{ with } |a \cap C_\sigma| = |b \cap C_\sigma| \text{ for all } \sigma\text{; then } a \equiv b \pmod{\mathbf{\Delta}}\,. \tag{1}$$

The number of sequences $\langle n_1, \ldots, n_p\rangle$ where $p \geqslant 1$ and $1 \leqslant n_i \leqslant n$ and $n_1 + \ldots + n_p = n$ is 2^{n-1}; list these as $\langle n_1(j), \ldots, n_{p(j)}(j)\rangle$ for j with $j < 2^{n-1}$. Define a partition

$$\mathbf{\Gamma} = \{\Gamma(k_j; j < 2^{n-1});\ k_j < \gamma \text{ for each } j\}$$

of $[\omega]^n$ as follows: if $x \in [\omega]^n$, say $x = \{\sigma_1, \ldots, \sigma_n\}_<$, then

$$x \in \Gamma(k_j; j < 2^{n-1}) \Leftrightarrow \text{for each } j \text{, whenever } a \in [\textstyle\bigcup \mathcal{C}]^n \text{, with}$$

$$|a \cap C_{\sigma_i}| = n_i(j) \text{ for } i \leqslant p(j) \text{ then } a \in \Delta_{k_j}.$$

By (1), this is a valid definition.

From Ramsey's Theorem, $\aleph_0 \to (\aleph_0)^n_\delta$ for δ finite; thus there is a set I in $[\omega]^{\aleph_0}$ such that for some sequence $\langle k_j; j < 2^{n-1}\rangle$,

$$[I]^n \subseteq \Gamma(k_j; j < 2^{n-1})\,.$$

Put $H = \bigcup\{C_\sigma; \sigma \in I\}$, then $|H| = \kappa$ and $[H]^n \subseteq \bigcup\{\Delta_{k_j}; j < 2^{n-1}\}$. Since $\gamma > 2^{n-1}$ there is some class of $\mathbf{\Delta}$ which has empty intersection with $[H]^n$. This proves the theorem.

In fact the theorem above is best possible in γ, by the following general remark.

Theorem 2.7.8. *If κ is singular and $n \geqslant 2$, then* $\kappa \nrightarrow [\kappa]^n_{2^{n-1}}$.

Proof. Let S be any set with $|S| = \kappa$, and write S as a disjoint union, $S = \bigcup\{S_\sigma; \sigma < \kappa'\}$ where always $|S_\sigma| < \kappa$. For each sequence $\langle n_1, ..., n_p\rangle$ where $p \geqslant 1$, $1 \leqslant n_1, ..., n_p \leqslant n$ and $n_1 + ... + n_p = n$, put

$$\Delta(n_1, ..., n_p) = \{a \in [S]^n;\ \text{there are } \sigma_1, ..., \sigma_p \text{ with } \sigma_1 < ... < \sigma_p < \kappa' \text{ and } |a \cap S_{\sigma_i}| = n_i\ (1 \leqslant i \leqslant p)\}\ .$$

This gives a partition of $[S]^n$ into 2^{n-1} disjoint classes. And for any H in $[S]^\kappa$ since $|H \cap S_\sigma| \geqslant n$ for at least n values of σ, it follows that $[H]^n$ meets every class of this partition.

We shall now make a couple of further isolated remarks about the case $n \geqslant 3$. From Theorem 2.6.2 we know $\kappa \nrightarrow (n+1, \kappa)^n$, and so $\kappa \nrightarrow [n+1, \kappa]^n$, for κ uncountable and not strongly inaccessible. In most cases there is a best-possible strengthening of this negative relation.

Theorem 2.7.9 (GCH). *Let $n \geqslant 3$ and suppose κ' is a successor cardinal. Then* $\kappa \nrightarrow [n+1, (\kappa)_{\kappa'}]^n$.

Proof. Suppose first that κ is regular, say $\kappa = \lambda^+$. The construction of an example is somewhat similar to that in Lemma 2.7.3, only rather simpler. Let $\langle A_\alpha; \alpha < \lambda^+\rangle$ be a well ordering of $[\lambda^+]^\lambda$, and for each τ with $\tau < \lambda^+$, put

$$\mathcal{A}_\tau = \{A_\alpha; \alpha < \tau \text{ and } A_\alpha \subseteq \tau\}\ .$$

For each ξ with $\xi < \lambda^+$, we seek to pick sets $a(\xi, \sigma, k)$ from $[A_\sigma]^{n-1}$ for A_σ in $\mathcal{A}_\xi$ and k with $k < \xi$, such that $a(\xi, \sigma, k) \cap a(\xi, \tau, l) = \emptyset$ if $\langle\sigma, k\rangle \neq \langle\tau, l\rangle$. Since $|\mathcal{A}_\xi \times \xi| \leqslant \lambda$, by working along a well ordering of $\mathcal{A}_\xi \times \xi$ of order type λ we can clearly do this. Now define subsets Δ_k of $[\lambda^+]^n$ for k with $k < \lambda^+$ as follows: if $\{\alpha_1, ..., \alpha_n\} \subseteq \lambda^+$ with $\alpha_1 < ... < \alpha_n$, then

$$\{\alpha_1, ..., \alpha_n\} \in \Delta_k \Leftrightarrow \{a_1, ..., \alpha_{n-1}\} = a(\alpha_n, \sigma, k) \text{ for some } \sigma\ .$$

Put $\Delta = [\lambda^+]^n - \bigcup\{\Delta_k; k < \lambda^+\}$. Then $\{\Delta\} \cup \{\Delta_k; k < \lambda^+\}$ is a disjoint partition of $[\lambda^+]^n$.

Suppose $H \in [\lambda^+]^{n+1}$, say $H = \{\alpha_0, ..., \alpha_n\}_<$, with $[H]^n \subseteq \bigcup\{\Delta_k; k < \lambda^+\}$. Then in particular there are σ, τ, k, l such that $\{\alpha_0, ..., \alpha_{n-2}, \alpha_n\} = a(\alpha_n, \sigma, k)$

and $\{\alpha_1, ..., \alpha_n\} = a(\alpha_n, \tau, l)$, and $\langle\sigma, k\rangle \neq \langle\tau, l\rangle$. But $a(\alpha_n, \sigma, k) \cap a(\alpha_n, \tau, l) = \emptyset$ and $\{\alpha_0, ..., \alpha_{n-2}, \alpha_n\} \cap \{\alpha_1, ..., \alpha_n\} \neq \emptyset$, so this is impossible. Hence if $H \in [\lambda^+]^{n+1}$ then $[H]^n \cap \Delta \neq \emptyset$.

Now take H in $[\lambda^+]^{\lambda^+}$. Choose A in $[H]^\lambda$, so for some σ with $\sigma < \lambda^+$ we have $A = A_\sigma$. Take any k with $k < \lambda^+$. Since $|H| = \lambda^+$, there is ξ in H with $\xi > \sigma, k$ and $A \in \mathscr{A}_\xi$. Then $a(\xi, \sigma, k) \in [A]^{n-1}$ so that $a(\xi, \sigma, k) \cup \{\xi\} \in \Delta_k$. Since $a(\xi, \sigma, k) \cup \{\xi\} \subseteq H$, this means $[H]^n \cap \Delta_k \neq \emptyset$. Thus the theorem is proved in this case.

Now suppose that κ is singular, with $\kappa' = \lambda^+$. Take any set S with $|S| = \kappa$, and write S as a disjoint union $S = \bigcup\{S_\sigma; \sigma < \lambda^+\}$ where always $|S_\sigma| < \kappa$. We shall use the partition of $[\lambda^+]^n$ constructed above to define a partition $\{\Gamma\} \cup \{\Gamma_k; k < \lambda^+\}$ of $[S]^n$. For k with $k < \lambda^+$ and $\{x_1, ..., x_n\}$ from $[S]^n$, define

$$\{x_1, ..., x_n\} \in \Gamma_k \Leftrightarrow \text{there is } \{\sigma_1, ..., \sigma_n\} \text{ in } \Delta_k \text{, such that}$$

$$x_1 \in S_{\sigma_1}, ..., x_n \in S_{\sigma_n},$$

and put

$$\Gamma = [S]^n - \bigcup\{\Gamma_k; k < \lambda^+\}.$$

Just as in the proof of Theorem 2.7.6, this partition has the required property.

To obtain positive relations when $n \geqslant 3$, there is a Stepping-up Lemma similar to that which holds for the ordinary partition relation.

Theorem 2.7.10. *Let κ and λ be infinite cardinals such that $\lambda < \kappa'$. Let $n \geqslant 1$ and suppose $\lambda \to [\eta_k; k < \gamma]^n$ where $|\gamma|^{|\sigma|} < \kappa'$ whenever $\sigma < \lambda$. Then $\kappa \to [\eta_k \dotplus 1; k < \gamma]^{n+1}$.*

The proof is entirely similar to that of Lemma 2.3.1.

As in the case of the ordinary partition relation, one can hope for a converse to Theorem 2.7.10, namely that under suitable conditions, if $\kappa \nrightarrow [\eta_k; k < \gamma]^n$ then $2^\kappa \nrightarrow [\eta_k \dotplus 1; k < \gamma]^{n+1}$. Very little along these lines is known. See Problem 17 in [24] and the discussion in [25]. Even the conjecture $\aleph_2 \nrightarrow [\aleph_1]^3_4$ is unsolved. Shore [83] has shown that if Gödel's Axiom of Constructibility ($V = L$) is true then $\kappa^{(n+)} \nrightarrow [\kappa^+]^{n+1}_{\kappa^+}$ (where $n \geqslant 2$), which corresponds to the negative relation $\kappa^+ \nrightarrow [\kappa^+]^2_{\kappa^+}$ in Theorem 2.7.4.

The results in this section have depended heavily on the GCH, and without GCH very little seems known about square bracket relations. For example, Galvin and Shelah [47], [48] have shown $2^{\aleph_0} \nrightarrow [2^{\aleph_0}]^2_{\aleph_0}$ and $\aleph_1 \nrightarrow [\aleph_1]^2_4$.

Shelah has established $\aleph_n \nrightarrow [\aleph_n]_n^{n+2}$ if $n \geqslant 1$. The relation $2^{\aleph_0} \rightarrow [\aleph_1]_3^2$ is open.

For the ordinary partition symbol, all relations with infinite exponent n are false. Little seems known about square bracket relations with infinite n. Erdös and Hajnal have shown the following.

Theorem 2.7.11. *Suppose κ and λ are infinite. Then $\kappa \nrightarrow [\lambda]_{2^\lambda}^\lambda$.*

Proof. This is trivial if $\kappa < \lambda$, so suppose $\kappa \geqslant \lambda$. Take any set S with $|S| = \kappa$; we have to find a disjoint partition $\mathbf{\Delta} = \{\Delta_\beta; \beta < 2^\lambda\}$ of $[S]^\lambda$ such that $[H]^\lambda \cap \Delta_\beta \neq \emptyset$ for every β and every H in $[S]^\lambda$.

Suppose first that $\lambda = \kappa$. Then we can well order $[S]^\lambda$ with order type 2^λ, say $[S]^\lambda = \{A_\alpha; \alpha < 2^\lambda\}$. Take any one-to-one and onto map p: $2^\lambda \times 2^\lambda \rightarrow 2^\lambda$. Choose sets $A_{\alpha\beta}$ from $[A_\alpha]^\lambda$ for α, β with $\alpha, \beta < 2^\lambda$ by induction on $p(\alpha, \beta)$ so that $A_{\alpha\beta} \neq A_{\gamma\delta}$ whenever $p(\gamma, \delta) < p(\alpha, \beta)$. Put $\Gamma_\beta = \{A_{\alpha\beta}; \alpha < 2^\lambda\}$, so $\{\Gamma_\beta; \beta < 2^\lambda\}$ is pairwise disjoint, and take any disjoint partition $\mathbf{\Delta} = \{\Delta_\beta; \beta < 2^\lambda\}$ of $[S]^\lambda$ with $\Gamma_\beta \subseteq \Delta_\beta$ for each β. Then if $H \in [S]^\lambda$, for some α we have $H = A_\alpha$ and so $A_{\alpha\beta} \in [H]^\lambda \cap \Delta_\beta$. Thus $\mathbf{\Delta}$ is as required.

Now suppose $\kappa > \lambda$. By virtue of Zorn's Lemma, there is a maximal almost disjoint family $\mathcal{M}$ with $\mathcal{M} \subseteq [S]^\lambda$. For each M in $\mathcal{M}$ we have just shown that there is a disjoint partition $\mathbf{\Delta}(M) = \{\Delta_\beta(M); \beta < 2^\lambda\}$ of $[M]^\lambda$ such that $[H]^\lambda \cap \Delta_\beta(M) \neq \emptyset$ whenever $H \in [M]^\lambda$, $\beta < 2^\lambda$. Put $\Gamma_\beta = \bigcup\{\Delta_\beta(M); M \in \mathcal{M}\}$. By the choice of $\mathbf{\Delta}(M)$ and since $|M \cap N| < \lambda$ for distinct M, N in $\mathcal{M}$, certainly $\Gamma_\beta \cap \Gamma_\gamma = \emptyset$ when $\beta \neq \gamma$. Take any disjoint partition $\mathbf{\Delta} = \{\Delta_\beta; \beta < 2^\lambda\}$ of $[S]^\lambda$ with $\Gamma_\beta \subseteq \Delta_\beta$ for each β. If $H \in [S]^\lambda$, by the maximality of $\mathcal{M}$ there is M in $\mathcal{M}$ with $|H \cap M| = \lambda$. Then $[H \cap M]^\lambda \cap \Delta_\beta(M) \neq \emptyset$ for every β, and so $[H]^\lambda \cap \Delta_\beta \neq \emptyset$. This completes the proof.

Theorem 2.7.11 leaves unsettled most of the cases of the following (Problem 14 in [24]).

Problem 2.7.12. If $\kappa, \lambda \geqslant \mu \geqslant \aleph_0$, is it ture that $\kappa \nrightarrow [\lambda]_{\lambda^\mu}^\mu$?

If the Axiom of Constructibility is true then Problem 2.7.12 is answered positively in Shore [83]. He shows that if $V = L$ then $\kappa \nrightarrow [\lambda^+]_{\lambda^+}^{\aleph_0}$ for all κ and λ.

§8. Partitions of all the finite subsets

The partition symbol $\kappa \rightarrow (\eta)_\gamma^{<\omega}$ represents an extension of the symbol $\kappa \rightarrow (\eta)_\gamma^n$.

Definition 2.8.1. The partition relation $\kappa \to (\eta)_\gamma^{<\omega}$ means the following: given any set S with $|S| = \kappa$, for all partitions $\mathbf{\Delta} = \{\Delta_k; k < \gamma\}$ of $[S]^{<\aleph_0}$ into γ parts there is a subset H of S of power η which is homogeneous for $\mathbf{\Delta}$, in the sense that for each n with $n < \omega$ there is $k(n)$ with $k(n) < \gamma$ such that $[H]^n \subseteq \Delta_{k(n)}$.

From the earlier results of this chapter, it is easy to see that given η, γ and finite n, there is κ so large that κ has the property $\kappa \to (\eta)_\gamma^{<n}$ (with the obvious meaning of this symbol) and moreover questions concerning this relation easily reduce to questions about the ordinary partition symbol. The relation $\kappa \to (\eta)_\gamma^{<\omega}$ is thus seen to be a natural extension of the ordinary partition symbol. Clearly we must suppose η to be infinite, for otherwise $\kappa \to (\eta)_\gamma^{<\omega}$ is equivalent to $\kappa \to (\eta)_\gamma^{<\eta}$, which reduces to an ordinary partition relation.

The results to be presented in this section will show that the least cardinal κ with any of these properties is strongly inaccessible, and so these partition relations are "large cardinal" properties. We shall pursue the matter no further. For more information, see for example Drake [10].

Our first result is due to Rowbottom [81].

Lemma 2.8.1. *Let η be infinite and suppose $\kappa \to (\eta)_2^{<\omega}$. Then $\kappa \to (\eta)_\gamma^{<\omega}$ for all γ with $\gamma \leqslant 2^{\aleph_0}$.*

Proof. Let any disjoint partition $\mathbf{\Delta}$ of $[\kappa]^{<\aleph_0}$ into $2^{\aleph_0}$ classes be given, and use the functions in ${}^\omega 2$ to index the classes in $\mathbf{\Delta}$, so $\mathbf{\Delta} = \{\Delta_f; f \in {}^\omega 2\}$. Take any one-to-one and onto map $p\colon \omega \times \omega \to \omega$ with the property that always $l, m \leqslant p(l, m)$. Define a partition $\mathbf{\Gamma} = \{\Gamma_0, \Gamma_1\}$ of $[\kappa]^{<\aleph_0}$ as follows: if $\alpha_1 < \alpha_2 < \ldots < \alpha_n < \kappa$ and $n = p(l, m)$ then for i with $i = 0,1$

$$\{\alpha_1, \ldots, \alpha_n\} \in \Gamma_i \Leftrightarrow \exists\, f \in {}^\omega 2 (f(l) = i \text{ and } \{\alpha_1, \ldots, \alpha_m\} \in \Delta_f)\,. \tag{1}$$

Take H from $[\kappa]^\eta$ such that H is homogeneous for $\mathbf{\Gamma}$. Then H is also homogeneous for $\mathbf{\Delta}$. For take any sets $\{\alpha_1, \ldots, \alpha_m\}_<$, $\{\beta_1, \ldots, \beta_m\}_<$ from $[H]^m$. Suppose $\{\alpha_1, \ldots, \alpha_m\} \in \Delta_f$ and $\{\beta_1, \ldots, \beta_m\} \in \Delta_g$. Take any l with $l < \omega$ and put $n = p(l, m)$. Choose $\gamma_{m+1}, \ldots, \gamma_n$ from H with $\alpha_m, \beta_m < \gamma_{m+1} < \ldots < \gamma_n$; since H is infinite this is certainly possible. Then since $\{\alpha_1, \ldots, \alpha_m\} \in \Delta_f$, we have $\{\alpha_1, \ldots, \alpha_m, \gamma_{m+1}, \ldots, \gamma_n\} \in \Gamma_{f(l)}$ from (1). Then $\{\beta_1, \ldots, \beta_m, \gamma_{m+1}, \ldots, \gamma_n\} \in \Gamma_{f(l)}$ since H is homogeneous for Γ, and so $g(l) = f(l)$. Since this is true for every l, we have $f = g$ and so H is indeed homogeneous for $\mathbf{\Delta}$.

The following two theorems for $\eta = \aleph_0$ follow from methods credited by Erdös and Hajnal [18, Thm 9b] to G. Fodor. We shall follow the proofs in Morley [74].

Theorem 2.8.2. *Let η be infinite. If κ is the least cardinal such that $\kappa \to (\eta)_2^{<\omega}$, then κ is regular.*

Proof. Let κ be least such that $\kappa \to (\eta)_2^{<\omega}$ and suppose in fact κ is singular. Take a pairwise disjoint decomposition $\kappa = \bigcup\{K_\sigma; \sigma < \kappa'\}$ where always $|K_\sigma| < \kappa$. By the choice of κ, there are partitions $\mathbf{\Gamma}_\sigma = \{\Gamma_{\sigma 0}, \Gamma_{\sigma 1}\}$ of $[K_\sigma]^{<\aleph_0}$ and a partition $\mathbf{\Gamma} = \{\Gamma_0, \Gamma_1\}$ of $[\kappa']^{<\aleph_0}$, none of which have a homogeneous set of size κ. Define a partition $\mathbf{\Delta} = \{\Delta_{00}, \Delta_{01}, \Delta_{10}, \Delta_{11}\}$ of $[\kappa]^{<\aleph_0}$ as follows: if $a \in [\kappa]^{<\aleph_0}$ and $i = 0,1$

$$a \in \Delta_{0i} \Leftrightarrow \exists\ \sigma < \kappa'\ (a \subseteq K_\sigma \cap \Gamma_{\sigma i}),$$

$$a \in \Delta_{1i} \Leftrightarrow [|a| = 1 \text{ or } \forall \sigma < \kappa'\ (a \not\subseteq K_\sigma)]$$

and

$$\{\sigma < \kappa'; a \cap K_\sigma \neq \emptyset\} \in \Gamma_i.$$

Let H be a subset of κ homogeneous for $\mathbf{\Delta}$. Either there is σ such that $H \subseteq K_\sigma$, or else $|H \cap K_\sigma| \leq 1$ for every σ. In the first case, H is homogeneous for Γ_σ; in the second, $\{\sigma < \kappa'; |H \cap K_\sigma| = 1\}$ is homogeneous for Γ. In either event, $|H| < \eta$. Thus $\mathbf{\Delta}$ has no homogeneous set of size η. In view of Lemma 2.8.1, this contradicts that $\kappa \to (\eta)_2^{<\omega}$. Hence κ must be regular.

Theorem 2.8.3. *Let η be infinite. If $\kappa \nrightarrow (\eta)_2^{<\omega}$ then $2^\kappa \nrightarrow (\eta)_2^{<\omega}$.*

Proof. Suppose $\kappa \nrightarrow (\eta)_2^{<\omega}$ and take a partition $\mathbf{\Gamma} = \{\Gamma_0, \Gamma_1\}$ of $[\kappa]^{<\aleph_0}$ which has no homogeneous set of size η. We shall construct a partition $\mathbf{\Delta} = \{\Delta_k; k < \omega\}$ of $[{}^\kappa 2]^{<\aleph_0}$ which has no homogeneous set of size η; by Lemma 2.8.1 this will suffice to prove the theorem.

Extend $\mathbf{\Gamma}$ to a partition $\mathbf{\Gamma}^* = \{\Gamma_k^*; k < \omega\}$ of the finite sequences from κ so that $\langle\alpha_1, ..., \alpha_n\rangle$ and $\langle\beta_1, ..., \beta_n\rangle$ are in the same class of $\mathbf{\Gamma}^*$ if and only if $\{\alpha_1, ..., \alpha_n\}$ and $\{\beta_1, ..., \beta_n\}$ are in the same class of $\mathbf{\Gamma}$ and moreover,

$$\alpha_i < \alpha_j \Leftrightarrow \beta_i < \beta_j \qquad (i, j = 1, ..., n).$$

Let $\prec$ be the lexicographic ordering of ${}^\kappa 2$, and choose a partition $\mathbf{\Delta} = \{\Delta_k; k < \omega\}$ of ${}^\kappa 2$ such that if $f_0, ..., f_n \in {}^\kappa 2$ with $f_0 \prec f_1 \prec ... \prec f_n$ then

$$\{f_0, ..., f_n\} \in \Delta_k \Leftrightarrow \langle\delta(f_0, f_1), ..., \delta(f_{n-1}, f_n)\rangle \in \Gamma_k^*,$$

where, as in §6, $\delta(f, g)$ is the least α where $f(\alpha) \neq g(\alpha)$.

Let H be a subset of ${}^\kappa 2$ homogeneous for $\mathbf{\Delta}$. Suppose $f, g, h \in H$ with $f \prec g \prec h$. Then $\delta(f, g) \neq \delta(g, h)$ so either $\delta(f, g) < \delta(g, h)$ or $\delta(f, g) > \delta(g, h)$, and moreover this order is the same for all triples from H. Suppose in fact $\delta(f, g) < \delta(g, h)$; the argument is similar in the other case. Thus whenever $f, g, h \in H$ with $f \prec g \prec h$ we have $\delta(f, g) = \delta(f, h)$. Put

$$I = \{\delta(f, g); f, g \in H\} ,$$

then $|I| = |H|$. Further, I is easily seen to be homogeneous for $\mathbf{\Gamma}$, so $|I| < \eta$. Thus $|H| < \eta$, and so $\mathbf{\Delta}$ has no homogeneous set of size η.

Corollary 2.8.4. *If κ is the least cardinal such that $\kappa \to (\eta)_2^{<\omega}$ where η is infinite, then κ is strongly inaccessible.*

We shall conclude this section by showing that increasing the value of η in the relation $\kappa \to (\eta)_2^{<\omega}$ represents a genuine strengthening of the relation. In this respect, to relate these properties to the ordinary partition symbol studied before, we should remark that Silver [90] has shown that if κ is the least cardinal with even the property $\kappa \to (\aleph_0)_2^{<\omega}$, then there is a cardinal λ with $\lambda < \kappa$ such that $\lambda \to (\lambda)_2^2$. Silver's proof of this result depends on methods from mathematical logic. A combinatorial proof has since been given by Henle and Kleinberg [58].

Theorem 2.8.5. *Let η_1, η_2 be infinite cardinals with $\eta_1 < \eta_2$. If κ_1 and κ_2 are the least cardinals with the properties $\kappa \to (\eta_1)_2^{<\omega}$ and $\kappa \to (\eta_2)_2^{<\omega}$ respectively, then $\kappa_1 < \kappa_2$.*

Proof. Clearly $\kappa_1 \leqslant \kappa_2$, so suppose in fact $\kappa_1 = \kappa_2 = \kappa$ say. Then for each cardinal α with $\alpha < \kappa$, we have $|\alpha| < \kappa$ so there is a partition $\mathbf{\Delta}(\alpha) = \{\Delta_0(\alpha), \Delta_1(\alpha)\}$ of $[\alpha]^{<\aleph_0}$ which has no homogeneous subset of power η_1. Define a partition $\mathbf{\Delta} = \{\Delta_0, \Delta_1\}$ of $[\kappa]^{<\aleph_0}$ as follows: if $\alpha_1, ..., \alpha_n \in \kappa$ with $\alpha_1 < ... < \alpha_n$, when $i = 0,1$

$$\{\alpha_1, ..., \alpha_n\} \in \Delta_i \Leftrightarrow \{\alpha_1, ..., \alpha_{n-1}\} \in \Delta_i(\alpha_n) .$$

Let H be a subset of κ homogeneous for $\mathbf{\Delta}$. Then for each α in H, it follows that $H \cap \alpha$ is homogeneous for $\mathbf{\Delta}(\alpha)$, so $|H \cap \alpha| < \eta_1$. Hence $|H| \leqslant \eta_1 < \eta_2$, so $\mathbf{\Delta}$ has no homogeneous set of size η_2, contrary to the relation $\kappa \to (\eta_2)_2^{<\omega}$.

CHAPTER 3

SET MAPPINGS

§1. Set mappings of small order

Let S be an arbitrary set. By a *set-mapping* on S is meant a function f mapping S into the powerset $\mathfrak{P}S$ of S such that $x \notin f(x)$ for each x in S. The set map f is said to be of *order* λ if λ is the least cardinal such that $|f(x)| < \lambda$ for each x in S. A subset S' of S is said to be *free* for f if $S' \cap \bigcup f[S'] = \emptyset$. Equivalently then, S' is free for f if for all pairs x, y from S' both $x \notin f(y)$ and $y \notin f(x)$.

We shall investigate in this chapter some conditions on the set map f under which there will be a large free subset. Apparently P. Turán (in about 1930) was the first to ask questions of this nature – they arose in connnection with a problem on interpolation. Specifically, he asked if when S is of the power of the continuum and each $f(x)$ is finite, must there be an infinite set? Ruziewicz [82] generalized the problem to the following: if S is infinite with $|S| = \kappa$ and f is a set map on S of order λ where $\lambda < \kappa$, will there be a free set of power κ? Solutions to special cases of this problem are in Lázár [66], Sierpinski [86], Piccard [76], [77] and Fodor [40]. In [14] Erdös proved, under the assumption of the Generalized Continuum Hypothesis, that there always is such a free set. Finally Hajnal [53] proved this result without appeal to GCH.

It is easily seen that if the set map f has order λ with $\lambda \geqslant \kappa$ then there need be no free set even of size 2. For let f be the set map on κ where $f(\alpha) = \{\beta \in \kappa;\ \beta < \alpha\}$. Clearly f has order κ, and there is no free pair for f.

We shall now give Hajnal's proof.

Theorem 3.1.1. *Let S be a set with $|S| = \kappa$ and let f be a set map on S of order λ where $\lambda < \kappa$. Then there is a free set of size κ for f.*

Proof. Suppose first that κ is regular. (The proof in this case is due to D. Lázár.) We use transfinite induction and choose subsets S_α of S for α with $\alpha < \lambda$ so that S_α is a maximal free subset of $S - \bigcup\{S_\beta; \beta < \alpha\}$. Since being a free subset is a property of finite character, such a maximal set exists. If for any α it happens that $|S_\alpha| = \kappa$ then there is nothing more to do. So suppose that $|S_\alpha| < \kappa$ whenever $\alpha < \lambda$. We shall obtain a contradiction.

Since always $|f(x)| < \lambda$ where $\lambda < \kappa$, it follows that for any α we have $|\bigcup f[S_\alpha]| < \kappa$ and consequently, from the regularity of κ, if $S^* = \bigcup\{S_\alpha \cup \bigcup f[S_\alpha]; \alpha < \lambda\}$ then $|S^*| < \kappa$. Thus $S - S^* \neq \emptyset$. Choose x from $S - S^*$. Then $x \notin S_\alpha \cup \bigcup f[S_\alpha]$ for each α. Since S_α is a maximal free set, there must be x_α in S_α such that $x_\alpha \in f(x)$, for otherwise $S_\alpha \cup \{x\}$ contradicts the maximality of S_α. Now $x_\alpha \neq x_\beta$ when $\alpha \neq \beta$ since by construction $S_\alpha \cap S_\beta = \emptyset$, and $\{x_\alpha; \alpha < \lambda\} \subseteq f(x)$ so $|f(x)| \geqslant \lambda$. This contradicts that f has order λ; and this case is proved.

Now suppose that κ is singular. Choose a regular cardinal θ such that $\kappa', \lambda < \theta < \kappa$. Take a strictly increasing sequence $\langle \kappa_\sigma; \sigma < \kappa' \rangle$ of regular cardinals all less than κ with $\kappa = \Sigma(\kappa_\sigma; \sigma < \kappa')$, where $\theta \leqslant \kappa_0$. Write S as a disjoint union, $S = \bigcup\{S_\sigma; \sigma < \kappa'\}$ where $|S_\sigma| = \kappa_\sigma$. Then f induces set maps $f_\sigma : S_\sigma \to \mathfrak{P} S_\sigma$ where $f_\sigma(x) = f(x) \cap S_\sigma$, and f_σ has order at most λ. Since $|S_\sigma| = \kappa_\sigma$ where κ_σ is regular with $\lambda < \kappa_\sigma$, there is a set S'_σ in $[S_\sigma]^{\kappa_\sigma}$ free for f_σ, and S'_σ is then also free for f. Put

$$S^*_\sigma = S'_\sigma - \bigcup\{\bigcup f[S_\tau]; \tau < \sigma\},$$

then by the regularity of κ_σ we have $|S^*_\sigma| = \kappa_\sigma$. Also

$$f[S^*_\sigma] \cap \bigcup\{S^*_\tau; \tau \geqslant \sigma\} = \emptyset .$$

We shall seek sets T^*_σ with $T^*_\sigma \in [S^*_\sigma]^{\kappa_\sigma}$ such that

$$f[T^*_\sigma] \cap \bigcup\{T^*_\tau; \tau < \sigma\} = \emptyset , \tag{1}$$

for then if $T' = \bigcup\{T^*_\sigma; \sigma < \kappa'\}$ clearly T' is free for f, and $|T'| = \kappa$.

We start by finding sets T_σ in $[S^*_\sigma]^{\kappa_\sigma}$ such that if $Z_\sigma = \bigcup\{T_\tau; \tau < \sigma\}$ then for a particular pairwise disjoint decomposition $\{Z_{\sigma\alpha}; \alpha < \theta\}$ of Z_σ, there is an ordinal $\alpha(\sigma)$ with $\alpha(\sigma) < \theta$ such that

$$Z_\sigma \cap \bigcup f[T_\sigma] \subseteq \bigcup\{Z_{\sigma\alpha}; \alpha < \alpha(\sigma)\} . \tag{2}$$

Use induction on σ with $\sigma < \kappa'$ to define sets T_σ with $|T_\sigma| = \kappa_\sigma$ together with pairwise disjoint decompositions $T_\sigma = \bigcup\{T_{\sigma\alpha}; \alpha < \theta\}$ of T_σ where always $|T_{\sigma\alpha}| = \kappa_\sigma$. Suppose that σ is given with $\sigma < \kappa'$, and that this has already been done for all τ where $\tau < \sigma$. Put $Z_\sigma = \bigcup\{T_\tau; \tau < \sigma\}$ and $Z_{\sigma\alpha} = \bigcup\{T_{\tau\alpha}; \tau < \sigma\}$, so that Z_σ is a disjoint union, $Z_\sigma = \bigcup\{Z_{\sigma\alpha}; \alpha < \theta\}$.

For any x in S^*_σ,

$$f(x) \cap Z_\sigma = \bigcup\{f(x) \cap Z_{\sigma\alpha}; \alpha < \theta\} .$$

Since $|f(x)| < \lambda < \theta$ and θ is regular there is an ordinal $\alpha(x, \sigma)$ with $\alpha(x, \sigma) < \theta$ such that $f(x) \cap Z_{\sigma\alpha} = \emptyset$ whenever $\alpha \geqslant \alpha(x, \sigma)$. For each β with $\beta < \theta$ put

$$S^*_{\sigma\beta} = \{x \in S^*_\sigma; \alpha(x, \sigma) \leqslant \beta\}\ ,$$

so $S^*_{\sigma\beta} \subseteq S^*_{\sigma\gamma}$ if $\beta < \gamma$ and $S^*_\sigma = \bigcup\{S^*_{\sigma\beta}; \beta < \theta\}$. Since $|S^*_\sigma| = \kappa_\sigma$ and $\theta < \kappa'_\sigma = \kappa_\sigma$, there is $\alpha(\sigma)$ with $\alpha(\sigma) < \theta$ for which $|S^*_{\sigma\alpha(\sigma)}| = \kappa_\sigma$. Put $T_\sigma = S^*_{\sigma\alpha(\sigma)}$. Since $\theta < \kappa_\sigma$ there is a pairwise disjoint decomposition $T_\sigma = \bigcup\{T_{\sigma\alpha}; \sigma < \theta\}$ where always $|T_{\sigma\alpha}| = \kappa_\sigma$. This completes the definition of T_σ and $T_{\sigma\alpha}$. Notice that for x in T_σ we have $\alpha(x, \sigma) \leqslant \alpha(\sigma)$ so that $f(x) \cap Z_{\sigma\alpha} = \emptyset$ whenever $\alpha \geqslant \alpha(\sigma)$, and thus (2) holds.

Put $\alpha^* = \sup\ \{\alpha(\sigma); \sigma < \kappa'\}$. Since θ is regular with $\kappa' < \theta$, it follows that $\alpha^* < \theta$. Put

$$T^*_\sigma = \bigcup\{T_{\sigma\alpha}; \alpha^* < \alpha < \theta\}\ ,$$

so $|T^*_\sigma| = \kappa_\sigma$. Then if $y \in T^*_\tau$ and $x \in T^*_\sigma$ where $\tau < \sigma$ we have $y \in T_{\tau\alpha} \subseteq Z_{\sigma\alpha}$ for some α with $\alpha > \alpha^* \geqslant \alpha(\sigma)$. From (2) it follows that $f(x) \cap Z_{\sigma\alpha} = \emptyset$ for such an α; hence $y \notin f(x)$. This establishes (1) and completes the proof.

There are a couple of interesting extensions of Theorem 3.1.1. The first one is taken from Erdös and Fodor [16].

Theorem 3.1.2. *Let S be a set with $|S| = \kappa$ and suppose a family $\{R_\gamma; \gamma < \eta\}$ of κ-size subsets of S is given, where η is any cardinal with $\eta < \kappa$. Let f be a set mapping on S of order λ where $\lambda < \kappa$. Then there is a subset S' of S free for f, and moreover $|S' \cap R_\gamma| = \kappa$ for each γ.*

Proof. By Lemma 2.7.2, we may suppose that the sets R_γ are pairwise disjoint. First suppose κ is regular. Let $p : \kappa \to \eta$ be any onto map with the property that for each γ with $\gamma < \eta$ we have $|\{\alpha < \kappa; p(\alpha) = \gamma\}| = \kappa$. Let $\boldsymbol{S}$ be the set of all sequences $\boldsymbol{s} = \langle s_\nu; \nu < \varphi\rangle$ where $\varphi \leqslant \kappa$, $s_\nu \in R_{p(\nu)}$ for each ν, $s_\mu \neq s_\nu$ if $\mu \neq \nu$, and ran $(\boldsymbol{s}) = \{s_\nu; \nu < \varphi\}$ is free for f.

Inductively, define sequences $\boldsymbol{s}_\alpha = \langle s_{\alpha\nu}; \nu < \varphi_\alpha\rangle$ in $\boldsymbol{S}$ for α with $\alpha < \kappa$ as follows. Choose $\boldsymbol{s}_\alpha$ as a sequence in $\boldsymbol{S}$ all the entries for which are in $S - \bigcup\{\text{ran}(\boldsymbol{s}_\beta); \beta < \alpha\}$, and which has maximal length amongst all such sequences from $\boldsymbol{S}$. If any $\boldsymbol{s}_\alpha$ has length κ, then ran$(\boldsymbol{s}_\alpha)$ is a free set which has the required property, so for a contradiction suppose $\varphi_\alpha < \kappa$ for all α.

Consider the values $p(\varphi_\alpha)$ where $\alpha < \kappa$. Since p maps into η and $\lambda, \eta < \kappa$, there must be A in $[\kappa]^\lambda$ and γ with $\gamma < \eta$ such that $p(\varphi_\alpha) = \gamma$ for all α in A. Put

$$S^* = \bigcup\{\text{ran}(\boldsymbol{s}_\alpha) \cup \bigcup f[\text{ran}(\boldsymbol{s}_\alpha)]; \alpha \in A\}\ .$$

§2. Set mappings of large order

Let S be a set of power κ and let $f : S \to \mathfrak{P}S$ be a set map on S of order κ. We shall investigate conditions on f under which there will be a large subset of S which is free for f. An example was given in the last section of a set map on κ of order κ with no free set of size 2, so clearly some restriction on f is required. The conditions we shall impose are of the following type. Let η, θ be cardinals; we demand that whenever X is an η-size subset of S then the intersection of all the $f(x)$ for x in X has power less than θ. Questions of this type were first raised by Erdös and Fodor [16] and discussed in detail in Erdös, Hajnal and Rado [38].

The following notation is convenient to summarize the results. (This corresponds to the symbol $\kappa \to [[\kappa, \eta, \theta, \lambda]]$ of [38].)

Definition 3.2.1. The relation $(\kappa, \eta, \theta) \to \lambda$ means the following. Whenever f is a set map of order κ on a set S with $|S| = \kappa$ such that

$$X \in [S]^{\eta} \Rightarrow |\bigcap f[X]| < \theta ,$$

then there is a subset S' of S free for f with $|S'| = \lambda$.

We start by noting a couple of lemmas that extend some of the results from Chapter 1, §1 on almost disjoint families to families with the present intersection property.

Lemma 3.2.2. *If a set S with $|S| = \kappa$ can be decomposed into a family $\mathcal{A}$ with $|\mathcal{A}| = \lambda$ such that $|\bigcap\mathcal{F}| < \theta$ whenever $\mathcal{F} \in [\mathcal{A}]^{\eta^+}$ then $\lambda \leqslant \eta \cdot \kappa^{\theta}$.*

Proof. Suppose $S = \bigcup\mathcal{A}$, where $\mathcal{A}$ is a family with the properties in the statement of the lemma. For each A in $\mathcal{A}$ with $|A| \geqslant \theta$, choose A^* in $[A]^{\theta}$. Then for any B in $[S]^{\theta}$, we have $|\{A \in \mathcal{A}; A^* = B|\} \leqslant \eta$. Thus $\{A \in \mathcal{A}\,; |A| \geqslant \theta\}$ has power at most $\eta \cdot |[S]^{\theta}|$, and so $|\mathcal{A}| \leqslant \eta \cdot |[S]^{\theta}| + |[S]^{<\theta}| = \eta \cdot \kappa^{\theta}$.

One proves the following lemma and its corollary from Lemma 3.2.2 in much the same way that, in Chapter 1, Theorem 1.1.6 and Corollary 1.1.7 were proved from Theorem 1.1.2.

Lemma 3.2.3 (GCH). *Suppose κ and θ are infinite cardinals with $\kappa' \neq \theta'$. Then there is no decomposition $\mathcal{A}$ of a set of power κ into κ^+ subsets each of power at least θ such that $|\bigcap\mathcal{F}| < \theta$ for every family $\mathcal{F}$ from $[\mathcal{A}]^{\kappa^+}$.*

Clearly $\{x_\alpha; \alpha < \kappa\}$ is the required free set.

Thus Theorem 3.1.3 will follow from the following theorem. (The proof requires the GCH only when κ is singular.)

Theorem 3.1.4 (GCH). *Let f be a set map of order λ on a set S where $|S| = \kappa$ with $\lambda < \kappa$. Then there is A in $[S]^\kappa$ with $|\Pi A| < \kappa$.*

Proof. Suppose first that κ is regular, and let the set map $f: S \to \mathfrak{P}S$ be given. We shall show that for any T from $[S]^\kappa$ there is A in $[T]^\kappa$ with $|\Pi A| < \kappa$. For a contradiction, suppose for some particular T from $[S]^\kappa$ that this is false; thus whenever $A \subseteq T$ and $|\Pi A| < \kappa$ then $|A| < \kappa$.

Use induction to define sets A_α in $[T]^{<\kappa}$ for α with $\alpha < \lambda$ as follows. Put $A_\alpha^* = \bigcup\{\bigcup f[A_\beta]; \beta < \alpha\}$ so $|A_\alpha^*| < \kappa$ by the regularity of κ. Use Zorn's Lemma to choose for A_α a subset of $T - \bigcup\{A_\beta; \beta < \alpha\}$ maximal with the property $\Pi A_\alpha \subseteq A_\alpha^*$. Then $A_\alpha \neq \emptyset$ and $|A_\alpha| < \kappa$ by our assumption.

Put $A = \bigcup\{A_\alpha; \alpha < \lambda\}$ so $|A| < \kappa$. Thus $T - A \neq \emptyset$, so choose x from $T - A$. For each α there must be y_α in S_α such that $f(x) \cap f(y_\alpha) \not\subseteq A_\alpha^*$, for otherwise $S_\alpha \cup \{x\}$ contradicts the maximality of S_α. Choose x_α with $x_\alpha \in (f(x) \cap f(y_\alpha)) - A_\alpha^*$. Then $x_\alpha \neq x_\beta$ whenever $\alpha \neq \beta$, for if say $\beta < \alpha$ then $x_\alpha \in f(y_\alpha) - f(y_\beta)$ yet $x_\beta \in f(y_\beta)$. Thus since $\{x_\alpha; \alpha < \lambda\} \subseteq f(x)$ we have $|f(x)| \geqslant \lambda$, contradicting that f has order λ. This completes this part of the proof.

If in fact $\lambda^+ < \kappa$, by assuming GCH the result above can be improved as follows. Given any subset T of S with $|T| = \theta^+$ where θ is a cardinal with $\lambda < \theta' \leqslant \theta < \kappa$, then there is a subset B of T with $|B| = \theta^+$ and $|\Pi B| < \lambda$. To see this, note that by the construction above there is a subset A of T with $|A| = \theta^+$ and $|\Pi A| < \theta^+$. Look at $A' = \{x \in A; f(x) \cap \Pi A \neq \emptyset\}$. The sets $f(x)$ for x in $A - A'$ are pairwise disjoint, so if $|A'| < \theta^+$ then $B = A - A'$ has the required property. So we may suppose $|A'| = \theta^+$. For each x in A', the set $f(x) \cap \Pi A$ is a subset of ΠA of power less than λ. The number of such subsets is at most θ^λ, and $\theta^\lambda = \theta$ by GCH, since $\lambda < \theta'$. Thus there must be a subset B of A' with $|B| = \theta^+$ such that for all x in B the set $f(x) \cap \Pi A$ is the same. But then $\Pi B = f(x) \cap \Pi A$ for any x in B, so $|\Pi B| \not< \lambda$, and the claim is established.

We are now ready to prove the theorem for singular κ. Take a strictly increasing sequence $\langle \kappa_\sigma; \sigma < \kappa' \rangle$ where each κ_σ is the successor of a regular cardinal, with $\kappa', \lambda^+ < \kappa_\sigma < \kappa$ and $\kappa = \Sigma(\kappa_\sigma; \sigma < \kappa')$. Write S as a disjoint union, $S = \bigcup\{S_\sigma; \sigma < \kappa'\}$ where always $|S_\sigma| = \kappa_\sigma$. By the remark above, for each σ there are subsets B_σ of S_σ with $|B_\sigma| = \kappa_\sigma$ such that $|\Pi B_\sigma| < \lambda$. Put $B_\sigma^* = \bigcup\{\bigcup f[B_\tau]; \tau < \sigma\}$, so $|B_\sigma^*| \leqslant \lambda \cdot \Sigma(\kappa_\tau; \tau < \sigma) < \kappa_\sigma$, noting that κ_σ (being a successor cardinal) is regular. Since the sets $f(x) - \Pi B_\sigma$ for x in B_σ are pairwise disjoint, the number of x in B_σ for which $f(x) - \Pi B_\sigma$ meets B_σ^* is at most $|B_\sigma^*|$. Hence if

$$A_\sigma = \{x \in B_\sigma; (f(x) - \Pi B_\sigma) \cap B_\sigma^* = \emptyset\}$$

then $|A_\sigma| = \kappa_\sigma$. Put $A = \bigcup\{A_\sigma; \sigma < \kappa'\}$, so $|A| = \kappa$. Moreover, $\Pi A \subseteq \bigcup\{\Pi B_\sigma; \sigma < \kappa'\}$. For take x, y in A, so suppose $x \in A_\sigma$, $y \in A_\tau$. If $\sigma = \tau$ certainly $f(x) \cap f(y) \subseteq \Pi B_\sigma$, so suppose without loss of generality that $\tau < \sigma$. Then $y \in A_\tau \subseteq B_\tau$ so $f(y) \subseteq B_\sigma^*$. By the definition of A_σ then $f(x) \cap f(y) \subseteq \Pi B_\sigma$. Thus indeed $\Pi A \subseteq \bigcup\{\Pi B_\sigma; \sigma < \kappa'\}$, and hence $|\Pi A| \leqslant \lambda \cdot \kappa' < \kappa$. This constructs A with the desired property, and completes the proof.

We conclude this section with a theorem of Fodor [40] concerning the decomposition of a set into a union of free sets.

Theorem 3.1.5. *Let S be a set of power κ and let f be a set mapping on S of order λ where $\aleph_0 \leqslant \lambda < \kappa$. Then there is a family $\mathcal{H}$ with $|\mathcal{H}| \leqslant \lambda$ of subsets of S each free for f, such that $S = \bigcup\mathcal{H}$.*

Proof. Given X in $[S]^\lambda$, define X_n where $n < \omega$ by induction on n as follows: $X_0 = X$, $X_{n+1} = \bigcup f[X_n]$. Put $X_\omega = \bigcup\{X_n; n < \omega\}$. Then $|X_\omega| = \lambda$ and X_ω is closed under f, in the sense that whenever $x \in X_\omega$ then $f(x) \subseteq X_\omega$. So clearly S may be written as a union (probably not disjoint) of κ sets each of power λ and each closed under f, say $S = \bigcup\{S_\alpha; \alpha < \kappa\}$.

Using such a decomposition of S, we shall show the following: For any non-empty subset T of S, there is a family $\mathcal{H}(T) = \{T_\nu; \nu < \lambda\}$ of λ subsets of T, each free for f, such that

$$y \in T - \bigcup\mathcal{H}(T) \Rightarrow \exists\, x \in \bigcup\mathcal{H}(T)(x \in f(y)) . \tag{1}$$

Start by putting, for α with $\alpha < \kappa$,

$$R_\alpha = T \cap (S_\alpha - \bigcup\{S_\beta; \beta < \alpha\}) .$$

Discard any empty R_α. Then always $1 \leqslant |R_\alpha| \leqslant \lambda$, and so we may write $R_\alpha = \{s_{\alpha\nu}; \nu < \lambda\}$. Define subsets T_ν of T by induction on ν (where $\nu < \lambda$) so that $T_\nu \subseteq \{s_{\alpha\nu}; \alpha < \kappa\}$ as follows. Suppose for a particular value of ν that the sets T_μ for μ with $\mu < \nu$ have already been defined. We determine whether or not $s_{\alpha\nu}$ is in T_ν using induction on α, by specifying that $s_{\alpha\nu} \in T_\nu$ if and only if the following two conditions are met:

(i) $\forall x \in \bigcup\{T_\mu; \mu < \nu\}\ (x \notin f(s_{\alpha\nu}))$,

(ii) $\forall \beta < \alpha(s_{\beta\nu} \in T_\nu \Rightarrow s_{\beta\nu} \notin f(s_{\alpha\nu}))$.

This defines T_ν. Put $\mathcal{H}(T) = \{T_\nu; \nu < \lambda\}$; we show (1) holds. Take y with $y \in T - \bigcup \mathcal{H}(T)$. Since $T = \bigcup\{R_\alpha; \alpha < \kappa\}$ there is α such that $y \in R_\alpha$, and so $y = s_{\alpha\nu}$ for some ν. Since $y \notin \bigcup \mathcal{H}(T)$, in particular $y \notin T_\nu$. From the definition of T_ν, either there is x in $\bigcup\{T_\mu; \mu < \nu\}$ such that $x \in f(y)$ or else there is $s_{\beta\nu}$ in T_ν such that $s_{\beta\nu} \in f(y)$; in any event there is x in $\bigcup \mathcal{H}(T)$ such that $x \in f(y)$. Thus (1) holds. Further, each T_ν is free for f. For take $s_{\alpha\nu}$, $s_{\beta\nu}$ from T_ν and suppose say $\beta < \alpha$. Since $f(s_{\beta\nu}) \subseteq S_\beta$ and $s_{\alpha\nu} \in S_\alpha - \bigcup\{S_\gamma; \gamma < \alpha\}$, certainly $s_{\alpha\nu} \notin f(s_{\beta\nu})$. And condition (ii) in the definition of T_ν ensures that $s_{\beta\nu} \notin f(s_{\alpha\nu})$. Hence T_ν is indeed free for f.

We are now ready to construct the family $\mathcal{H}$ of free subsets of S. Inductively define families $\mathcal{H}_\nu$ of subsets of S as follows. Put $U_\nu = S - \bigcup\{\bigcup \mathcal{H}_\mu; \mu < \nu\}$, then $\mathcal{H}_\nu = \mathcal{H}(U_\nu)$ if $U_\nu \neq \emptyset$; $\mathcal{H}_\nu = \emptyset$ if $U_\nu = \emptyset$. Now define $\mathcal{H} = \bigcup\{\mathcal{H}_\nu; \nu < \lambda\}$. Then certainly $\mathcal{H}$ is a family of free subsets of S, with $|\mathcal{H}| \leqslant \lambda$.

Suppose for some ν that $U_\nu \neq \emptyset$, and choose y from U_ν. Then $y \in U_\mu$ for each μ with $\mu < \nu$ so $y \in U_\mu - \bigcup \mathcal{H}_\mu = U_\mu - \bigcup \mathcal{H}(U_\mu)$. Thus by the property (1) for $\mathcal{H}(U_\mu)$, we can choose x_μ from $\bigcup \mathcal{H}_\mu$ such that $x_\mu \in f(y)$. Now $U_\mu \cap \bigcup \mathcal{H}_\xi = \emptyset$ whenever $\xi < \mu$; since $\bigcup \mathcal{H}_\mu \subseteq U_\mu$ it follows that $\bigcup \mathcal{H}_\xi \cap \bigcup \mathcal{H}_\mu = \emptyset$ whenever $\xi \neq \mu$. Hence $x_\xi \neq x_\mu$ whenever $\xi \neq \mu$. Since always $|f(y)| < \lambda$ and $\{x_\mu; \mu < \nu\} \subseteq f(y)$ it follows that $|\nu| < \lambda$ and so $\nu < \lambda$. Thus if $U_\nu \neq \emptyset$ we must have $\nu < \lambda$. We are now set to show that $S = \bigcup \mathcal{H}$. For take any y in S. There must be ν with $\nu < \lambda$ for which $y \in \bigcup \mathcal{H}_\nu$, for otherwise $y \in S - \bigcup\{\bigcup \mathcal{H}_\nu; \nu < \lambda\} = U_\lambda$, and we have just remarked that $U_\lambda = \emptyset$. Since $\mathcal{H}_\nu \subseteq \mathcal{H}$, this shows $y \in \bigcup \mathcal{H}$, and consequently $S = \bigcup \mathcal{H}$. This completes the proof.

There are various ways in which the problems investigated in this section can be extended. The original question of Ruziewicz can be phrased in terms of the order type of the sets involved, rather than their cardinality. Questions of this nature have been investigated by Erdös, Hajnal and Milner in [37], for countable order types. On the other hand, one can consider set mappings defined on a topological space, and replace the cardinality requirements by conditions of a topological nature. As an example, we mention the following result of Bagemihl [2]: If $f: \boldsymbol{R} \to \mathfrak{P}\boldsymbol{R}$ is a set mapping on the set $\boldsymbol{R}$ of real numbers such that always $f(x)$ is nowhere dense, then there is an everywhere dense free set for f.

Since f has order λ and always $|\mathrm{ran}(s_\alpha)| < \kappa$, it follows from the regularity of κ that $|S^*| < \kappa$. Thus $R_\gamma - S^* \neq \emptyset$. Choose x from $R_\gamma - S^*$. Then since $x \notin S^*$, there must be for each α in A an element $s_{\alpha\nu}$ in $\mathrm{ran}(s_\alpha)$ such that $s_{\alpha\nu} \in f(x)$, for otherwise the sequence $s_\alpha \cup \{\langle\gamma, x\rangle\}$ obtained by extending s_α one place further with the value x would contradict the maximal length of s_α. Now by construction $s_{\alpha\nu} \neq s_{\beta\mu}$ if $\alpha \neq \beta$ so since $\{s_{\alpha\nu}; \alpha \in A\} \subseteq f(x)$ it follows that $|f(x)| \geqslant \lambda$, contradicting that f has order λ. This completes the proof in this case.

The proof when κ is singular is very similar to that of Theorem 3.1.1, so we don't give full details. Choose θ regular so that $\kappa', \lambda, \eta < \theta < \kappa$. In the notation of the last proof, ensure that the sets S_σ are chosen so that $|S_\sigma \cap R_\gamma| = \kappa_\sigma$ for $\gamma < \eta$ and $\sigma < \kappa'$, and take a free set S'_σ such that $|S'_\sigma \cap R_\gamma| = \kappa_\sigma$ for all γ. Rather than constructing the one set T^*_σ, construct sets $T^*_{\sigma\gamma}$ for all γ with $\gamma < \eta$, so that

$$f[T^*_{\sigma\gamma}] \cap \bigcup\{T^*_{\tau\delta}\,; \tau < \sigma \text{ and } \delta < \eta\} = \emptyset\ .$$

The set $\bigcup\{T^*_{\sigma\gamma}; \sigma < \kappa' \text{ and } \gamma < \eta\}$ is the required free set.

It is easy to see that the result in Theorem 3.1.2 is the best possible, in the sense that the family $\{R_\gamma; \gamma < \eta\}$ cannot be replaced by a family $\{R_\gamma; \gamma < \kappa\}$ of κ pairwise disjoint sets from $[S]^\kappa$. For write κ as a disjoint union $\kappa = \bigcup\{R_\gamma; \gamma < \kappa\}$ where always $|R_\gamma| = \kappa$ and ensure that $\gamma \notin R_\gamma$; consider the set map $f : \kappa \to \mathfrak{P}\kappa$ where $f(\alpha) = \{\gamma\}$ just when $\alpha \in R_\gamma$. If S' meets all the R_γ then $f[S'] = \kappa$ and so S' is not free for f.

The second extension of Theorem 3.1.1 is from Fodor [41].

Theorem 3.1.3 (GCH). *Let f be a set map of order λ on a set S where $|S| = \kappa$ with $\lambda < \kappa$. Then there is a free subset S' of S with $|S'| = \kappa$ such that*

$$|\bigcup\{f(x) \cap f(y); x, y \in S' \text{ and } x \neq y\}| < \kappa\ .$$

With the set map f on S assumed given, for a subset A of S let us put

$$\Pi A = \bigcup\{f(x) \cap f(y); x, y \in A \text{ and } x \neq y\}\ .$$

If we could find A in $[S]^\kappa$ with $|\Pi A| < \kappa$, then we could obtain a free set S' in $[A]^\kappa$ (so necessarily $|\Pi S'| < \kappa$) as follows. Note that for x in $A - \Pi A$, for at most one y in A can $x \in f(y)$ hold – write $y(x)$ for this y, when it exists. Inductively choose x_α from $A - \Pi A$ for α with $\alpha < \kappa$ so that

$$x_\alpha \in A - (\Pi A \cup \{x_\beta; \beta < \alpha\} \cup \{y(x_\beta); \beta < \alpha\} \cup \bigcup\{f(x_\beta); \beta < \alpha\})\ .$$

Since $|\bigcup\{f(x_\beta); \beta < \alpha\}| \leqslant \lambda \cdot |\alpha| < \kappa$, the choice of x_α is always possible.

Corollary 3.2.4 (GCH). *Suppose $\theta^+ \leqslant \kappa$. Then there is no decomposition $\mathcal{A}$ of a set of power κ into κ^+ subsets each of power at least θ^+ with $|\bigcap \mathcal{F}| < \theta$ whenever $\mathcal{F}$ is a family from $[\mathcal{A}]^{\kappa^+}$. (If θ is finite, GCH is not needed.)*

We are now in a position to prove the following positive result.

Theorem 3.2.5 (GCH). *Suppose κ is regular and $\theta^+ < \kappa$. Then $(\kappa, \kappa, \theta) \to \kappa$.*

Proof. Suppose the theorem is false, and let f be a set map on a set S where $|S| = \kappa$ such that

$$X \in [S]^\kappa \Rightarrow |\bigcap f[X]| < \theta \ , \tag{1}$$

and yet there is no free set for f of power κ. We start much as in the proof of Theorem 3.1.1. Choose by transfinite induction sets S_α for α with $\alpha < \theta^+$ such that S_α is a maximal free subset of $S - \bigcup\{S_\beta; \beta < \alpha\}$, so always $|S_\alpha| < \kappa$. Since κ is regular with $\theta^+ < \kappa$ and always $|f(x)| < \kappa$ it follows that if $S^* = \bigcup\{S_\alpha \cup \bigcup f[S_\alpha]; \alpha < \theta^+\}$ then $|S^*| < \kappa$. Hence $|S - S^*| = \kappa$. Let $x \in S - S^*$. By the maximality of S_α, there must be x_α in S_α such that $x_\alpha \in f(x)$, and $x_\alpha \neq x_\beta$ whenever $\alpha \neq \beta$. Put $T = \bigcup\{S_\alpha; \alpha < \theta^+\}$, then it follows that $|T \cap f(x)| \geqslant \theta^+$. Suppose $|T| = \lambda$.

Assume $\lambda^+ < \kappa$. Then since

$$|\{T \cap f(x); x \in S - S^*\}| \leqslant |\mathfrak{P}T| = \lambda^+ < \kappa \ ,$$

by the regularity of κ there must be X in $[S - S^*]^\kappa$ such that $T \cap f(x)$ is the same for all x in X. But then $|\bigcap f[X]| \geqslant \theta^+$, contradicting (1).

The only other possibility is that $\lambda^+ = \kappa$. Consider the family $\mathcal{A} = \{T \cap f(x); x \in S - S^*\}$. Then $|A| \geqslant \theta^+$ for each A in $\mathcal{A}$, and $|\bigcap \mathcal{F}| < \theta$ whenever $\mathcal{F} \in [\mathcal{A}]^{\lambda^+}$ by (1). Since $|S - S^*| = \kappa = \lambda^+$, this contradicts Corollary 3.2.4.

Thus in either case a contradiction is reached. This suffices to prove the theorem.

The symbol $(\kappa, \eta, \theta) \to \lambda$ is fairly readily discussed in the case that κ is singular. The relation is fully covered by the following theorem.

Theorem 3.2.6 (GCH). *Let κ be singular and suppose that $\eta, \theta < \kappa$. Then*

(i) $(\kappa, \eta, \theta) \to \kappa$,

(ii) $(\kappa, \kappa, \theta) \to \kappa'$,

(iii) $(\kappa, \kappa, 1) \not\to (\kappa')^+$.

Proof. Take an increasing sequence $\langle \kappa_\sigma; \sigma < \kappa' \rangle$ of cardinals such that $\kappa = \Sigma(\kappa_\sigma; \sigma < \kappa')$ and always $\eta, \theta < \kappa_\sigma < \kappa$.

To establish (iii), take any set S with $|S| = \kappa$ and write S as a disjoint union, $S = \bigcup\{S_\sigma; \sigma < \kappa'\}$ where $|S_\sigma| = \kappa_\sigma$. Define $f : S \to \mathfrak{P}S$ by

$$f(x) = S_\sigma - \{x\} \quad \text{where } x \in S_\sigma \,.$$

Then f is a set map on S of order κ. If $X \in [S]^\kappa$ then X must meet more than one of the sets S_σ and so $\bigcap f[X] = \emptyset$. And if S' is a subset of S free for f, for each σ we must have $|S' \cap S_\sigma| \leqslant 1$, so $|S'| \leqslant \kappa'$. Thus f is an example which proves (iii).

To show (ii), let f be any set map of order κ on a set S with $|S| = \kappa$, such that $|\bigcap F[X]| < \theta$ whenever $X \in [S]^\kappa$. We may suppose $S = \kappa$, and write $S = \bigcup\{S_\sigma; \sigma < \kappa'\}$ where $|S_\sigma| = \kappa_\sigma$ and $S_\sigma < S_\tau$ whenever $\sigma < \tau$. Define a disjoint partition $\mathbf{\Delta} = \{\Delta_1, \Delta_2, \Delta_3\}$ of $[\kappa]^2$ as follows: if $\alpha, \beta < \kappa$ with $\alpha < \beta$ then

$$\{\alpha, \beta\} \in \Delta_1 \Leftrightarrow \beta \in f(\alpha) \,,$$

$$\{\alpha, \beta\} \in \Delta_2 \Leftrightarrow \alpha \in f(\beta) \text{ and } \beta \notin f(\alpha) \,,$$

$$\{\alpha, \beta\} \in \Delta_3 \Leftrightarrow \{\alpha, \beta\} \notin \Delta_1 \cup \Delta_2 \,.$$

By the Canonization Lemma, Lemma 2.4.2, there is a family $\mathcal{C} = \{C_\sigma; \sigma < \kappa'\}$ with $|C_\sigma| = \kappa_\sigma$ such that $\mathbf{\Delta}$ is canonical in $\mathcal{C}$. Moreover (see the remarks after the proof of this lemma) we may suppose that $C_\sigma < C_\tau$ whenever $\sigma < \tau$. So for all pairs a, b from $\bigcup\mathcal{C}$ we have

$$\forall\sigma(|a \cap C_\sigma| = |b \cap C_\sigma|) \Rightarrow a \equiv b \ (\text{mod } \mathbf{\Delta}) \,. \tag{1}$$

For each σ with $\sigma < \kappa'$ choose α_σ from C_σ and put

$$X_\sigma = \{\tau < \kappa'; \sigma < \tau \text{ and } \exists\beta \in C_\tau(\{\alpha_\sigma, \beta\} \in \Delta_1)\} \,.$$

By (1) and the definition of Δ_1, whenever $\tau \in X_\sigma$ then $C_\tau \subseteq f(\alpha_\sigma)$. Consequently always $|X_\sigma| < \kappa'$, for otherwise $|f(\alpha_\sigma)| = \kappa$. Similarly, put

$$Y_\sigma = \{\tau < \kappa'; \sigma < \tau \text{ and } \exists\beta \in C_\tau(\{\alpha_\sigma, \beta\} \in \Delta_2)\} \,.$$

By (1) and the definition of Δ_2, if $\tau \in Y_\sigma$ then $C_\sigma \subseteq \bigcap f[C_\tau]$, and so $|\bigcap f[\bigcup\{C_\tau; \tau \in Y_\sigma\}]| \geqslant |C_\sigma| = \kappa_\sigma > \theta$. Hence $|Y_\sigma| < \kappa'$, for otherwise the intersection property of f is violated.

Inductively choose ordinals $\sigma(\xi)$ where $\xi < \kappa'$ so that

$$\sigma(\xi) \in \kappa' - (\{\sigma(\zeta); \zeta < \xi\} \cup \bigcup\{X_{\sigma(\zeta)} \cup Y_{\sigma(\zeta)}; \zeta < \xi\}) \,;$$

by the regularity of κ' this is always possible. Put $S' = \{\alpha_{\sigma(\xi)}; \xi < \kappa'\}$ so $|S'| = \kappa'$. And if $\zeta < \xi$ then $\sigma(\xi) \notin X_{\sigma(\zeta)} \cup Y_{\sigma(\zeta)}$ so that $\{\alpha_{\sigma(\zeta)}, \alpha_{\sigma(\xi)}\} \in \Delta_3$.

Hence $\alpha_{\sigma(\zeta)} \notin f(\alpha_{\sigma(\xi)})$ and $\alpha_{\sigma(\xi)} \notin f(\alpha_{\sigma(\zeta)})$ so that S' is free for f. This proves (ii).

Finally, to settle (i), let f be a set map on S as in the proof of (ii), except now with the intersection property

$$X \in [S]^{\eta} \Rightarrow |\bigcap f[X]| < \theta \, . \tag{2}$$

Define the ordinals $\sigma(\xi)$ for ξ with $\xi < \kappa'$ and the set S' as above, so that by (1) whenever $\alpha \in C_{\sigma(\zeta)}$, $\beta \in C_{\sigma(\xi)}$ where $\zeta \neq \xi$ then $\alpha \notin f(\beta)$ and $\beta \notin f(\alpha)$. Moreover, in this case each C_σ is free for f. Once this is established, it follows that if $S'' = \bigcup\{C_{\sigma(\xi)}; \xi < \kappa'\}$ then S'' is free for f. Since $|S''| = \kappa$, (i) would hold. To show that C_σ is free, note that since $\eta, \theta < \kappa_\sigma$ there are subsets A_σ, B_σ of C_σ with $|A_\sigma| = |B_\sigma| = \eta \dotplus \theta$ and $A_\sigma < B_\sigma$. By (1), C_σ is homogeneous for $\mathbf{\Delta}$. However, if $[C_\sigma]^2 \subseteq \Delta_1$ then $B_\sigma \subseteq \bigcap f[A_\sigma]$ and if $[C_\sigma]^2 \subseteq \Delta_2$ then $A_\sigma \subseteq \bigcap f[B_\sigma]$. Either of these contradicts (2), and so it must be that $[C_\sigma]^2 \subseteq \Delta_3$. Hence C_σ is free for f, and the proof is complete.

Let us return to problems involving regular cardinals. Here a few unsolved problems remain. Theorem 3.2.5 gives the best possible result for inaccessible κ, so only the case when κ is a successor cardinal need be considered. If $\kappa = \lambda^+$ where λ is regular, the following strong negative result of Hajnal [52] shows that the restriction $\theta^+ < \kappa$ in Theorem 3.2.5 cannot be relaxed.

Theorem 3.2.7 (GCH). *Let λ be regular and let S be any set with $|S| = \lambda^+$. Then there is an almost disjoint family $\mathcal{B}^*$ with $|\mathcal{B}^*| = \lambda^+$ of the λ-size subsets of S such that $|S - \bigcup\mathcal{B}| \leqslant \lambda$ for any subfamily $\mathcal{B}$ of $\mathcal{B}^*$ with power λ^+.*

Corollary 3.2.8 (GCH). *Let λ be regular. Then $(\lambda^+, 2, \lambda) \nrightarrow \lambda^+$.*

Proof of Corollary 3.2.8. Let S be any set with $|S| = \lambda^+$, and let $\mathcal{B}^*$ be a family on S with the properties given by Theorem 3.2.7. Take well orderings $\langle x_\alpha; \alpha < \lambda^+ \rangle$ and $\langle B_\alpha; \alpha < \lambda^+ \rangle$ of S and of $\mathcal{B}^*$. Define sets $f(x_\alpha)$ where $\alpha < \lambda^+$ by induction on α as follows. Let γ be least such that $f(x_\beta) \neq B_\gamma$ for all β with $\beta < \alpha$; then

$$f(x_\alpha) = \begin{cases} B_\gamma \text{ if } x_\alpha \notin B_\gamma \\ \emptyset \text{ otherwise} \, . \end{cases}$$

It is clear that if $S^* = \{x \in S; f(x) \neq \emptyset\}$ then $|S^*| = \lambda^+$.

Define $f^* : S^* \to \mathfrak{P}S^*$ by $f^*(x) = S^* \cap f(x)$. Then f^* is a set map on S^* of order at most λ^+, and for distinct x, y from S^* we have $|f^*(x) \cap f^*(y)| \leqslant |f(x) \cap f(y)| < \lambda$ since the family $\mathcal{B}^*$ is almost disjoint. Further, f^* has no free set of power λ^+. For take S' with $S' \subseteq S^*$ and $|S'| = \lambda^+$. Then $f[S'] \subseteq \mathcal{B}^*$ and $|f[S']| = \lambda^+$ so that $|S' - \bigcup f[S']| \leqslant \lambda$ by the properties of $\mathcal{B}^*$. Since $|S'| = \lambda^+$ this means $S' \cap \bigcup f[S'] \neq \emptyset$. Since $S' \subseteq S^*$, we have $S' \cap \bigcup f[S'] = S' \cap \bigcup f^*[S']$, and consequently S' is not free for f^*. Thus f^* is a set map which illustrates $(\lambda^+, 2, \lambda) \nrightarrow \lambda^+$.

Proof of Theorem 3.2.7. Take the set S with $|S| = \lambda^+$. By GCH, $|[S]^\lambda| = \lambda^+$, so there is a well ordering $\langle A_\alpha; \alpha < \lambda^+ \rangle$ of $[S]^\lambda$. Put $\mathcal{A}_\alpha = \{A_\beta; \beta < \alpha\}$. Use transfinite induction to define sets B_α in $[S]^\lambda$ for α with $\alpha < \lambda^+$ as follows. Put $\mathcal{B}_\alpha = \{B_\beta; \beta < \alpha\}$, and let γ be the least ordinal with $\gamma \geqslant \alpha$ for which there is A in $\mathcal{A}_\gamma$ with $|A - \bigcup \mathcal{B}| = \lambda$ for all $\mathcal{B}$ from $[\mathcal{B}_\alpha]^{<\lambda}$. Since $|\bigcup \mathcal{B}_\alpha| \leqslant \lambda$ whereas $|S| = \lambda^+$ there certainly is such an ordinal γ. Put

$$\mathcal{A}^*_\alpha = \{A \in \mathcal{A}_\gamma; \forall\, \mathcal{B} \in [\mathcal{B}_\alpha]^{<\lambda}(|A - \bigcup \mathcal{B}| = \lambda)\} ,$$

so $1 \leqslant |\mathcal{A}^*_\alpha| \leqslant \lambda$. Write $\mathcal{A}^*_\alpha = \{A_{\alpha\nu}; \nu < \lambda\}$ and $\mathcal{B}_\alpha = \{B_{\alpha\nu}; \nu < \lambda\}$, listed with repetitions if necessary. Choose elements $x_{\alpha\nu}$ for ν with $\nu < \lambda$ so that

$$x_{\alpha\nu} \in A_{\alpha\nu} - (\bigcup \{B_{\alpha\mu}; \mu < \nu\} \cup \{x_{\alpha\mu}; \mu < \nu\}) ;$$

since $A_{\alpha\nu} \in \mathcal{A}^*_\alpha$ such a choice is always possible. Put $B_\alpha = \{x_{\alpha\nu}; \nu < \lambda\}$. This completes the definition of B_α.

Then always $B_\alpha \in [S]^\lambda$. And $|B_\alpha \cap B_\beta| < \lambda$ if $\alpha \neq \beta$. For if say $\beta < \alpha$ then $B_\beta \in \mathcal{B}_\alpha$ so $B_\beta = B_{\alpha\nu}$ for some ν with $\nu < \lambda$. Then

$$B_\beta \cap B_\alpha \subseteq \{x_{\alpha\mu}; \mu \leqslant \nu\} ,$$

so $|B_\alpha \cap B_\beta| < \lambda$. Thus if

$$\mathcal{B}^* = \{B_\alpha; \alpha < \lambda^+\} ,$$

then $\mathcal{B}^* \subseteq [S]^\lambda$, $|\mathcal{B}^*| = \lambda^+$ and $\mathcal{B}^*$ is an almost disjoint family. In fact $\mathcal{B}^*$ has the property asserted in the statement of the theorem. This we show now.

$$\text{Put } \mathcal{A}^* = \{A \in [S]^\lambda; \forall\, \mathcal{B} \in [\mathcal{B}^*]^{<\lambda}(|A - \bigcup \mathcal{B}| = \lambda)\} .$$

Then we have

$$A_\beta \in \mathcal{A}^* \text{ and } \beta < \alpha < \lambda^+ \Rightarrow A_\beta \cap B_\alpha \neq \emptyset . \tag{1}$$

For if $A_\beta \in \mathcal{A}^*$ certainly $A_\beta \in \mathcal{A}^*_\alpha$ so that $A_\beta = A_{\alpha\nu}$ for some ν. Then $x_{\alpha\nu} \in A_\beta \cap B_\alpha$. Also we have

$$S' \subseteq S \text{ and } |S'| = \lambda^+ \Rightarrow \exists\, A \in \mathcal{A}^* \, (A \subseteq S') . \tag{2}$$

For let S' from $[S]^{\lambda^+}$ be given, and put

$$\mathcal{B}' = \{B \in \mathcal{B}^*; |B \cap S'| = \lambda\} .$$

Consider two cases.

Case 1. $|\mathcal{B}'| \leqslant \lambda$. Since $|S' - \bigcup\mathcal{B}'| = \lambda^+$, we may choose A from $[S' - \bigcup\mathcal{B}']^\lambda$. Then $|A \cap B| < \lambda$ for every B in $\mathcal{B}^*$. By the regularity of λ, then $|A - \bigcup\mathcal{B}| = \lambda$ for every $\mathcal{B}$ from $[\mathcal{B}^*]^{<\lambda}$, and so $A \in \mathcal{A}^*$.

Case 2. $|\mathcal{B}'| > \lambda$. Choose $\mathcal{B}''$ from $[\mathcal{B}']^\lambda$, and put $A = S' \cap \bigcup\mathcal{B}''$. Take any $\mathcal{B}$ from $[\mathcal{B}^*]^{<\lambda}$. Then there is B'' in $\mathcal{B}'' - \mathcal{B}$. Since the sets in $\mathcal{B}^*$ are almost disjoint, $|(B'' \cap S') \cap B| < \lambda$ for each B in $\mathcal{B}$. Again by the regularity of λ, it follows that $|(B'' \cap S') \cap \bigcup\mathcal{B}| < \lambda$. Since $|B'' \cap S'| = \lambda$, this means that $|(B'' \cap S') - \bigcup\mathcal{B}| = \lambda$, and so $|A - \bigcup\mathcal{B}| = \lambda$. Thus $A \in \mathcal{A}^*$, and (2) is established.

We can now see that $\mathcal{B}^*$ has the property claimed. For take any family $\mathcal{B}$ from $[\mathcal{B}^*]^{\lambda^+}$; we show $|S - \bigcup\mathcal{B}| \leqslant \lambda$. For if not, $|S - \bigcup\mathcal{B}| = \lambda^+$ and so by (2) there is A in A^* with $A \subseteq S - \bigcup\mathcal{B}$. Now $A = A_\beta$ for some β. Since $|\mathcal{B}| = \lambda^+$, there is α with $\beta < \alpha < \lambda^+$ such that $B_\alpha \in \mathcal{B}$. Then by (1), $A \cap B_\alpha \neq \emptyset$ which is impossible with $A \subseteq S - \bigcup\mathcal{B}$. Thus $|S - \bigcup\mathcal{B}| \leqslant \lambda$, and the proof is complete.

In view of the result $(\lambda^+, 2, \lambda) \nrightarrow \lambda^+$ for λ regular from Corollary 3.2.8, the strongest positive result for regular λ would be $(\lambda^+, \lambda^+, \lambda) \to \lambda$. In fact this result holds, as is shown by the following theorem.

Theorem 3.2.9 (GCH). *If λ is infinite then* $(\lambda^+, \lambda^+, \lambda) \to \lambda'$.

Proof. Take S with $|S| = \lambda^+$ and suppose $f: S \to \mathfrak{P}S$ is a set map of order λ with $|\bigcap f[X]| < \lambda$ whenever $X \in [S]^{\lambda^+}$. Choose elements x_α from S for α with $\alpha < \lambda^+$ by induction so that

$$x_\alpha \in S - (\{x_\beta; \beta < \alpha\} \cup \bigcup\{f(x_\beta); \beta < \alpha\}) .$$

Then $x_\alpha \notin f(x_\beta)$ whenever $\beta < \alpha$. Put $T = \{x_\alpha; \alpha < \lambda^+\}$, so $|T| = \lambda^+$. Define a partition $\{\Delta_0, \Delta_1\}$ of $[T]^2$ by, if $\beta < \alpha$ then

$$\{x_\alpha, x_\beta\} \in \Delta_0 \Leftrightarrow x_\beta \in f(x_\alpha) ,$$

$$\{x_\alpha, x_\beta\} \in \Delta_1 \Leftrightarrow x_\beta \notin f(x_\alpha) .$$

By Corollary 2.2.7, $\lambda^+ \to (\lambda^+, \lambda')^2_2$, so either there is H in $[T]^{\lambda^+}$ with $[H]^2 \subseteq \Delta_0$ or there is H in $[T]^{\lambda'}$ with $[H]^2 \subseteq \Delta_1$. Suppose there is H in $[T]^{\lambda^+}$ with $[H]^2 \subseteq \Delta_0$. Let $H = \{x_{\alpha(\gamma)}; \gamma < \varphi\}$ where $\gamma_1 < \gamma_2 \Rightarrow \alpha(\gamma_1) < \alpha(\gamma_2)$, (so $\varphi \geqslant \lambda^+$). But then $\{x_{\alpha(\gamma)}; \gamma < \lambda\} \subseteq \bigcap\{f(x_{\alpha(\gamma)}); \gamma \geqslant \lambda\}$, contradicting the choice of f. Thus there must be H in $[T]^{\lambda'}$ with $[H]^2 \subseteq \Delta_1$. But then H is free for f, and the theorem is proved.

The results for λ regular can be summarized as follows: $(\lambda^+, \lambda^+, \theta) \to \lambda^+$ if $\theta < \lambda$ (Theorem 3.2.5), and $(\lambda^+, \lambda^+, \lambda) \to \lambda$ (Theorem 3.2.9), yet $(\lambda^+, 2, \lambda) \nrightarrow \lambda^+$ (Corollary 3.2.8).

There remains the possibility that $|S| = \lambda^+$ where λ is singular. The positive result of Theorem 3.2.9 cannot be greatly strengthened even in this case, as is shown by the following theorem. (If λ is regular, the theorem is trivial, but then Corollary 3.2.8 gives a strong result.)

Theorem 3.2.10 (GCH). *If κ is infinite then $(\kappa^+, \kappa^+, \kappa) \nrightarrow (\kappa')^{++}$.*

Proof. We may suppose that κ is singular, and choose an increasing sequence $\langle \kappa_\sigma; \sigma < \kappa' \rangle$ of cardinals with always $\kappa_\sigma < \kappa$ and $\kappa = \Sigma(\kappa_\sigma; \sigma < \kappa')$. Write $[\kappa^+]^\kappa = \{A_\alpha; \alpha < \kappa^+\}$, put $\mathcal{A}_\alpha = \{A_\beta; \beta < \alpha \text{ and } A_\beta \subseteq \alpha\}$ and choose subfamilies $\mathcal{A}_{\alpha\sigma}$ of $\mathcal{A}_\alpha$ with $|\mathcal{A}_{\alpha\sigma}| \leqslant \kappa_\sigma$ and $\mathcal{A}_\alpha = \bigcup\{\mathcal{A}_{\alpha\sigma}; \sigma < \kappa'\}$. As in the proof of Theorem 2.5.9, define subsets $f_\sigma(\alpha)$ of α with $|f_\sigma(\alpha)| < \kappa_{\sigma+1}$ such that $A \cap f_\sigma(\alpha) \neq \emptyset$ for each A in $\mathcal{A}_{\alpha\sigma}$ and $\gamma \notin f_\sigma(\beta)$ for all β, γ with $\beta, \gamma \in f_\sigma(\alpha)$. Define a function $g : \kappa^+ \to \mathfrak{P}\kappa^+$ by

$$g(\alpha) = \alpha - \bigcup\{f_\sigma(\alpha); \sigma < \kappa'\}\,.$$

Then g is a set mapping on κ^+ of order at most κ^+. I claim that g has the properties required to prove the theorem.

Take X in $[\kappa^+]^{\kappa^+}$; we show $|\bigcap g[X]| < \kappa$. For suppose on the contrary that $|\bigcap g[X]| \geqslant \kappa$, and choose A from $[\bigcap g[X]]^\kappa$, so $A = A_\beta$ for some β. Since $|X| = \kappa^+$ we can choose α from X with $\beta < \alpha$ and $A \subseteq \alpha$. Then $A \in \mathcal{A}_{\alpha\sigma}$ for some σ with $\sigma < \kappa'$, and consequently there is γ with $\gamma \in A \cap f_\sigma(\alpha)$. Thus $\gamma \notin g(\alpha)$, and so $\gamma \notin \bigcap g[X]$. Yet $\gamma \in A$ and $A \subseteq \bigcap g[X]$. This contradiction shows that in fact $|\bigcap g[X]| < \kappa$.

Now suppose S is a subset of κ^+ free for g. We need to show that $|S| < (\kappa')^{++}$. Since S is free, if we take α, β from S, say where $\beta < \alpha$, then $\beta \notin g(\alpha)$ so $\beta \in \bigcup\{f_\sigma(\alpha); \beta < \kappa'\}$. Thus we may define a partition $[S]^2 = \bigcup\{\Delta_\sigma; \sigma < \kappa'\}$ where, if $\sigma < \kappa'$,

$$\Delta_\sigma = \{\{\beta, \alpha\}_< \in [S]^2; \beta \in f_\sigma(\alpha)\}\,.$$

As in the proof of Theorem 2.5.9, this partition has no homogeneous set of size 3. However, from Theorem 2.2.4 the relation $(\kappa')^{++} \to (3)^2_{\kappa'}$ holds, so if $|S| \geqslant (\kappa')^{++}$ there would have to be a homogeneous triple. Hence $|S| < (\kappa')^{++}$, and the theorem is proved.

Let us review the situation with regard to the relation $(\lambda^+, \eta, \theta) \to \iota$ when λ is singular. By Theorem 3.2.5, if $\theta < \lambda$ then the best possible result $(\lambda^+, \lambda^+, \theta)$

$\to \lambda^+$ holds, so we need only consider the relation $(\lambda^+, \eta, \lambda) \to \iota$. When $\eta = \lambda^+$, from Theorems 3.2.9 and 3.2.10 the relation is true for $\iota = \lambda'$ and false for $\iota = (\lambda')^{++}$. The truth of the relation $(\lambda^+, \lambda^+, \lambda) \to (\lambda')^+$ is an open problem. A couple of results for the case $\eta \leqslant \lambda$ are announced in Erdös, Hajnal and Rado [38], but a complete discussion is lacking. Among the simplest unsolved problems are the following. Is $(\aleph_{\omega+1}, \eta, \aleph_\omega) \to \aleph_{\omega+1}$ true when $2 \leqslant \eta \leqslant \aleph_1$, and is $(\aleph_{\omega+1}, \aleph_\omega, \aleph_\omega) \to \iota$ true when $\aleph_1 \leqslant \iota \leqslant \aleph_\omega$?

§3. Set mappings of higher type

So far in this chapter we have considered maps $f : S \to \mathfrak{P}S$. The concept of a set mapping can be extended by considering functions $f : [S]^\eta \to \mathfrak{P}S$ for any cardinal η. Such a mapping will be said to have *type* η. The maps of type 1 can be identified with the set maps of the earlier sections.

Definition 3.3.1. A set mapping on S of *type* η is a function $f : [S]^\eta \to \mathfrak{P}S$ such that $X \cap f(X) = \emptyset$ for each X in $[S]^\eta$. A subset S' of S is said to be *free* for f if $S' \cap \bigcup f[[S']^\eta] = \emptyset$.

Set mappings of type greater than 1 were first discussed by Erdös and Hajnal in [18].

The first result shows that if we are seeking a free set, only set maps of finite type need be considered, for if η is infinite then there is a set map of type η of order 2 which has no free set of size η. (Trivially, any set of size less than η is free.)

Theorem 3.3.2. *Let η be infinite and let S be a set with $|S| \geqslant \eta$. Then there is a set map f on S of type η and order 2 which has no free set of power η.*

Proof. Consider first the case where $|S| = \eta$. Write $[S]^\eta = \{A_\alpha; \alpha < 2^\eta\}$. Choose by induction sets B_α from $[A_\alpha]^\eta$ for α with $\alpha < 2^\eta$ so that $B_\alpha \neq A_\alpha$ and $B_\alpha \neq B_\beta$ for any β with $\beta < \alpha$. Since $|[A_\alpha]^\eta| = 2^\eta$, such a choice of B_α is always possible. Choose x_α from $A_\alpha - B_\alpha$. Now define a function $f : [S]^\eta \to \mathfrak{P}S$ as follows:

$$f(X) = \begin{cases} \{x_\alpha\} \text{ if } X = B_\alpha , \\ \emptyset \text{ otherwise.} \end{cases}$$

Then always $X \cap f(X) = \emptyset$, so f is a set map of type η, and order 2, on S. Let S' be any subset of S of power η, so $S' = A_\alpha$ for some α. Then $B_\alpha \in [S']^\eta$ and $x_\alpha \in S' \cap f(B_\alpha)$. Thus $x_\alpha \in S' \cap \bigcup f[[S']^\eta]$ and so S' is not free for f. Thus f has no free set of size η, and the theorem holds in this case.

Now suppose $|S| > \eta$. By Zorn's Lemma, there is a maximal almost disjoint family $\mathfrak{M}$ with $\mathfrak{M} \subseteq [S]^\eta$. For each M in $\mathfrak{M}$, we have just shown that there is a set map $f_M : [M]^\eta \to \mathfrak{P}M$ of type η and order 2 on M, which has no free set of size η. By the maximality of $\mathfrak{M}$, for each X in $[S]^\eta$ there is M in $\mathfrak{M}$ such that $|X \cap M| = \eta$. Choose such an M, and call it $M(X)$. Note that if $Y \in [X \cap M(X)]^\eta$ then necessarily $M(Y) = M(X)$, by the almost disjoint property of the family $\mathfrak{M}$. Define a function $f : [S]^\eta \to \mathfrak{P}S$ by

$$f(X) = f_{M(X)}(X \cap M(X)) .$$

Since $f_{M(X)}(X \cap M(X)) \subseteq M(X) - (X \cap M(X))$, certainly $X \cap f(X) = \emptyset$. Thus f is a set mapping on S of type η and order 2. Suppose there is S' in $[S]^\eta$ with S' free for f; thus $S' \cap \bigcup f[[S']^\eta] = \emptyset$. In particular, if $S'' = S' \cap M(S')$ then $S'' \cap \bigcup f_{M(S')}[[S'']^\eta] = \emptyset$, since for any Y in $[S'']^\eta$,

$$f(Y) = f_{M(Y)}(Y \cap M(Y)) = f_{M(S')}(Y \cap M(S')) = f_{M(S')}(Y) .$$

Thus S'' would be a set of power η free for $f_{M(S')}$, which is impossible. Thus f has no free set of size η, and the proof is complete.

In view of Theorem 3.3.2 then, we need only consider the case of set maps of finite type. Just as a set map of type 1 and order κ on a set of size κ need not have a free pair, also a set map of type n and order κ on a set of size $\kappa^{(n-1)+}$ need not have a free set of size $n + 1$.

Theorem 3.3.3. *Let n be a positive integer and let S be a set with $|S| = \kappa^{(n-1)+}$. Then there is a set mapping on S of type n and order κ which has no free set of power $n + 1$.*

Proof. By induction on n. When $n = 1$, we noted in §1 that this is so – identify S with κ, and for α in κ put $f(\alpha) = \alpha = \{\beta; \beta < \alpha\}$.

Now suppose the theorem holds for a particular value of n, and take S with $|S| = \kappa^{(n+)}$. We seek a set map $F : [S]^{n+1} \to [S]^{<\kappa}$ with no free set of size $n + 2$. Identify S with $\{\alpha; \kappa^{(n-1)+} \leqslant \alpha < \kappa^{(n+)}\}$, so by the inductive assumption, for each α in S there is a set map $f_\alpha : [\alpha]^n \to [\alpha]^{<\kappa}$ with no free set of size $n + 1$. Define $F : [S]^{n+1} \to \mathfrak{P}S$ by, if $a \in [S]^{n+1}$ then

$$F(a) = f_\alpha(a - \{\alpha\}) \text{ where } \alpha = \max(a) .$$

Then F is a set map on S of type $n + 1$ and order at most κ. And clearly a free set X for F, with $|X| = n + 2$, would give a free set $X - \{\alpha\}$ of size $n + 1$ for f_α, where $\alpha = \max X$, so that F can have no free set of size $n + 2$. This completes the proof.

The value of $|S|$ in Theorem 3.3.3 is the maximum possible if there is to be no non-trivial free set (at least if the GCH is assumed), as is shown by the next theorem.

Theorem 3.3.4 (GCH). *Let n be a positive integer and let S be a set with $|S| = \kappa^{(n+)}$. Then every set mapping on S of type n and order κ has a free set of power κ^+.*

Proof. If $n = 1$ this follows from Theorem 3.1.1, so we shall suppose henceforth that $n \geqslant 2$. Let $f : [S]^n \to [S]^{<\kappa}$ be a set map on S, where $|S| = \kappa^{(n+)}$. For each a in $[S]^{n-1}$, define an (ordinary) set mapping f_a of order κ on $S-a$ by putting, for x in $S-a$,

$$f_a(x) = f(a \cup \{x\}) .$$

By Theorem 3.1.5, always $S-a$ can be written as a union of at most κ sets each free for f_a, say $S-a = \bigcup\{H_\alpha(a); \alpha < \kappa\}$.

Take a well ordering, $<$ say, of S. Define a partition $\mathbf{\Delta}$ of $[S]^n$ as follows: if $a, b \in [S]^n$ with $a = \{x_0, ..., x_{n-1}\}_<$ and $b = \{y_0, ..., y_{n-1}\}_<$, then

$$a \equiv b \pmod{\mathbf{\Delta}} \Leftrightarrow (\forall i < n)(\exists \alpha < \kappa)\, [x_i \in H_\alpha(a - \{x_i\}) \text{ and }$$
$$y_i \in H_\alpha(a - \{y_i\}))] .$$

Then clearly $\mathbf{\Delta}$ has at most κ classes. Since $|S| = \kappa^{(n+)}$, by the partition relation $\kappa^{(n+)} \to (\kappa^+)^n_\kappa$ from Theorem 2.3.2, there is a set H in $[S]^{\kappa^+}$ which is homogeneous for $\mathbf{\Delta}$. But then H is free for f. For take a from $[H]^n$ and x in $H - a$; it suffices to show that $x \notin f(a)$. Define an element y in a as follows. If $x < u$ for some u in a, then y is the least such u; otherwise y is the largest element of a. Thus y is the i-th element of a if and only if x is the i-element of $a \cup \{x\} - \{y\}$. Since H is homogeneous for $\mathbf{\Delta}$, we have $a \equiv a \cup \{x\} - \{y\}$ (mod $\mathbf{\Delta}$). The definition of $\mathbf{\Delta}$ shows that for some α, both $y \in H_\alpha(a - \{y\})$ and $x \in H_\alpha(a - \{y\})$. Since $H_\alpha(a - \{y\})$ is free for $f_{a-\{y\}}$, then $x \notin f_{a-\{y\}}(y)$; thus $x \notin f(a)$. So H is indeed free for f, and the theorem is proved.

One can ask if, in the situation of Theorem 3.3.4, the size κ^+ of the free set is the largest possible. For $n = 1$ trivially this is so, for $n = 2$ we shall show in Corollary 3.3.6 below that again κ^+ is the best possible, but for $n \geqslant 3$ the problem is open. An upper bound for the size of a free set is given by Corollary 3.3.6. (Compare with Problem 34B in Erdös and Hajnal [24].) First we prove the following result, which should anyway be compared with Theorem 3.3.3.

Theorem 3.3.5 (GCH). *Let $n = 2, 3, \ldots$ and let S be a set with $|S| = \kappa^{(n-1)+}$. There is a set mapping on S of type n and order 2 which has no free set of power $\kappa^{(n-1)+}$.*

Proof. By induction on n. If $n = 2$, take S with $|S| = \kappa^+$, and identify S with κ^+. From Lemma 2.7.3, there is a disjoint partition $[S]^2 = \bigcup\{\Gamma_\gamma; \gamma < \kappa^+\}$ such that whenever $A \in [S]^\kappa$, $B \in [S]^{\kappa^+}$ and $\gamma \in S$ then there are α in A and β in B for which $\{\alpha, \beta\} \in \Gamma_\gamma$. Define $f: [S]^2 \to [S]^{\leqslant 1}$ by:

$$f(\{\alpha, \beta\}) = \begin{cases} \{\gamma\} & \text{if } \{\alpha,\beta\} \in \Gamma_\gamma \text{ and } \gamma \neq \alpha, \beta\,, \\ \emptyset & \text{otherwise.} \end{cases}$$

Then f is a set mapping on S of type 2 and order 2. Moreover, f can have no free set of size κ^+. For take any subset S' of S with $|S'| = \kappa^+$. Choose γ from S' and A from $[S']^\kappa$, B from $[S']^{\kappa^+}$ with $\gamma \notin A \cup B$. By the choice of the partition, there are α, β with $\alpha \in A$, $\beta \in B$ and $\{\alpha, \beta\} \in \Gamma_\gamma$. Thus $f(\{\alpha,\beta\}) = \{\gamma\}$ so $\gamma \in f(\{\alpha,\beta\})$, and hence S' is not free. This completes the case $n = 2$.

For the inductive step, proceed much as in the proof of Theorem 3.3.3. Suppose the theorem holds for a particular value of n, and take S with $|S| = \kappa^{(n+)}$. Identify S with $\{\alpha; \kappa^{(n-1)+} \leqslant \alpha < \kappa^{(n+)}\}$ and for each α in S let $f_\alpha: [\alpha]^n \to [\alpha]^{<2}$ be a set map on α of type n and order 2 with no free set of power $\kappa^{(n-1)+}$. Define $F: [S]^{n+1} \to \mathfrak{P}S$ by, if $a \in [S]^{n+1}$ then

$$F(a) = f_\alpha(a - \{\alpha\}) \text{ where } \alpha = \max(a)\,,$$

so F is a set map on S of type $n + 1$ and order 2. Suppose S' is a subset of S with $|S'| = \kappa^{(n+)}$ which is free for F. Take α in S' such that α has $\kappa^{(n-1)+}$ predecessors in S'. Then clearly $S' \cap \alpha$ is a set of power $\kappa^{(n-1)+}$ which is free for f_α, contrary to the choice of f_α. Thus F has no free set of size $\kappa^{(n+)}$. This completes the proof.

Replacing κ by κ^+ in Theorem 3.3.5 gives the following.

Corollary 3.3.6 (GCH). *Let $n = 2, 3, \ldots$ and let S be a set with $|S| = \kappa^{(n+)}$. There is a set mapping on S of type n and order 2 which has no free set of power $\kappa^{(n+)}$.*

The result in Theorem 3.3.5 shows that a set map on a set S of type n ($n \geqslant 2$) and order λ where $\lambda < |S|$ need not necessarily have a free set of size $|S|$, in contrast to the result in Theorem 3.1.1 for set maps of type 1. However, if $|S|$ is singular, then the exact analogue of Theorem 3.1.1 holds.

Theorem 3.3.7 (GCH). *Let S be a set with $|S| = \kappa$ where κ is singular, and let f be a set map on S of type n and order λ where $\lambda < \kappa$. Then there is a subset S' of S with $|S'| = \kappa$ which is free for f.*

Proof. By Theorem 3.1.1, we need only be concerned with n where $n \geqslant 2$. Identify S with κ. Choose a disjoint partition $\mathbf{\Delta} = \{\Delta_i; i \leqslant n+1\}$ of $[S]^{n+1}$ such that for a in $[S]^{n+1}$, say $a = \{\alpha_0, ..., \alpha_n\}_<$,

$$a \in \Delta_i \Rightarrow \alpha_i \in f(a - \{\alpha_i\}), \text{ for } i \text{ with } i \leqslant n$$

$$a \in \Delta_{n+1} \Rightarrow \forall i \leqslant n\ (\alpha_i \notin f(a - \{\alpha_i\}))\ .$$

From Lemma 4.2.2 (the Canonization Lemma) there is a pairwise disjoint family $\mathcal{C} = \{C_\sigma; \sigma < \kappa'\}$ with $|\bigcup \mathcal{C}| = \kappa$ such that $\mathbf{\Delta}$ is canonical in $\mathcal{C}$ and, by the remarks after the proof of that lemma, we may suppose that always $|C_\sigma| > \lambda$ and $C_\sigma < C_\tau$ whenever $\sigma < \tau$. Thus, whenever $a, b \in [\bigcup \mathcal{C}]^{n+1}$,

$$\forall \sigma\ (|a \cap C_\sigma| = |b \cap C_\sigma|) \Rightarrow a \equiv b\ (\text{mod } \mathbf{\Delta})\ . \tag{1}$$

I claim $[\bigcup \mathcal{C}]^{n+1} \subseteq \Delta_{n+1}$, and consequently $\bigcup \mathcal{C}$ is a set of power κ free for f, as needed. For take any set a from $[\bigcup \mathcal{C}]^{n+1}$, and suppose on the contrary that $a \in \Delta_i$ for some i with $i \leqslant n$. Let the i-th element of a be in C_τ. Choose b from $[\bigcup \mathcal{C}]^n$ such that $b \cap C_\sigma = a \cap C_\sigma$ if $\sigma \neq \tau$, and if $b = \{\beta_0, ..., \beta_{n-1}\}_<$ then for $B = C_\tau \cap \{\beta; \beta_{i-1} < \beta < \beta_i\}$ we have $|B| = \lambda$. Since $|C_\tau| > \lambda$ such a choice of b is possible. Then for any β from B it follows that β is the i-th element of $b \cup \{\beta\}$. Moreover, for such β we have $|a \cap C_\sigma| = |(b \cup \{\beta\}) \cap C_\sigma|$ for all σ, so since $a \in \Delta_i$ it follows from (1) that $b \cup \{\beta\} \in \Delta_i$, that is, $\beta \in f(b)$. Hence $B \subseteq f(b)$. Since f has order λ, this is impossible. This shows that $[\bigcup \mathcal{C}]^{n+1} \subseteq \Delta_{n+1}$, and completes the proof.

CHAPTER 4

POLARIZED PARTITION RELATIONS

§ 1. Partitions of $\kappa \times \kappa$

The polarized partition relations that are to be studied in this chapter were first defined in Erdös and Rado [29] and the more tractable cases carefully investigated in Erdös, Hajnal and Rado [38]. Most of the results in this chapter first appeared in [38]. The polarized relations are an extension of the relation $\kappa \to (\eta_k; k < \gamma)^n$ from Chapter 2, an extension obtained by considering partitions of sequences of finite subsets, rather than partitions of just finite subsets.

Definition 4.1.1. The *polarized partition symbol*

$$\begin{bmatrix} \kappa_1 \\ \cdot \\ \cdot \\ \cdot \\ \kappa_m \end{bmatrix} \to \begin{bmatrix} \eta_{1k} \\ \cdot \\ \cdot \\ \cdot \\ \eta_{mk} \end{bmatrix}_{k<\gamma}^{n_1,\ldots,\, n_m}$$

means: given any sets $S_1, \ldots, S_m$ with $|S_i| = \kappa_i$ (where $i = 1, \ldots, m$), for all partitions $\mathbf{\Delta} = \{\Delta_k; k < \gamma\}$ of $[S_1]^{n_1} \times \ldots \times [S_m]^{n_m}$ into γ parts, there are k with $k < \gamma$ and subsets H_i of S_i with $|H_i| = \eta_{ik}$ (where $i = 1, \ldots, m$) such that $[H_1]^{n_1} \times \ldots \times [H_m]^{n_m} \subseteq \Delta_k$.

We shall adopt without specific mention conventions in the use of the polarized partition symbol similar to those used with the ordinary partition symbol. In particular, the sequence $H_1, \ldots, H_m$ in the above definition will be called a *homogeneous sequence* for the partition. The same simple remarks that were made at the time of the introduction of the ordinary partition symbol apply, mutatis mutandis, to the polarized symbol.

In the first two sections we shall be concerned with the special case where $m = 2, n_1 = n_2 = 1$, and usually $\gamma = 2$ as well. Thus we shall be investigating properties of the symbol

$$\begin{pmatrix} \kappa_1 \\ \kappa_2 \end{pmatrix} \to \begin{pmatrix} \eta_{10} & \eta_{11} \\ \eta_{20} & \eta_{21} \end{pmatrix}^{1,1},$$

that is, seeking homogeneous pairs for partitions of $[S_1]^1 \times [S_2]^1$. We shall always identify $[S]^1$ with S, and so we are dealing with partitions of the cartesian product $S_1 \times S_2$.

A few of the results that will be proved are stronger than can be expressed by the polarized partition symbol. To formulate these compactly, it is convenient to introduce the relation *with alternatives.*

Definition 4.1.2. The *polarized partition symbol with alternatives*

$$\begin{pmatrix} \kappa_1 \\ \kappa_2 \end{pmatrix} \to \begin{pmatrix} \eta_{10} \vee \lambda_{10} & \eta_{11} \vee \lambda_{11} \\ \eta_{20} \vee \lambda_{20} & \eta_{21} \vee \lambda_{21} \end{pmatrix}^{1,1}$$

means: given sets S_1, S_2 with $|S_1| = \kappa_1$ and $|S_2| = \kappa_2$, for all partitions $S_1 \times S_2 = \Delta_0 \cup \Delta_1$, there are k with $k = 0,1$ and subsets H_1 of S_1 and H_2 of S_2 with $H_1 \times H_2 \subseteq \Delta_k$ such that either $|H_1| = \eta_{1k}$ and $|H_2| = \eta_{2k}$, or else $|H_1| = \lambda_{1k}$ and $|H_2| = \lambda_{2k}$.

The definition above can be extended to other cases of the polarized partition symbol. This relation will not be investigated for its own sake. However, the methods that we shall use to establish results concerning the symbol without alternatives at times, without extra work, will lead to a stronger result that can be expressed using the symbol with alternatives.

In this first section, we shall concentrate on the relation with $m = 2$, $n_1 = n_2 = 1$ and further $\kappa_1 = \kappa_2 = \kappa$. We shall start by deducing several results using the theory of set mappings from the previous chapter.

Theorem 4.1.3. *Let κ be infinite with $\eta < \kappa$. Then*

$$\begin{pmatrix} \kappa \\ \kappa \end{pmatrix} \to \begin{pmatrix} \kappa & 1 \\ \kappa & \eta \end{pmatrix}^{1,1}$$

Proof. Let S be any set with $|S| = \kappa$, so we may identify S with κ. Let a partition

$$S \times S = \Delta_0 \cup \Delta_1$$

be given, and suppose $|\{\beta \in S; \langle\alpha, \beta\rangle \in \Delta_1\}| < \eta$ for each α in S. We need to find H_1, H_2 in $[S]^\kappa$ with $H_1 \times H_2 \subseteq \Delta_0$. Define a set mapping f on S as follows: for α in S, put

$$f(\alpha) = \{\beta \in S; \langle\alpha, \beta\rangle \in \Delta_1 \text{ and } \alpha \neq \beta\},$$

so f has order at most η. By Theorem 3.1.1 there is a set S' in $[S]^\kappa$ free for f. Choose disjoint sets H_1, H_2 from $[S']^\kappa$; then $H_1 \times H_2 \subseteq \Delta_0$.

The value of η in this last theorem cannot be increased to $\eta = \kappa$, as the following strong negative relation shows.

Theorem 4.1.4. *For all* κ,

$$\binom{\kappa}{\kappa} \nrightarrow \binom{\kappa \quad 1}{1 \quad \kappa}^{1,1}.$$

Proof. Consider the partition $\kappa \times \kappa = \Delta_0 \cup \Delta_1$ where

$$\langle\alpha, \beta\rangle \in \Delta_0 \Leftrightarrow \alpha \leqslant \beta; \langle\alpha, \beta\rangle \in \Delta_1 \Leftrightarrow \alpha > \beta.$$

Likewise, an attempt to improve Theorem 4.1.3 by increasing the "1" may not meet with success, although this is not as easy to show.

Theorem 4.1.5 (GCH). *Suppose* κ *is regular. Then*

$$\binom{\kappa^+}{\kappa^+} \nrightarrow \binom{\kappa^+ \quad 1 \vee 2}{\kappa^+ \quad \kappa^+ \vee \kappa}^{1,1}.$$

Proof. By virtue of Theorem 3.2.7 there is an almost disjoint family $\mathscr{A} = \{A_\alpha; \alpha < \kappa^+\}$ with always $A_\alpha \in [\kappa^+]^\kappa$ such that whenever $\mathfrak{B}$ is a family in $[\mathscr{A}]^{\kappa^+}$ then $|\kappa^+ - \bigcup\mathfrak{B}| \leqslant \kappa$. Take such a family $\mathscr{A}$, and define a disjoint partition $\kappa^+ \times \kappa^+ = \Delta_0 \cup \Delta_1$ by

$$\langle\alpha, \beta\rangle \in \Delta_1 \Leftrightarrow \beta \in A_\alpha.$$

This partition has the correct properties to prove the theorem. For each α, by definition $\{\beta; \langle\alpha, \beta\rangle \in \Delta_1\} = A_\alpha$, and always $|A_\alpha| < \kappa^+$. Also $|A_\alpha \cap A_\gamma| < \kappa$ if $\alpha \neq \gamma$, so that Δ_1 is as required. With respect to Δ_0, take H_1 from $[\kappa^+]^{\kappa^+}$ and H_2 from $\mathfrak{P}\kappa^+$ with $H_1 \times H_2 \subseteq \Delta_0$; we need to show that $|H_2| < \kappa^+$. However, if $\mathfrak{B} = \{A_\alpha; \alpha \in H_1\}$ then

$$\{\beta; \langle\alpha, \beta\rangle \in \Delta_1 \text{ for some } \alpha \text{ in } H_1\} = \bigcup\mathfrak{B}.$$

Thus $H_2 \subseteq \kappa^+ - \bigcup\mathcal{B}$, and so $|H_2| \leqslant \kappa$. This proves the theorem.

At this stage, we note another negative relation which follows easily from an example constructed in Chapter 2.

Theorem 4.1.6 (GCH). *For all infinite* κ,

$$\binom{\kappa^+}{\kappa^+} \nrightarrow \begin{pmatrix} \kappa \vee \kappa^+ & \kappa \vee \kappa^+ \\ \kappa^+ \vee \kappa & \kappa^+ \vee \kappa \end{pmatrix}^{1,1} .$$

Proof. By Lemma 2.7.3, there is a partition $[\kappa^+]^2 = \bigcup\{\Gamma_k; k < \kappa^+\}$ such that whenever $A \in [\kappa^+]^\kappa$, $B \in [\kappa^+]^{\kappa^+}$, $k \in \kappa^+$ then $\{a, b\} \in \Gamma_k$ for some $\{a, b\}$ where $a \in A$, $b \in B$. Using this partition of $[\kappa^+]^2$, define a disjoint partition $\kappa^+ \times \kappa^+ = \Delta_0 \cup \Delta_1$ by:

$$\langle \alpha, \beta \rangle \in \Delta_0 \Leftrightarrow \{\alpha, \beta\} \in \Gamma_0 .$$

Clearly this partition has the required property.

In the proof of the next theorem, we shall need a polarized canonization lemma, similar to Lemma 2.4.2. This can be proved quite easily from Lemma 2.4.2. We shall state and prove only the special case we need.

Lemma 4.1.7 (GCH). *Let* κ *be singular and let* $\langle \kappa_\sigma; \sigma < \kappa' \rangle$ *be an increasing sequence of cardinals below* κ *with* $\kappa = \Sigma(\kappa_\sigma; \sigma < \kappa')$. *Let a disjoint partition* $\kappa \times \kappa = \bigcup\{\Delta_k; k < \gamma\}$ *be given, where* $\gamma < \kappa$. *Then there are pairwise disjoint families* $\{A_\sigma; \sigma < \kappa'\}$ *and* $\{B_\sigma; \sigma < \kappa'\}$ *and a function* $h : \kappa' \times \kappa' \to \gamma$ *such that* $|A_\sigma| = |B_\sigma| = \kappa_\sigma$ *and*

$$\sigma, \tau < \kappa' \Rightarrow A_\sigma \times B_\tau \subseteq \Delta_{h(\sigma,\tau)} .$$

Proof. For k, l with $k, l < \gamma$ put

$$\Delta(k, l) = \{\{\alpha, \beta\}_< \in [\kappa]^2 ; \langle \alpha, \beta \rangle \in \Delta_k \text{ and } \langle \beta, \alpha \rangle \in \Delta_l\} .$$

This gives a disjoint partition $\mathbf{\Delta}$ of $[\kappa]^2$,

$$[\kappa]^2 = \bigcup\{\Delta(k, l); k, l < \gamma\} .$$

By Lemma 2.4.2, there is a family $\mathcal{C} = \{C_\sigma; \sigma < \kappa'\}$ where $C_\sigma < C_\tau$ whenever $\sigma < \tau$, such that $\mathbf{\Delta}$ is canonical in $\mathcal{C}$, and we may assume $|C_\sigma| \geqslant \kappa_\sigma$. Thus there are functions $h_0, h_1 : \kappa' \to \gamma$ such that

$$\sigma < \tau < \kappa' \text{ and } \alpha \in C_\sigma, \beta \in C_\tau \Rightarrow \{\alpha, \beta\} \in \Delta(h_0(\sigma, \tau), h_1(\sigma, \tau)) . \qquad (1)$$

For σ with $\sigma < \kappa'$, choose A_σ from $[C_{2\sigma}]^{\kappa\sigma}$ and B_σ from $[C_{2\sigma+1}]^{\kappa\sigma}$, so certainly the families $\{A_\sigma; \sigma < \kappa'\}$ and $\{B_\sigma; \sigma < \kappa'\}$ are both pairwise disjoint. Take σ, τ with $\sigma, \tau < \kappa'$, and consider α from A_σ and β from B_τ. If $\sigma \leqslant \tau$ then $2\sigma < 2\tau + 1$ so that $\alpha < \beta$ and by (1),

$$\{\alpha, \beta\} \in \Delta(h_0(2\sigma, 2\tau + 1), h_1(2\sigma, 2\tau + 1))$$

so that $\langle\alpha, \beta\rangle \in \Delta_{h_0(2\sigma, 2\tau+1)}$; whereas if $\tau < \sigma$ then $2\tau + 1 < 2\sigma$ so that $\beta < \alpha$ and by (1),

$$\{\alpha, \beta\} \in \Delta(h_0(2\tau + 1, 2\sigma), h_1(2\tau + 1, 2\sigma)) ,$$

so that $\langle\alpha, \beta\rangle \in \Delta_{h_1(2\tau+1, 2\sigma)}$.

Hence define $h : \kappa' \times \kappa' \to \gamma$ by

$$h(\sigma, \tau) = \begin{cases} h_0(2\sigma, 2\tau + 1) & \text{if } \sigma \leqslant \tau , \\ h_1(2\tau + 1, 2\sigma) & \text{if } \tau < \sigma . \end{cases}$$

Then always $A_\sigma \times B_\tau \subseteq \Delta_{h(\sigma,\tau)}$, and the lemma is proved.

Theorem 4.1.8 (GCH). *Let η_0, η_1 be cardinals with $\eta_0^+, \eta_1^+ < \kappa$. Then*

$$\binom{\kappa}{\kappa} \to \begin{pmatrix} \kappa & \kappa \vee \eta_0 \\ \kappa & \eta_1 \vee \kappa \end{pmatrix}^{1,1} . \qquad (1)$$

Proof. Let S be any set with $|S| = \kappa$, so we may identify S with κ. Let any disjoint partition $S \times S = \Delta_0 \cup \Delta_1$ be given.

Consider first the case that κ is regular. Define set mappings f_0, f_1 on S as follows:

$$f_0(\alpha) = \{\beta < \alpha; \langle\beta, \alpha\rangle \in \Delta_1\}, \quad f_1(\alpha) = \{\beta < \alpha; \langle\alpha, \beta\rangle \in \Delta_1\} .$$

Then both f_0, f_1 have order at most κ. If there is X in $[S]^\kappa$ with $|\bigcap f_0[X]| \geqslant \eta_0$, since $\bigcap f_0[X] \times X \subseteq \Delta_1$ there is nothing more to show. Similarly if there is X in $[S]^\kappa$ with $|\bigcap f_1[X]| \geqslant \eta_1$, since $X \times \bigcap f_1[X] \subseteq \Delta_1$ there is nothing more to do. So suppose that $|\bigcap f_0[X]| < \eta_0$ and $|\bigcap f_1[X]| < \eta_1$ whenever $X \in [S]^\kappa$. By Theorem 3.2.5, there is a set A in $[S]^\kappa$ which is free for f_0. Let $f_1' : A \to \mathfrak{P}A$ be the restriction of f_1 to A, that is $f_1'(\alpha) = f_1(\alpha) \cap A$. Again by Theorem 3.2.5 there is a set B in $[A]^\kappa$ which is free for f_1'. Take two disjoint sets H_1, H_2 from $[B]^\kappa$. Take α from H_1 and β from H_2. If $\alpha < \beta$, since A is free for f_0 we know $\alpha \notin f_0(\beta)$ and so $\langle \alpha, \beta \rangle \notin \Delta_1$. If $\beta < \alpha$, since B is free for f_1' we know $\beta \notin f_1(\alpha)$ and so $\langle \alpha, \beta \rangle \notin \Delta_1$. So in any event $\langle \alpha, \beta \rangle \in \Delta_0$ and hence $H_1 \times H_2 \subseteq \Delta_0$. This proves the relation (1) when κ is regular.

Now suppose that κ is singular. Choose an increasing sequence of cardinals $\langle \kappa_\sigma; \sigma < \kappa' \rangle$ with always $\eta_0, \eta_1 \leqslant \kappa_\sigma < \kappa$ and $\kappa = \Sigma(\kappa_\sigma; \sigma < \kappa')$. By Lemma 4.1.7 there are pairwise disjoint families $\{A_\sigma; \sigma < \kappa'\}$ and $\{B_\sigma; \sigma < \kappa'\}$ with $A_\sigma, B_\sigma \in [S]^{\kappa_\sigma}$, together with a function $h : \kappa' \times \kappa' \to \{0,1\}$ such that $A_\sigma \times B_\tau \subseteq \Delta_{h(\sigma,\tau)}$. Define a disjoint partition $\kappa' \times \kappa' = \Delta_0^* \cup \Delta_1^*$, where

$$\langle \sigma, \tau \rangle \in \Delta_i^* \Leftrightarrow h(\sigma, \tau) = i\,, \quad (i = 0,1)\,.$$

Since κ' is regular, the theorem applies to κ'. In particular

$$\begin{pmatrix} \kappa' \\ \kappa' \end{pmatrix} \to \begin{pmatrix} \kappa' & \kappa' \vee 1 \\ \kappa' & 1 \vee \kappa' \end{pmatrix}^{1,1}.$$

So one of the following three events happens. (1) There are I_1, I_2 from $[\kappa']^{\kappa'}$ such that $I_1 \times I_2 \subseteq \Delta_0^*$. If we put $H_1 = \bigcup\{A_\sigma; \sigma \in I_1\}$, $H_2 = \bigcup\{B_\tau; \tau \in I_2\}$ then $|H_1| = |H_2| = \kappa$ and $H_1 \times H_2 \subseteq \Delta_0$. (2) There are τ in κ' and I_1 in $[\kappa']^{\kappa'}$ such that $I_1 \times \{\tau\} \subseteq \Delta_1^*$. Put $H_1 = \bigcup\{A_\sigma; \sigma \in I_1\}$ so $|H_1| = \kappa$, and $H_1 \times B_\tau \subseteq \Delta_1$. (3) There are σ in κ' and I_2 in $[\kappa']^{\kappa'}$ such that $\{\sigma\} \times I_2 \subseteq \Delta_1^*$. Put $H_2 = \bigcup\{B_\tau; \tau \in I_2\}$ so $|H_2| = \kappa$, and $A_\sigma \times H_2 \subseteq \Delta_1$. Since $|B_\tau| = \kappa_\tau \geqslant \eta_1$ and $|A_\sigma| = \kappa_\sigma \geqslant \eta_0$, this proves that the relation (1) holds. The proof is complete.

Theorem 4.1.9 (GCH). *Let η_0, η_1 be cardinals with $\eta_0^+, \eta_1 < \kappa$. Then*

$$\begin{pmatrix} \kappa \\ \kappa \end{pmatrix} \to \begin{pmatrix} \kappa & \kappa \\ \eta_0 & \eta_1 \end{pmatrix}^{1,1}.$$

Proof. Let any disjoint partition $\kappa \times \kappa = \Delta_0 \cup \Delta_1$ be given. Suppose first that κ is regular. Define a set mapping f on κ as follows:

$$f(\alpha) = \{\beta < \alpha; \langle \alpha, \beta \rangle \in \Delta_0\}\,,$$

so f has order at most κ. If there is X in $[\kappa]^\kappa$ with $|\bigcap f[X]| \geqslant \eta_0$, since $X \times \bigcap f[X] \subseteq \Delta_0$ there is nothing more to show. Otherwise, for all X in $[\kappa]^\kappa$ we have $|\bigcap f[X]| < \eta_0$, so by Theorem 3.2.5 there must be S in $[\kappa]^\kappa$ which is free for f. Choose subsets H_1, H_2 of S with $|H_1| = \kappa$, $|H_2| = \eta_1$ and $H_2 < H_1$. Then if $\alpha \in H_1$ and $\beta \in H_2$ we know $\beta \notin f(\alpha)$ and so $\langle \alpha, \beta \rangle \notin \Delta_0$; hence $H_1 \times H_2 \subseteq \Delta_1$ and the result follows.

The argument when κ is singular is the same as that in Theorem 4.1.8.

We come now to one of the most difficult theorems involving the symbol

$$\begin{pmatrix} \kappa \\ \kappa \end{pmatrix} \to \begin{pmatrix} \eta_{10} & \eta_{11} \\ \eta_{20} & \eta_{21} \end{pmatrix}^{1,1} .$$

We shall need two lemmas first.

Lemma 4.1.10. *For each finite n,*

$$\begin{pmatrix} \kappa \\ \kappa^+ \end{pmatrix} \to \begin{pmatrix} \kappa & \kappa \\ \kappa^+ & n \end{pmatrix}^{1,1} . \tag{1}$$

Proof. Use induction on n, starting from the trivial relation with $n = 0$. So make the inductive hypothesis that (1) is true for a particular value of n. Take sets S, T with $|S| = \kappa$, $|T| = \kappa^+$ and let a disjoint partition $S \times T = \Delta_0 \cup \Delta_1$ be given. Suppose there are no sets H_1 in $[S]^\kappa$ and H_2 in $[T]^{\kappa^+}$ with $H_1 \times H_2 \subseteq \Delta_0$. By the inductive hypothesis, there must then be sets H_1 in $[S]^\kappa$ and H_2 in $[T]^n$ such that $H_1 \times H_2 \subseteq \Delta_1$.

For each y in T, put $Q_1(y) = \{x \in S; \langle x, y\rangle \in \Delta_1\}$. If there is y in $T - H_2$ for which $|H_1 \cap Q_1(y)| = \kappa$, then $(H_1 \cap Q_1(y)) \times (H_2 \cup \{y\}) \subseteq \Delta_1$ and the result follows. So suppose $|H_1 \cap Q_1(y)| < \kappa$ for all y in $T - H_2$. We shall show that a contradiction results.

For each cardinal λ with $\lambda < \kappa$, put

$$T_\lambda = \{y \in T - H_2 ; |H_1 \cap Q_1(y)| = \lambda\} .$$

There must be some λ for which $|T_\lambda| = \kappa^+$. For such a λ, write H_1 as a disjoint union, $H_1 = \bigcup\{H_{1\alpha}; \alpha < \lambda^+\}$ where always $|H_{1\alpha}| = \kappa$. If $y \in T_\lambda$ then $|H_1 \cap Q_1(y)| = \lambda < \lambda^+$ and hence there is $\alpha(y)$ with $\alpha(y) < \lambda^+$ for which $Q_1(y) \cap H_{1\alpha(y)} = \emptyset$. Since κ^+ is regular, there is Y in $[T_\lambda]^{\kappa^+}$ such that $\alpha(y)$ is constant for y in Y, say with value α. Thus $Q_1(y) \cap H_{1\alpha} = \emptyset$ for all y in Y. Hence $H_{1\alpha} \times Y \subseteq \Delta_0$. But $|H_{1\alpha}| = \kappa$ and $|Y| = \kappa^+$, so this contradicts the assumed property of the partition. This completes the inductive step, and the lemma is proved.

Lemma 4.1.11 (GCH). *Let κ be singular and suppose a partition $\kappa \times \kappa^+ = \Delta_0 \cup \Delta_1$ is given for which there are no sets H_1 in $[\kappa]^\kappa$ and H_2 in $[\kappa^+]^{\kappa^+}$ with $H_1 \times H_2 \subseteq \Delta_1$. Let λ be a regular cardinal with $\kappa' < \lambda < \kappa$. Take any set A in $[\kappa]^\kappa$ and subset B of κ^+. Then there is a family $\mathcal{A}$ of at most κ sets from $[A]^\lambda$ and a map $f: \mathcal{A} \to \mathfrak{P}B$ such that $|B - \bigcup f[\mathcal{A}]| \leqslant \kappa$ and $X \times f(X) \subseteq \Delta_0$ for all X in $\mathcal{A}$.*

Proof. Let A and B be given, with $A \in [\kappa]^\kappa$ and $B \subseteq \kappa^+$. Write $Q_0(\beta) = \{\alpha < \kappa; \langle\alpha, \beta\rangle \in \Delta_0\}$. Put $B_1 = \{\beta \in B; |A \cap Q_0(\beta)| < \lambda\}$ and $B_2 = B - B_1$. From lemma 4.1.10, the relation

$$\begin{pmatrix} \kappa \\ \kappa^+ \end{pmatrix} \to \begin{pmatrix} \kappa & \kappa \\ 1 & \kappa^+ \end{pmatrix}^{1,1}$$

holds. Since certainly $A \times B_1 \subseteq \Delta_0 \cup \Delta_1$, if $|B_1| = \kappa^+$ by this relation either there would be β in B_1 such that $|A \cap Q(\beta)| = \kappa$ (contrary to the definition of B_1) or else there would be H_1 in $[A]^\kappa$ and H_2 in $[B_1]^{\kappa^+}$ with $H_1 \times H_2 \subseteq \Delta_1$ (contrary to the choice of the partition). So we deduce that $|B_1| \leqslant \kappa$.

Let $\langle\kappa_\sigma; \sigma < \kappa'\rangle$ be a sequence of cardinals with $\kappa_\sigma < \kappa$ and $\kappa = \Sigma(\kappa_\sigma; \sigma < \kappa')$, and write A as a disjoint union $A = \bigcup\{A_\sigma; \sigma < \kappa'\}$ where always $|A_\sigma| = \kappa_\sigma$. Put $\mathscr{A} = \bigcup\{[A_\sigma]^\lambda; \sigma < \kappa'\}$, so $|\mathscr{A}| \leqslant \Sigma(\kappa_\sigma^\lambda; \sigma < \kappa') \leqslant \kappa$. For X in $\mathscr{A}$, define $f(X) = \{\beta \in B_2; X \subseteq A \cap Q_0(\beta)\}$, so f maps $\mathscr{A}$ into $\mathfrak{P}B$ and $X \times f(X) \subseteq \Delta_0$ for each X. Further, for each β in B_2 there is X in $\mathscr{A}$ for which $\beta \in f(X)$, since otherwise $|A_\sigma \cap Q_0(\beta)| < \lambda$ for all σ with $\sigma < \kappa'$ and so by the regularity of λ and the definition of B_2, we would have the contradiction

$$\lambda \leqslant |A \cap Q_0(\beta)| = \Sigma(|A_\sigma \cap Q_0(\beta)|; \sigma < \kappa'\} < \lambda\,.$$

Hence $B_2 = \bigcup f[\mathscr{A}]$. Thus $B - \bigcup f[\mathscr{A}] = B_1$, and the result follows.

We are now set for the major theorem.

Theorem 4.1.12 (GCH). *For all* κ,

$$\begin{pmatrix} \kappa^+ \\ \kappa^+ \end{pmatrix} \to \begin{pmatrix} \kappa & \kappa \vee \kappa^+ \\ \kappa & \kappa^+ \vee \kappa \end{pmatrix}^{1,1}.$$

Proof. Let S be any set with $|S| = \kappa^+$, and let a partition $S \times S = \Delta_0 \cup \Delta_1$ be given. Assume for all H_1, H_2 with $H_1 \in [S]^\kappa$, $H_2 \in [S]^{\kappa^+}$ that $H_1 \times H_2 \not\subseteq \Delta_1$ and $H_2 \times H_1 \not\subseteq \Delta_1$, so we seek H_1, H_2 in $[S]^\kappa$ such that $H_1 \times H_2 \subseteq \Delta_0$.

Identify S with κ^+. For α, β with $\alpha, \beta < \kappa^+$ define

$$P_0(\alpha) = \{\beta < \kappa^+; \langle\alpha, \beta\rangle \in \Delta_0\};\ Q_0(\beta) = \{\alpha < \kappa^+; \langle\alpha, \beta\rangle \in \Delta_0\}\,.$$

Then if $A \in [\kappa^+]^\kappa$ and $B \subseteq \kappa^+$, since

$$A \times (B - \bigcup\{P_0(\alpha); \alpha \in A\}) \subseteq \Delta_1$$

it follows that $|B - \bigcup\{P_0(\alpha); \alpha \in A\}| \leqslant \kappa$. Similarly, if $B \in [\kappa^+]^\kappa$ and $A \subseteq \kappa^+$ then $|A - \bigcup\{Q_0(\beta); \beta \in B\}| \leqslant \kappa$.

Suppose first that κ is regular. We shall define a ramification system $\mathfrak{R}$ on κ^+ of height κ so that reading along any branch of the ramification tree gives a chain of elements $\alpha_0, \beta_0, \alpha_1, \beta_1, \ldots$ such that always $\langle \alpha_\sigma, \beta_\tau \rangle \in \Delta_0$. The idea of the construction is roughly as follows (with the notation conventions of Chapter 2, §2). Suppose we have already constructed the ramification system up to level σ, so for a sequence $\mathbf{v}$ in $N \cap SEQ_\sigma$ we know $S(\mathbf{v})$. Choose A from $[S(\mathbf{v})]^\kappa$. By the remark above (with B replaced by $S(\mathbf{v})$), at most κ elements β of $S(\mathbf{v})$ fail to have a companion α in A such that $\langle \alpha, \beta \rangle \in \Delta_0$. Discard these elements by putting them in $F(\mathbf{v})$, split up the remaining elements of $S(\mathbf{v})$ according to their companion α in A and push them up to the next level. Do this alternately on the first component and the second component of the pairs $\langle \alpha, \beta \rangle$.

Formally, then, take σ with $\sigma < \kappa$ and a sequence $\mathbf{v}$ from SEQ_σ. Suppose $S(\mathbf{v})$ has already been defined. If $|S(\mathbf{v})| \leqslant \kappa$, put $F(\mathbf{v}) = S(\mathbf{v})$ and $n(\mathbf{v}) = 0$. If $|S(\mathbf{v})| = \kappa^+$, put $n(\mathbf{v}) = \kappa$. Choose $R(\mathbf{v})$ from $[S(\mathbf{v})]^\kappa$ and for $\boldsymbol{\mu}$ in $N \cap SEQ_{\sigma+1}$ with $\boldsymbol{\mu} \upharpoonright \sigma = \mathbf{v}$ let $\gamma(\boldsymbol{\mu})$ range over the elements of $R(\mathbf{v})$. If σ is even, put

$$F(\mathbf{v}) = R(\mathbf{v}) \cup (S(\mathbf{v}) - \bigcup \{P_0(\gamma); \gamma \in R(\mathbf{v})\})$$

and for $\mathbf{v}$ in $N \cap SEQ_{\sigma+1}$ with $\boldsymbol{\mu} \upharpoonright \sigma = \mathbf{v}$ put

$$S(\boldsymbol{\mu}) = (S(\mathbf{v}) \cap P_0(\gamma(\boldsymbol{\mu}))) - F(\mathbf{v}) .$$

If σ is odd, put

$$F(\mathbf{v}) = R(\mathbf{v}) \cup (S(\mathbf{v}) - \bigcup \{Q_0(\gamma); \gamma \in R(\mathbf{v})\})$$

and for $\mathbf{v}$ in $N \cap SEQ_{\sigma+1}$ with $\boldsymbol{\mu} \upharpoonright \sigma = \mathbf{v}$ put

$$S(\boldsymbol{\mu}) = (S(\mathbf{v}) \cap Q_0(\gamma(\boldsymbol{\mu}))) - F(\mathbf{v}) .$$

Thus in either case it follows that

$$S(\mathbf{v}) - F(\mathbf{v}) = \bigcup \{S(\boldsymbol{\mu}); \boldsymbol{\mu} \in N \cap SEQ_{\sigma+1} \text{ and } \boldsymbol{\mu} \upharpoonright \sigma = \mathbf{v}\} .$$

This completes the definition of the ramification system $\mathfrak{R}$.

By definition $|R(\mathbf{v})| = \kappa$, so from the remark above it follows that $|F(\mathbf{v})| \leqslant \kappa$. Hence the Ramification Lemma, Theorem 2.2.3(ii) applies to $\mathfrak{R}$. This yields a sequence $\mathbf{v}$ in $N \cap SEQ_\kappa$ such that $S(\mathbf{v}) \neq \emptyset$. Choose and fix such a sequence $\mathbf{v}$. Put $\alpha_\sigma = \gamma(\mathbf{v} \upharpoonright 2\sigma + 1)$ and $\beta_\sigma = \gamma(\mathbf{v} \upharpoonright 2\sigma + 2)$, for all σ with $\sigma < \kappa$. Write $H_1 = \{\alpha_\sigma; \sigma < \kappa\}$ and $H_2 = \{\beta_\sigma; \sigma < \kappa\}$. Since $\alpha_\sigma = \gamma(\mathbf{v} \upharpoonright 2\sigma + 1) \in F(\mathbf{v} \upharpoonright 2\sigma)$ and the $F(\mathbf{v} \upharpoonright 2\sigma)$ are pairwise disjoint (by Lemma 2.2.2(ii)) it follows that $|H_1| = \kappa$. Similarly $|H_2| = \kappa$. Moreover, $H_1 \times H_2 \subseteq \Delta_0$. For take σ, τ with

$\sigma, \tau < \kappa$; we want to show $\langle \alpha_\sigma, \beta_\tau \rangle \in \Delta_0$. Suppose $\sigma \leqslant \tau$. Then

$$\beta_\tau = \gamma(\mathbf{v} \restriction 2\tau + 2) \in R(\mathbf{v} \restriction 2\tau + 1) \subseteq S(\mathbf{v} \restriction 2\tau + 1) \subseteq S(\mathbf{v} \restriction 2\sigma + 1),$$

$$S(\mathbf{v} \restriction 2\sigma + 1) \subseteq P_0(\gamma(\mathbf{v} \restriction 2\sigma + 1)) = P_0(\alpha_\sigma),$$

and so $\langle \alpha_\sigma, \beta_\tau \rangle \in \Delta_0$. Similarly if $\tau < \sigma$ we find $\alpha_\sigma \in Q_0(\beta_\tau)$ and again $\langle \alpha_\sigma, \beta_\tau \rangle \in \Delta_0$. This proves the theorem if κ is regular.

Now suppose that κ is singular. Choose an increasing sequence of regular cardinals $\langle \kappa_\sigma; \sigma < \kappa' \rangle$ such that $\kappa' < \kappa_\sigma < \kappa$ and $\kappa = \Sigma(\kappa_\sigma; \sigma < \kappa')$. We shall define a ramification system $\mathfrak{R}$ on κ^+ of height κ' (so the Ramification Lemma will apply) in such a way that reading along any branch in the tree gives a sequence of sets $A_0, B_0, A_1, B_1, \ldots$ such that $|A_\sigma|, |B_\sigma| \geqslant \kappa_\sigma$ and always $A_\sigma \times B_\tau \subseteq \Delta_0$. The construction is similar to that used in the case above, except that Lemma 4.1.11 is applied to provide the appropriate splitting up of the set $S(\mathbf{v})$.

To define $\mathfrak{R}$ formally, suppose $S(\mathbf{v})$ is given for some $\mathbf{v}$ in $N \cap SEQ_\sigma$ where $\sigma < \kappa'$. If $|S(\mathbf{v})| \leqslant \kappa$, put $F(\mathbf{v}) = S(\mathbf{v})$ and $n(\mathbf{v}) = 0$. If $|S(\mathbf{v})| = \kappa^+$ put $n(\mathbf{v}) = \kappa$. Choose $R(\mathbf{v})$ from $[S(\mathbf{v})]^\kappa$. If σ is even, apply Lemma 4.1.11 with $\lambda = \kappa_\sigma$, $A = R(\mathbf{v})$ and $B = S(\mathbf{v})$ to obtain a subfamily $\mathcal{A}(\mathbf{v}) = \{A_\alpha(\mathbf{v}); \alpha < \kappa\}$ of $[R(\mathbf{v})]^{\kappa_\sigma}$ and a map $f_\mathbf{v}$ such that $A_\alpha(\mathbf{v}) \times f_\mathbf{v}(A_\alpha(\mathbf{v})) \subseteq \Delta_0$ and $|S(\mathbf{v}) - \bigcup f_\mathbf{v}[\mathcal{A}(\mathbf{v})]| \leqslant \kappa$. Put

$$F(\mathbf{v}) = R(\mathbf{v}) \cup (S(\mathbf{v}) - \bigcup f_\mathbf{v}[\mathcal{A}(\mathbf{v})]),$$

and for $\boldsymbol{\mu}$ in $N \cap SEQ_{\sigma+1}$ with $\boldsymbol{\mu} \restriction \sigma = \mathbf{v}$ define

$$S(\boldsymbol{\mu}) = f_\mathbf{v}(A_{\boldsymbol{\mu}(\sigma)}(\mathbf{v})) - F(\mathbf{v}).$$

If σ is odd, use the obvious modification to Lemma 4.1.11 to obtain a family $\mathcal{A}(\mathbf{v})$ and a map $f_\mathbf{v}$ as above, only that this time $f_\mathbf{v}(A_\alpha(\mathbf{v})) \times A_\alpha(\mathbf{v}) \subseteq \Delta_0$. Define $F(\mathbf{v})$ and $S(\boldsymbol{\mu})$ as above. This completes the definition of the Ramification system $\mathfrak{R}$.

Then, as in the regular case, there is a sequence $\mathbf{v}$ in $N \cap SEQ_{\kappa'}$ such that $S(\mathbf{v}) \neq \emptyset$. Put $A_\sigma = A_{\mathbf{v}(2\sigma)}(\mathbf{v} \restriction 2\sigma)$ and $B_\sigma = A_{\mathbf{v}(2\sigma+1)}(\mathbf{v} \restriction 2\sigma + 1)$, then $H_1 = \bigcup\{A_\sigma; \sigma < \kappa'\}$ and $H_2 = \bigcup\{B_\sigma; \sigma < \kappa'\}$. It follows that $|H_1| = |H_2| = \kappa$. And further, $H_1 \times H_2 \subseteq \Delta_0$. For take σ, τ with $\sigma, \tau < \kappa'$. We show $A_\sigma \times B_\tau \subseteq \Delta_0$. Suppose $\sigma \leqslant \tau$. Then

$$B_\tau = A_{\mathbf{v}(2\tau+1)}(\mathbf{v} \restriction 2\tau + 1) \subseteq R(\mathbf{v} \restriction 2\tau + 1) \subseteq S(\mathbf{v} \restriction 2\tau + 1),$$

$$S(\mathbf{v} \restriction 2\tau + 1) \subseteq S(\mathbf{v} \restriction 2\sigma + 1) \subseteq f_{\mathbf{v} \restriction 2\sigma}(A_{\mathbf{v}(2\sigma)}(\mathbf{v} \restriction 2\sigma)) = f_{\mathbf{v} \restriction 2\sigma}(A_\sigma),$$

and so $A_\sigma \times B_\tau \subseteq \Delta_0$. If $\sigma > \tau$, similarly $A_\sigma \subseteq f_{\mathbf{v} \restriction 2\tau+1}(B_\tau)$ so again $A_\sigma \times B_\tau$

$\subseteq \Delta_0$. This completes the proof for singular κ, and concludes the proof of Theorem 4.1.12.

We shall prove one more negative relation, and then we shall survey the results obtained so far.

Theorem 4.1.13 (GCH). *If* $\kappa' > \aleph_0$ *then*

$$\begin{pmatrix} \kappa^+ \\ \kappa^+ \end{pmatrix} \nrightarrow \begin{pmatrix} \kappa & \kappa \vee \kappa^+ \\ \kappa^+ & \aleph_0 \vee 1 \end{pmatrix}^{1,1},$$

and if $\kappa' = \aleph_0$ then

$$\begin{pmatrix} \kappa^+ \\ \kappa^+ \end{pmatrix} \nrightarrow \begin{pmatrix} \kappa & \kappa \vee \kappa^+ \\ \kappa^+ & \aleph_2 \vee 1 \end{pmatrix}^{1,1}.$$

Proof. We shall construct a partition $\kappa^+ \times \kappa^+ = \Delta_0 \cup \Delta_1$ which demonstrates the negative relation. As before, to ensure that there are not H_1 in $[\kappa^+]^\kappa$ and H_2 in $[\kappa^+]^{\kappa^+}$ with $H_1 \times H_2 \subseteq \Delta_0$, for each H_1 in $[\kappa^+]^\kappa$ there must be γ such that for all μ with $\mu \geqslant \gamma$ there is some chosen element $x(\mu)$ in H_1 such that $\langle x(\mu), \mu \rangle \in \Delta_1$. In order not to make Δ_1 too large, we want that any particular element in κ^+ be not chosen as $x(\mu)$ for too many μ. Since $|H_1| = \kappa$, when choosing $x(\mu)$ we can safely avoid only fewer than κ of the earlier $x(\nu)$. The formal construction is as follows.

Let $[\kappa^+]^\kappa = \{A_\alpha; \alpha < \kappa^+\}$. For each β with $0 < \beta < \kappa^+$, since $1 \leqslant |\beta| \leqslant \kappa$ we may write $\beta = \{\beta(\nu); \nu < \kappa\}$ and $\{A_\alpha; \alpha < \beta\} = \{B_{\beta\nu}; \nu < \kappa\}$. Use transfinite induction on β to choose elements $x(\beta, \nu)$ (where $\beta < \kappa^+$ and $\nu < \kappa$) such that

$$x(\beta, \nu) \in B_{\beta\nu} - \{x(\beta(\mu), \sigma); \mu < \nu \text{ and } \sigma \leqslant \nu\}.$$

Since $|\{\langle \mu, \sigma \rangle; \mu < \nu \text{ and } \sigma \leqslant \nu\}| < \kappa$ this is certainly possible. Define a partition $\kappa^+ \times \kappa^+ = \Delta_0 \cup \Delta_1$ as follows:

$$\langle \alpha, \beta \rangle \in \Delta_1 \Leftrightarrow \exists\, \nu < \kappa\ (\alpha = x(\beta, \nu)),$$

$$\langle \alpha, \beta \rangle \in \Delta_0 \Leftrightarrow \langle \alpha, \beta \rangle \notin \Delta_1.$$

Then if $H_1 \in [\kappa^+]^\kappa$, $H_2 \in [\kappa^+]^{\kappa^+}$, certainly $H_1 = A_\alpha$ for some α and there is β in H_2 with $\alpha < \beta$. Thus $H_1 = B_{\beta\nu}$ for some ν with $\nu < \kappa$. Then $x(\beta, \nu) \in B_{\beta\nu} = H_1$ and $\langle x(\beta, \nu), \beta \rangle \in \Delta_1$ so that $H_1 \times H_2 \nsubseteq \Delta_0$. Also for any β, we have $|\{\alpha; \langle \alpha, \beta \rangle \in \Delta_1\}| = |\{x(\beta, \nu); \nu < \kappa\}| \leqslant \kappa$ and so if $H_1 \in [\kappa^+]^{\kappa^+}$, $\beta \in \kappa^+$ we have $H_1 \times \{\beta\} \nsubseteq \Delta_1$.

Note that the choice of $x(\beta, \nu)$ ensures

$$\mu < \nu < \kappa \text{ and } \alpha = \beta(\mu) \text{ and } x(\alpha, \sigma) = x(\beta, \nu) \Rightarrow \nu < \sigma \,. \tag{1}$$

Take any infinite set B from $[\kappa^+]^{<\kappa}$ and suppose there is ρ with $\rho < \kappa$ such that

$$\rho > \sup\{\mu < \kappa;\ \exists\ \alpha, \beta \in B\ (\alpha = \beta(\mu))\}\,.$$

Put $Q_1(B) = \{\alpha;\ \forall \beta \in B(\langle\alpha, \beta\rangle \in \Delta_1)\}$; I claim $|Q_1(B)| < \kappa$. For take x from $Q_1(B)$. Then for each β in B there must be $\nu(\beta)$ such that $x = x(\beta, \nu(\beta))$. Suppose $\nu(\beta) \geqslant \rho$ for all β in B. Then if we take β_1, β_2 in B with $\beta_1 < \beta_2$, we have $\beta_1 = \beta_2(\mu)$ for some μ, and $\mu < \rho \leqslant \nu(\beta_2)$ so that by (1), $\nu(\beta_2) < \nu(\beta_1)$. Since B is infinite this leads to an infinite descending chain of ordinals, which is impossible. Consequently $\nu(\beta) < \rho$ for some β in B. Hence $Q_1(B) \subseteq \{x(\beta, \nu);\ \nu < \rho \text{ and } \beta \in B\}$ so that $|Q_1(B)| < \kappa$ as claimed.

Suppose $\kappa' > \aleph_0$. Take any B in $[\kappa^+]^{\aleph_0}$. Certainly there is ρ with $\rho < \kappa$ and $\rho > \sup\{\mu;\ \exists\ \alpha, \beta \in B(\alpha = \beta(\mu))\}$, so that $|Q_1(B)| < \kappa$. But if $A \times B \subseteq \Delta_1$ then $A \subseteq Q_1(B)$, so there are not A in $[\kappa^+]^\kappa$, B in $[\kappa^0]^{\aleph_0}$ with $A \times B \subseteq \Delta_1$.

Suppose $\kappa' = \aleph_0$, and let $\langle\kappa_n;\ n < \omega\rangle$ be a strictly increasing sequence of cardinals with $\kappa = \Sigma(\kappa_n;\ n < \omega)$. Take any set B in $[\kappa^+]^{\aleph_2}$, and form the partition $[B]^2 = \bigcup\{\Gamma_n;\ n < \omega\}$ where, in the notation above, for α, β in B with $\alpha < \beta$,

$$\{\alpha, \beta\} \in \Gamma_n \Leftrightarrow \exists\ \mu < \kappa_n\ (\alpha = \beta(\mu))\,.$$

By the relation $\aleph_2 \to (\aleph_0)^2_{\aleph_0}$ there is an infinite subset B^* of B such that $[B^*]^2 \subseteq \Gamma_n$ for some n. Thus

$$\kappa_{n+1} > \kappa_n \geqslant \sup\{\mu;\ \exists\ \alpha, \beta \in B^*\ (\alpha = \beta(\mu))\}\,,$$

so that $|Q_1(B^*)| < \kappa$ and hence $|Q_1(B)| < \kappa$. Thus there are not A in $[\kappa^+]^\kappa$, B in $[\kappa^+]^{\aleph_2}$ such that $A \times B \subseteq \Delta_1$.

Thus the partition $\{\Delta_0, \Delta_1\}$ provides an example which proves the theorem.

We shall now summarize the results concerning the symbol

$$\begin{pmatrix}\kappa\\ \kappa\end{pmatrix} \to \begin{pmatrix}\eta_0 & \eta_1\\ \eta_2 & \eta_3\end{pmatrix}^{1,1}$$

where $0 < \eta_0, \ldots, \eta_3 \leqslant \kappa$, assuming GCH throughout. It follows from Theorem 4.1.12 that the relation is true if $\eta_0, \ldots, \eta_3 < \kappa$. If any three of $\eta_0, \ldots, \eta_3$ are equal to κ, the relation is false (from Theorem 4.1.4).

By Theorem 4.1.9 the relation

$$\binom{\kappa}{\kappa} \to \begin{pmatrix} \kappa & \kappa \\ \eta_2 & \eta_3 \end{pmatrix}^{1,1}$$

is true if $\eta_2^+, \eta_3 < \kappa$. If κ is a successor cardinal, say $\kappa = \lambda^+$, a stronger relation would be

$$\binom{\kappa}{\kappa} \to \begin{pmatrix} \kappa & \kappa \\ \lambda & \lambda \end{pmatrix}^{1,1} ;$$

however it follows from Theorem 4.1.6 that this relation is false. If we now turn to the relation

$$\binom{\kappa}{\kappa} \to \begin{pmatrix} \kappa & \eta_1 \\ \kappa & \eta_3 \end{pmatrix}^{1,1} ,$$

by Theorem 4.1.8 this holds if $\eta_1^+, \eta_3^+ < \kappa$. If κ is a successor cardinal, $\kappa = \lambda^+$, we can ask for stronger relations of the form

$$\binom{\kappa}{\kappa} \to \begin{pmatrix} \kappa & \lambda \\ \kappa & \eta_3 \end{pmatrix}^{1,1} . \tag{1}$$

This relation holds if $\eta_3 = 1$ (Theorem (4.1.3) and if λ is regular it is false with $\eta_3 \geqslant 2$ (Theorem 4.1.5). When λ is singular, it follows from Theorem 4.1.13 that (1) is false with $\eta_3 \geqslant \aleph_2$ or $\eta_3 \geqslant \aleph_0$, depending on whether λ has cofinality $\aleph_0$ or greater. The situation when $1 < \eta_3 < \aleph_2$ or $1 < \eta_3 < \aleph_0$ (respectively) is an open problem.

Finally, we consider the relation

$$\binom{\kappa}{\kappa} \to \begin{pmatrix} \kappa & \eta_1 \\ \eta_2 & \eta_3 \end{pmatrix}^{1,1}$$

where $\eta_1, \eta_2, \eta_3 < \kappa$. The relation is true if either $\eta_2^+ < \kappa$ or $\eta_3^+ < \kappa$ (Theorem 4.1.9), or if $\eta_1^+, \eta_3^+ < \kappa$ (Theorem 4.1.8). This leaves the possibility $\kappa = \lambda^+$ and relations of the form

$$\binom{\kappa}{\kappa} \to \begin{pmatrix} \kappa & \eta_1 \\ \lambda & \lambda \end{pmatrix}^{1,1}$$

to consider. If η_1 is finite, the relation is true (Lemma 4.1.10). If $\lambda' > \aleph_0$ the relation is false for η_1 infinite (Theorem 4.1.13). When $\lambda' = \aleph_0$, the relation is true for $\eta_1 = \aleph_0$ (from Theorem 4.2.4 of the next section) and is false for $\eta_1 \geqslant \aleph_2$ (Theorem 4.1.13). For $\lambda = \aleph_0$ and $\eta_1 = \aleph_1$ it is false (from Theorem 4.1.6 with $\kappa = \aleph_0$). The case $\lambda > \lambda' = \aleph_0$, $\eta_1 = \aleph_1$ is an unsolved problem.

§2. Partitions of $\kappa \times \kappa^+$

We shall now seek results concerning the relation

$$\begin{pmatrix} \kappa \\ \kappa^+ \end{pmatrix} \to \begin{pmatrix} \eta_0 & \eta_1 \\ \eta_2 & \eta_3 \end{pmatrix}^{1,1} .$$

As in the last section, we shall soon run against unsolved problems. We start with the following easy lemma. (Compare with Lemma 4.1.10.)

Lemma 4.2.1. *For each finite n,*

$$\begin{pmatrix} \kappa \\ \kappa^+ \end{pmatrix} \to \begin{pmatrix} \kappa & n \\ \kappa^+ & \kappa^+ \end{pmatrix}^{1,1} . \tag{1}$$

Proof. By induction on n. The case $n = 0$ is trivial, so make the inductive assumption that the relation (1) is true for a particular value of n. Take a disjoint partition

$$\kappa \times \kappa^+ = \Delta_0 \cup \Delta_1$$

and suppose that there are no sets H_1 in $[\kappa]^\kappa$ and H_2 in $[\kappa^+]^{\kappa^+}$ such that $H_1 \times H_2 \subseteq \Delta_0$. By the inductive hypothesis, there must then be sets H_1 in $[\kappa]^n$ and H_2 in $[\kappa^+]^{\kappa^+}$ for which $H_1 \times H_2 \subseteq \Delta_1$.

For each α in κ, put $P_1(\alpha) = \{\beta < \kappa^+; \langle\alpha, \beta\rangle \in \Delta_1\}$. Then since $|\kappa - H_1| = \kappa$ and

$$(\kappa - H_1) \times (H_2 - \bigcup\{P_1(\alpha); \alpha \in \kappa - H_1\}) \subseteq \Delta_0 ,$$

by the choice of the partition $|H_2 - \bigcup\{P_1(\alpha); \alpha \in \kappa - H_1\}| < \kappa^+$. Thus $|H_2 \cap \bigcup\{P_1(\alpha); \alpha \in \kappa - H_1\}| = \kappa^+$ and so there must be α_0 in $\kappa - H_1$ such that $|H_2 \cap P_1(\alpha_0)| = \kappa^+$. But

$$(H_1 \cup \{\alpha_0\}) \times (H_2 \cap P_1(\alpha_0)) \subseteq \Delta_1 ,$$

and so since $|H_1 \cup \{\alpha_0\}| = n + 1$, the induction step is complete and the lemma proved.

Theorem 4.2.2 (GCH). *For any infinite* κ, *if* $\eta < \kappa$, *then*

$$\begin{pmatrix}\kappa\\ \kappa^+\end{pmatrix} \to \begin{pmatrix}\kappa & \eta\\ \kappa^+ & \kappa^+\end{pmatrix}^{1,1}.$$

Proof. The case $\kappa = \aleph_0$ is immediate from Lemma 4.2.1, so suppose $\kappa > \aleph_0$. Define λ as follows: if κ is regular, take $\lambda = \eta$, otherwise let λ be the larger of $(\kappa')^+, \eta^+$. In either case, $\eta \leqslant \lambda < \kappa$ and $\lambda' \neq \kappa'$.

Take any partition $\kappa \times \kappa^+ = \Delta_0 \cup \Delta_1$. To prove the theorem, we must either find sets H_1 in $[\kappa]^\kappa$ and H_2 in $[\kappa^+]^{\kappa^+}$ with $H_1 \times H_2 \subseteq \Delta_0$, or else find sets H_1 in $[\kappa]^\eta$ and H_2 in $[\kappa^+]^{\kappa^+}$ such that $H_1 \times H_2 \subseteq \Delta_1$.

Put $T = \{\beta < \kappa^+; |Q_1(\beta)| \geqslant \lambda\}$, where $Q_1(\beta) = \{\alpha < \kappa; \langle\alpha, \beta\rangle \in \Delta_1\}$. Suppose $|T| = \kappa^+$, and consider the family $\{Q_1(\beta); \beta \in T\}$. By Theorem 3.2.3, there must be T' in $[T]^{\kappa^+}$ such that $|\bigcap\{Q_1(\beta); \beta \in T'\}| \geqslant \lambda \geqslant \eta$. Since $\bigcap\{Q_1(\beta); \beta \in T'\} \times T' \subseteq \Delta_1$, there is nothing more to be shown. So consider the case when $|T| < \kappa^+$. Then $|\kappa^+ - T| = \kappa^+$. Now $\lambda^+ \leqslant \kappa$, so there is a pairwise disjoint decomposition $\kappa = \bigcup\{S_\mu; \mu < \lambda^+\}$ where always $|S_\mu| = \kappa$. Since $|Q_1(\beta)| < \lambda$ for β in $\kappa^+ - T$, for all such β there is $\mu(\beta)$ for which $S_{\mu(\beta)} \cap Q_1(\beta) = \emptyset$. There must be T_1 in $[T]^{\kappa^+}$ such that $\mu(\beta)$ is constant for β in T_1, say with value μ. Thus $S_\mu \cap Q_1(\beta) = \emptyset$ for all β in T_1 so that $S_\mu \times T_1 \subseteq \Delta_0$. Since $|S_\mu| = \kappa$ and $|T_1| = \kappa^+$, this proves the theorem.

Theorem 4.2.3 (GCH). *If* κ *is singular and* $\eta < \kappa$, *then*

$$\begin{pmatrix}\kappa\\ \kappa^+\end{pmatrix} \to \begin{pmatrix}\kappa & \kappa\\ \kappa & \eta\end{pmatrix}^{1,1}.$$

Proof. We may suppose that $\kappa' < \eta < \kappa$ and that η is regular. Choose an increasing sequence $\langle\kappa_\sigma; \sigma < \kappa'\rangle$ of regular cardinals such that $\eta < \kappa_\sigma < \kappa$ and $\kappa = \Sigma(\kappa_\sigma; \sigma < \kappa')$. Take any disjoint partition

$$\kappa \times \kappa^+ = \Delta_0 \cup \Delta_1,$$

and suppose there are no sets H_1 in $[\kappa]^\kappa$ and H_2 in $[\kappa^+]^\eta$ such that $H_1 \times H_2 \subseteq \Delta_1$. We must find H_1 in $[\kappa]^\kappa$ and H_2 in $[\kappa^+]^\kappa$ with $H_1 \times H_2 \subseteq \Delta_0$. We shall start by defining a ramification system $\mathfrak{R}$ on κ^+ of height κ' in such a way that associated with each branch in the ramification tree there are sets A_σ in $[\kappa]^{\kappa_\sigma}$ and B_σ in $[\kappa^+]^{\kappa_\sigma}$ such that $A_\sigma \times B_\tau \subseteq \Delta_0$ whenever $\sigma < \tau$. The construction is similar to that used in the proof of Theorem 4.1.12.

To define $\mathfrak{R}$, suppose $S(\mathbf{v})$ is given for some $\mathbf{v}$ in $N \cap SEQ_\sigma$ where $\sigma < \kappa'$. If $|S(\mathbf{v})| \leqslant \kappa$, put $F(\mathbf{v}) = S(\mathbf{v})$ and $n(\mathbf{v}) = 0$. If $|S(\mathbf{v})| = \kappa^+$ put $n(\mathbf{v}) = \kappa$. Choose

$B(\mathbf{v})$ from $[S(\mathbf{v})]^{\kappa_\sigma}$. Apply Lemma 4.1.11 with $\lambda = \kappa_\sigma$, $A = \kappa$ and $B = S(\mathbf{v}) - B(\mathbf{v})$ to obtain a family $\mathcal{A}(\mathbf{v}) = \{A_\alpha(\mathbf{v}); \alpha < \kappa\}$ of sets from $[\kappa]^{\kappa_\sigma}$ and a map $f_{\mathbf{v}}$ such that $|S(\mathbf{v}) - \bigcup f_{\mathbf{v}}[\mathcal{A}(\mathbf{v})]| \leqslant \kappa$ and $A_\alpha(\mathbf{v}) \times f_{\mathbf{v}}(A_\alpha(\mathbf{v})) \subseteq \Delta_0$. Put

$$F(\mathbf{v}) = B(\mathbf{v}) \cup (S(\mathbf{v}) - \bigcup f_{\mathbf{v}}[\mathcal{A}(\mathbf{v})]),$$

and for $\boldsymbol{\mu}$ in $N \cap SEQ_{\sigma+1}$ with $\boldsymbol{\mu} \upharpoonright \sigma = \mathbf{v}$ define

$$S(\boldsymbol{\mu}) = f_{\mathbf{v}}(A_{\boldsymbol{\mu}(\sigma)}(\mathbf{v})) - F(\mathbf{v}).$$

The Ramification Lemma (Theorem 2.2.3(ii)) applies to $\mathfrak{R}$, so we may choose a sequence $\mathbf{v}$ from $N \cap SEQ_{\kappa'}$ for which $S(\mathbf{v}) \neq \emptyset$. Put $A_\sigma = A_{\mathbf{v}(\sigma)}(\mathbf{v} \upharpoonright \sigma)$ and $B_\sigma = B(\mathbf{v} \upharpoonright \sigma)$. Then $|A_\sigma| = |B_\sigma| = \kappa_\sigma$. Take σ, τ with $\sigma < \tau < \kappa'$. Then

$$B_\tau = B(\mathbf{v} \upharpoonright \tau) \subseteq S(\mathbf{v} \upharpoonright \tau) \subseteq S(\mathbf{v} \upharpoonright \sigma + 1) \subseteq f_{\mathbf{v} \upharpoonright \sigma}(A_{\mathbf{v}(\sigma)}(\mathbf{v} \upharpoonright \sigma)) = f_{\mathbf{v} \upharpoonright \sigma}(A_\sigma)$$

and hence

$$\sigma < \tau < \kappa' \Rightarrow A_\sigma \times B_\tau \subseteq \Delta_0 . \tag{1}$$

Put $A = \bigcup\{A_\sigma; \sigma < \kappa'\}$ and $B = \bigcup\{B_\sigma; \sigma < \kappa'\}$ so $|A| = |B| = \kappa$. The original partition restricts to give a partition of $A \times B$. Hence by Lemma 4.1.7 there are sets C_σ in $[A]^{\kappa_\sigma}$ and D_σ in $[B]^{\kappa_\sigma}$ and a function $h : \kappa' \times \kappa' \to \{0, 1\}$ such that

$$\sigma, \tau < \kappa' \Rightarrow C_\sigma \times D_\tau \subseteq \Delta_{h(\sigma,\tau)} .$$

For i with $i = 0, 1$ put $\Delta_i^* = \{\langle\sigma, \tau\rangle \in \kappa' \times \kappa'; h(\sigma, \tau) = i\}$, so that $\kappa' \times \kappa' = \Delta_0^* \cup \Delta_1^*$. By Theorem 4.1.8, the relation

$$\begin{pmatrix} \kappa' \\ \kappa' \end{pmatrix} \to \begin{pmatrix} \kappa' & \kappa \vee 1 \\ \kappa' & 1 \vee \kappa \end{pmatrix}^{1,1}$$

holds. By applying it to the present partition of $\kappa' \times \kappa'$, there are three cases to consider.

Case 1. There are H in $[\kappa']^{\kappa'}$ and τ with $\tau < \kappa'$ such that $h(\sigma, \tau) = 1$ for all σ in H. Put $C = \bigcup\{C_\sigma; \sigma \in H\}$; then $|C| = \kappa$ and $C \times D_\tau \subseteq \Delta_1$. Since $|D_\tau| = \kappa_\tau > \eta$, this contradicts the original choice of the partition of $\kappa \times \kappa'$.

Case 2. There are H in $[\kappa']^{\kappa'}$ and σ with $\sigma < \kappa'$ such that $h(\sigma, \tau) = 1$ for all τ in H. Put $D = \bigcup\{D_\tau; \tau \in H\}$; then $|D| = \kappa$ and $C_\sigma \times D \subseteq \Delta_1$. Choose α in C_σ. Then there is ρ with $\rho < \kappa'$ such that $\alpha \in A_\rho$. It follows from (1) that we must have $D \subseteq \bigcup\{B_\tau; \tau \leqslant \rho\}$. However, this yields the contradiction

$$\kappa = |D| \leqslant \Sigma(|B_\tau|; \tau \leqslant \rho) = \Sigma(\kappa_\tau; \tau \leqslant \rho) < \kappa .$$

Case 3. This case must prevail. There are K_1, K_2 in $[\kappa']^{\kappa'}$ such that

$h(\sigma, \tau) = 0$ whenever $\sigma \in K_1, \tau \in K_2$. Put $H_1 = \bigcup\{C_\sigma; \sigma \in K_1\}$ and $H_2 = \bigcup\{D_\tau; \tau \in K_2\}$. Then $H_1 \in [\kappa]^\kappa$, $H_2 \in [\kappa^+]^\kappa$ and $H_1 \times H_2 \subseteq \Delta_0$. This completes the proof of the theorem.

In the case of cardinals κ with $\kappa' = \aleph_0$, there is a special method available which enables us to prove the following two theorems.

Theorem 4.2.4 (GCH). *If* $\kappa' = \aleph_0$ *then*

$$\binom{\kappa}{\kappa^+} \to \binom{\kappa \quad \kappa}{\kappa^+ \quad \aleph_0}^{1,1}.$$

Proof. Take a partition $\kappa \times \kappa^+ = \Delta_0 \cup \Delta_1$ and suppose there are no sets H_1 in $[\kappa]^\kappa$ and H_2 in $[\kappa^+]^{\kappa^+}$ such that $H_1 \times H_2 \subseteq \Delta_0$. We must find sets H_1 in $[\kappa]^\kappa$ and H_2 in $[\kappa^+]^{\aleph_0}$ such that $H_1 \times H_2 \subseteq \Delta_1$.

Write $\kappa = \Sigma(\kappa_n; n < \omega)$ where $\kappa_m < \kappa_n < \kappa$ when $m < n < \omega$. We shall define sets A_n from $[\kappa]^{\kappa_n}$ and elements y_n from κ^+ such that always $A_m \times \{y_n\} \subseteq \Delta_1$. By Theorem 4.2.2 and Lemma 4.1.10, the following two relations hold:

$$\binom{\kappa}{\kappa^+} \to \binom{\kappa \quad \kappa_n}{\kappa^+ \quad \kappa^+}^{1,1}, \quad \binom{\kappa}{\kappa^+} \to \binom{\kappa \quad \kappa}{\kappa^+ \quad 1}^{1,1}.$$

Use these relations to choose inductively A_n, B_n, C_n, y_n so that $A_0 \in [\kappa]^{\kappa_0}$, $B_0 \in [\kappa^+]^{\kappa^+}$ with $A_0 \times B_0 \subseteq \Delta_1$ and $C_0 \in [\kappa]^\kappa$, $y_0 \in B_0$ with $C_0 \times \{y_0\} \subseteq \Delta_1$. For n with $n > 0$, we want A_n in $[C_{n-1}]^{\kappa_n}$, B_n in $[B_{n-1}]^{\kappa^+}$ with $A_n \times B_n \subseteq \Delta_1$ and C_n in $[C_{n-1}]^\kappa$, y_n in $B_n - \{y_m; m < n\}$ with $C_n \times \{y_n\} \subseteq \Delta_1$.

Note that if $m \leqslant n$ then

$$A_m \times \{y_n\} \subseteq A_m \times B_n \subseteq A_m \times B_m \subseteq \Delta_1,$$

and if $m > n$ then

$$A_m \times \{y_n\} \subseteq C_{m-1} \times \{y_n\} \subseteq C_n \times \{y_n\} \subseteq \Delta_1.$$

Put $H_1 = \bigcup\{A_n; n < \omega\}$ and $H_2 = \{y_n; n < \omega\}$ so $|H_1| = \kappa$, $|H_2| = \aleph_0$ and $H_1 \times H_2 \subseteq \Delta_1$. Thus the theorem is proved.

Theorem 4.2.5 (GCH). *If* $\kappa' = \aleph_0$ *then*

$$\binom{\kappa}{\kappa^+} \to \binom{\kappa \quad \kappa}{\kappa \quad \kappa}^{1,1}.$$

Proof. The proof is very similar to that of the preceeding theorem, making use of Theorem 4.2.3 rather than Lemma 4.1.10. Take a partition $\kappa \times \kappa^+ = \Delta_0 \cup \Delta_1$ and suppose this time that there are no sets H_1 in $[\kappa]^\kappa$ and H_2 in $[\kappa^+]^\kappa$ with $H_1 \times H_2 \subseteq \Delta_0$; we must find such sets with $H_1 \times H_2 \subseteq \Delta_1$.

Again write $\kappa = \Sigma(\kappa_n; n<\omega)$ where $\kappa_m < \kappa_n < \kappa$ if $m<n<\omega$. We shall define sets A_n from $[\kappa]^{\kappa_n}$ and D_n from $[\kappa^+]^{\kappa_n}$ such that always $A_m \times D_n \subseteq \Delta_1$. If we then put $H_1 = \bigcup\{A_n; n<\omega\}$ and $H_2 = \bigcup\{D_n; n<\omega\}$ it follows that $|H_1| = |H_2| = \kappa$ and $H_1 \times H_2 \subseteq \Delta_1$, so this would suffice to prove the theorem.

We have the following relations at our disposal (from Theorems 4.2.2 and 4.2.3):

$$\begin{pmatrix} \kappa \\ \kappa^+ \end{pmatrix} \to \begin{pmatrix} \kappa & \kappa_n \\ \kappa & \kappa^+ \end{pmatrix}^{1,1}, \quad \begin{pmatrix} \kappa \\ \kappa^+ \end{pmatrix} \to \begin{pmatrix} \kappa & \kappa \\ \kappa & \kappa_n \end{pmatrix}^{1,1}.$$

Use these relations to choose inductively sets A_n, B_n, C_n, D_n as follows. Take A_0 from $[\kappa]^{\kappa_0}$, B_0 from $[\kappa^+]^{\kappa^+}$ such that $A_0 \times B_0 \subseteq \Delta_1$ and C_0 from $[\kappa]^\kappa$, D_0 from $[B_0]^{\kappa_0}$ such that $C_0 \times D_0 \subseteq \Delta_1$. For n with $n>0$, choose A_n from $[C_{n-1}]^{\kappa_n}$ and B_n from $[B_{n-1}]^{\kappa^+}$ with $A_n \times B_n \subseteq \Delta_1$ and C_n from $[C_{n-1}]^\kappa$, D_n from $[B_n]^{\kappa_n}$ with $C_n \times D_n \subseteq \Delta_1$. It follows that if $m \leqslant n$ then

$$A_m \times D_n \subseteq A_m \times B_n \subseteq A_m \times B_m \subseteq \Delta_1 ,$$

and if $m>n$ then

$$A_m \times D_n \subseteq C_{m-1} \times D_n \subseteq C_n \times D_n \subseteq \Delta_1 .$$

Thus $A_m \times D_n \subseteq \Delta_1$ for all m, n and the theorem is proved.

This completes the list of special results available. We shall now summarize (assuming GCH) the results concerning the relation

$$\begin{pmatrix} \kappa \\ \kappa^+ \end{pmatrix} \to \begin{pmatrix} \eta_0 & \eta_1 \\ \eta_2 & \eta_3 \end{pmatrix}^{1,1}.$$

By Theorem 4.2.2, the relation is true if either $\eta_0 < \kappa$ or $\eta_1 < \kappa$, so we need consider only the case $\eta_0 = \eta_1 = \kappa$. It follows from Theorem 4.1.6 that the strongest possible relation

$$\begin{pmatrix} \kappa \\ \kappa^+ \end{pmatrix} \to \begin{pmatrix} \kappa & \kappa \\ \kappa^+ & \kappa^+ \end{pmatrix}^{1,1}$$

is always false.

For the relation

$$\begin{pmatrix}\kappa\\ \kappa^+\end{pmatrix} \to \begin{pmatrix}\kappa & \kappa\\ \kappa^+ & \eta_3\end{pmatrix}^{1,1}$$

there are two cases to consider. If $\kappa' > \aleph_0$ then the relation is true if η_3 is finite (Lemma 4.1.10) and is otherwise false (from Theorem 4.1.13). If $\kappa' = \aleph_0$, this relation holds when $\eta_3 \leqslant \aleph_0$ (Theorem 4.2.4) and fails when $\eta_3 \geqslant \aleph_2$ (from Theorem 4.1.13). The situation when $\eta_3 = \aleph_1$ is an unsolved problem.

There are many unsolved problems concerning the relation

$$\begin{pmatrix}\kappa\\ \kappa^+\end{pmatrix} \to \begin{pmatrix}\kappa & \kappa\\ \eta_2 & \eta_3\end{pmatrix}^{1,1} \tag{1}$$

where $\eta_2, \eta_3 \leqslant \kappa$. From Theorem 4.2.5, the best possible such relation with $\eta_2 = \eta_3 = \kappa$ is true if $\kappa' = \aleph_0$. If κ is singular with $\kappa' > \aleph_0$ then (1) is true with $\eta_2 = \kappa$, $\eta_3 < \kappa$ (Theorem 4.2.3), and it is not known if this can be strengthened to have $\eta_3 = \kappa$. If κ is regular and uncountable, the only positive results known are those which follow trivially from partitions of $\kappa \times \lambda$ where $\lambda \leqslant \kappa$. Some of the simplest unsolved problems are (see [38, Problem 12]):

$$\begin{pmatrix}\aleph_1\\ \aleph_2\end{pmatrix} \overset{?}{\to} \begin{pmatrix}\aleph_1 & \aleph_1\\ \aleph_0 & \aleph_0\end{pmatrix}^{1,1}, \qquad \begin{pmatrix}\aleph_1\\ \aleph_2\end{pmatrix} \overset{?}{\to} \begin{pmatrix}\aleph_1 & \aleph_1\\ \aleph_1 & \aleph_1\end{pmatrix}^{1,1},$$

$$\begin{pmatrix}\aleph_2\\ \aleph_3\end{pmatrix} \overset{?}{\to} \begin{pmatrix}\aleph_2 & \aleph_2\\ \aleph_1 & \aleph_1\end{pmatrix}^{1,1}, \qquad \begin{pmatrix}\aleph_2\\ \aleph_3\end{pmatrix} \overset{?}{\to} \begin{pmatrix}\aleph_2 & \aleph_2\\ \aleph_2 & \aleph_0\end{pmatrix}^{1,1}.$$

Concerning the first two of these relations, Prikry [78] has shown that the relation

$$\begin{pmatrix}\aleph_1\\ \aleph_2\end{pmatrix} \nrightarrow \begin{pmatrix}\aleph_1 & \aleph_1\\ \aleph_0 & \aleph_0\end{pmatrix}^{1,1}$$

is consistent with the axioms of set theory that we have been using, so there is no hope of being able to prove the positive relations.

One can now consider the general relation

$$\begin{pmatrix}\kappa\\ \lambda\end{pmatrix} \to \begin{pmatrix}\eta_0 & \eta_1\\ \eta_2 & \eta_3\end{pmatrix}^{1,1}$$

where $\lambda > \kappa^+$. It turns out that the results we have already obtained are sufficient for a full discussion to be given; the only unsolved problems that arise are equivalent to unsolved problems we have already mentioned. We shall give the discussion for the special case

$$\binom{\kappa}{\lambda} \to \begin{pmatrix} \kappa & \kappa \\ \lambda & \lambda \end{pmatrix}^{1,1},$$

which will have applications in Chapter 6, but otherwise the reader is referred to Erdös, Hajnal and Rado [38, pp 188–193] for details. There are two lemmas first.

Lemma 4.2.6. *Let κ and λ be infinite with $2^\kappa < \lambda'$. Then*

$$\binom{\kappa}{\lambda} \to \begin{pmatrix} \kappa & \kappa \\ \lambda & \lambda \end{pmatrix}^{1,1}. \tag{1}$$

Proof. Take any disjoint partition $\kappa \times \lambda = \Delta_0 \cup \Delta_1$. For each subset A of κ, let $B(A) = \{\beta < \lambda;\ \{\alpha < \kappa; \langle\alpha, \beta\rangle \in \Delta_0\} = A\}$, so $\lambda = \bigcup\{B(A); A \in \mathfrak{P}\kappa\}$. Since $2^\kappa < \lambda'$, there is A_0 in $\mathfrak{P}\kappa$ with $|B(A_0)| = \lambda$. Now $A_0 \times B(A_0) \subseteq \Delta_0$ and $(\kappa - A_0) \times B(A_0) \subseteq \Delta_1$; since either $|A_0| = \kappa$ or $|\kappa - A_0| = \kappa$ the relation (1) holds.

Lemma 4.2.7. *Let κ and λ be infinite with $2^\kappa < \lambda$. Then the two relations*

$$\binom{\kappa}{\lambda} \to \begin{pmatrix} \kappa & \kappa \\ \lambda & \lambda \end{pmatrix}^{1,1} \tag{1}$$

and

$$\binom{\kappa}{\lambda'} \to \begin{pmatrix} \kappa & \kappa \\ \lambda' & \lambda' \end{pmatrix}^{1,1} \tag{2}$$

are equivalent.

Proof. If λ is regular, this is trivial, so suppose λ is singular. Let $\langle\lambda_\sigma; \sigma < \lambda'\rangle$ be an increasing sequence of regular cardinals with always $2^\kappa < \lambda_\sigma < \lambda$ for which $\lambda = \Sigma(\lambda_\sigma; \sigma < \lambda')$.

Suppose that the relation (1) holds. Take any partition $\kappa \times \lambda' = \Gamma_0 \cup \Gamma_1$. Take any pairwise disjoint decomposition $\lambda = \bigcup\{B_\sigma; \sigma < \lambda'\}$ where always $|B_\sigma| = \lambda_\sigma$. Consider the partition $\kappa \times \lambda = \Delta_0 \cup \Delta_1$ where for $\langle\alpha, \beta\rangle$ from $\kappa \times \lambda$ and $i = 0,1$

$$\langle\alpha, \beta\rangle \in \Delta_i \Leftrightarrow \beta \in B_\sigma \text{ and } \langle\alpha, \sigma\rangle \in \Gamma_i\,.$$

By assumption there are H_0 in $[\kappa]^\kappa$ and H_1 in $[\lambda]^\lambda$ such that $H_0 \times H_1 \subseteq \Delta_i$ where either $i = 0$ or $i = 1$. Put $H_1' = \{\sigma < \lambda'; B_\sigma \cap H_1 \neq \emptyset\}$; then $|H_1'| = \lambda'$ and $H_0 \times H_1' \subseteq \Gamma_i$. Thus the relation (2) holds.

Now suppose that (2) is in force. Let any partition $\kappa \times \lambda = \Delta_0 \cup \Delta_1$ be given. For each β, put $Q_0(\beta) = \{\alpha < \kappa; \langle \alpha, \beta \rangle \in \Delta_0\}$.For any σ with $\sigma < \lambda'$,

$$|(Q_0(\beta); \beta < \lambda_\sigma\}| \leqslant 2^\kappa < \lambda_\sigma .$$

Since λ_σ is regular, there is C_σ in $[\lambda_\sigma]^{\lambda_\sigma}$ such that $Q_0(\beta)$ is constant for β in C_σ, say $Q_0(\beta) = Q(\sigma)$. Define a disjoint partition $\kappa \times \lambda' = \Gamma_0 \cup \Gamma_1$ by, if $\langle \alpha, \sigma \rangle \in \kappa \times \lambda'$ then

$$\langle \alpha, \sigma \rangle \in \Gamma_0 \Leftrightarrow \alpha \in Q(\sigma) .$$

By the relation (2), there are H_0 in $[\kappa]^\kappa$, H_1 in $[\lambda']^{\lambda'}$ and i in $\{0,1\}$ such that $H_0 \times H_1 \subseteq \Gamma_i$. Put $H_2 = \bigcup\{C_\sigma; \sigma \in H_1\}$. Then $H_2 \in [\lambda]^\lambda$ and it is easy to see that $H_0 \times H_2 \subseteq \Delta_i$. Thus (1) holds.

Theorem 4.2.8 (GCH). *For an infinite cardinal κ, put $Z(\kappa) = \{\kappa, \kappa^+, \kappa', (\kappa')^+\}$. Then the relation*

$$\begin{pmatrix}\kappa\\ \lambda\end{pmatrix} \to \begin{pmatrix}\kappa & \kappa\\ \lambda & \lambda\end{pmatrix}^{1,1} \tag{1}$$

holds if and only if $Z(\kappa) \cap Z(\lambda) = \emptyset$.

Proof. We may assume throughout the proof that $\kappa \leqslant \lambda$. For brevity, write $R(\kappa, \lambda)$ to indicate that the relation (1) holds for the cardinals κ and λ.

Suppose first that $Z(\kappa) \cap Z(\lambda) = \emptyset$, so in fact $\kappa^+ < \lambda$. If $\kappa^+ < \lambda'$, by Lemma 4.2.6 $R(\kappa, \lambda)$ holds. If $\lambda' \leqslant \kappa^+$ we must have $(\lambda')^+ < \kappa$. Since either $(\lambda')^+ < \kappa'$ or $(\kappa')^+ < \lambda'$ and both κ' and λ' are regular, by Lemma 4.2.6 again $R(\kappa', \lambda')$ holds. Since $(\lambda')^+ < \kappa$, by Lemma 4.2.7 $R(\kappa, \lambda')$ holds, and since $\kappa^+ < \lambda$ again by Lemma 4.2.7 $R(\kappa, \lambda)$ is true. Thus in all cases, if $Z(\kappa) \cap Z(\lambda) = \emptyset$ then $R(\kappa, \lambda)$ holds.

Now suppose $Z(\kappa) \cap Z(\lambda) \neq \emptyset$. From Theorems 4.1.4 and 4.1.6, $R(\kappa, \lambda)$ is false if $\kappa = \lambda$ or $\kappa^+ = \lambda$, so we suppose henceforth that $\kappa^+ < \lambda$. This means that λ must be singular and either $\lambda' \in Z(\kappa)$ or $(\lambda')^+ \in Z(\kappa)$. If $\lambda' = \kappa$ (or equivalently $(\lambda')^+ = \kappa^+$) since $R(\kappa, \kappa)$ is false, by Lemma 4.2.7 also $R(\kappa, \lambda)$ is false. If $\lambda' = \kappa^+$, since $R(\kappa, \kappa^+)$ is false, also $R(\kappa, \lambda)$ is false. If $(\lambda')^+ = \kappa$, since $R((\lambda')^+, \lambda')$ is false again Lemma 4.2.7 shows that $R(\kappa, \lambda)$ is false. The remaining possibilities are that $(\lambda')^+ = \kappa'$, $(\lambda')^+ = (\kappa')^+$, $\lambda' = \kappa'$ or $\lambda' = (\kappa')^+$. In each case by either Theorem 4.1.4 or Theorem 4.1.6 the relation $R(\kappa', \lambda')$ is false.

If in fact $(\lambda')^+ < \kappa$, it follows that $R(\kappa, \lambda')$ is false and then that $R(\kappa, \lambda)$ is false, by two applications of Lemma 4.2.7. On the other hand, if $\kappa \leqslant (\lambda')^+$ this means either $\kappa \leqslant \kappa'$, $\kappa \leqslant (\kappa')^+$ or $\kappa \leqslant (\kappa')^{++}$; in any event κ is regular and from $R(\kappa', \lambda')$ being false it follows by Lemma 4.2.7 that $R(\kappa, \lambda)$ is false. Thus in all possible cases, we conclude that $R(\kappa, \lambda)$ is false. The theorem is proved.

We shall end this section with a few brief comments on the relation

$$\binom{\kappa}{\lambda} \to \binom{\eta_{1k}}{\eta_{2k}}^{1,1}_{k<\gamma}$$

where $\gamma > 2$. Few results have appeared in the literature, particularly for infinite γ. One can obtain, in trivial way, results from the ordinary partition relation, but this surely does not lead to best possible results.

Theorem 4.2.9. *Suppose that* $\kappa \to (\eta_k; k < \gamma)^2$ *and that* $\eta_{1k} + \eta_{2k} < \eta_k$ *for each k. Then*

$$\binom{\kappa}{\kappa} \to \binom{\eta_{1k}}{\eta_{2k}}^{1,1}_{k<\gamma} .$$

Proof. Given a partition $\boldsymbol{\Delta} = \{\Delta_k; k < \gamma\}$ of $\kappa \times \kappa$, define a partition $\boldsymbol{\Gamma} = \{\Gamma_k; k < \gamma\}$ of $[\kappa]^2$ as follows: if $\alpha, \beta \in \kappa$ with $\alpha < \beta$, then

$$\{\alpha, \beta\} \in \Gamma_k \Leftrightarrow \langle \alpha, \beta \rangle \in \Delta_k .$$

By the relation $\kappa \to (\eta_k; k < \gamma)^2$, there are k with $k < \gamma$ and H in $[\kappa]^{\eta_k}$ with $[H]^2 \subseteq \Delta_k$. Take H_1 from $[H]^{\eta_{1k}}$ and H_2 from $[H]^{\eta_{2k}}$ such that $H_1 < H_2$. Then $H_1 \times H_2 \subseteq \Delta_k$.

The positive results in §1 are nearly all proved by methods that are special to the case $\gamma = 2$, and don't generalize. The methods of §2, on the other hand, can be extended to finite γ (see [95]). In particular if $\kappa' = \aleph_0$ then (assuming GCH) for any finite n

$$\binom{\kappa}{\kappa^+} \to \binom{\kappa}{\kappa}^{1,1}_{n} .$$

This is the best possible, for if $\kappa' = \aleph_0$ then trivially

$$\binom{\kappa}{\kappa^+} \nrightarrow \binom{\kappa}{1}^{1,1}_{\aleph_0} .$$

Hajnal has proved (see [25]) assuming GCH that

$$\binom{\aleph_2}{\aleph_2} \to \binom{\aleph_1}{\aleph_1}^{1,1}_{3} .$$

This is the strongest relation of this type which is provable, for a generalization of the method of Prikry [78] shows that the relations

$$\binom{\aleph_2}{\aleph_2} \not\to \binom{\aleph_1}{\aleph_1}^{1,1}_{4} , \quad \binom{\aleph_2}{\aleph_2} \not\to \begin{pmatrix} \aleph_2 \vee \aleph_1 & \aleph_1 \; \aleph_1 \\ \aleph_1 \vee \aleph_2 & \aleph_1 \; \aleph_1 \end{pmatrix}^{1,1}$$

are consistent with GCH.

§3. Larger polarized relations

In this section, we shall consider a few further results concerning the polarized partition symbol. We first note two forms of the Stepping-up Lemma (Theorem 2.3.1) appropriate for the polarized relation.

Theorem 4.3.1. *Let $m, n_1, ..., n_m \geqslant 1$. Let λ be a cardinal such that*

$$\begin{bmatrix} \lambda \\ \cdot \\ \cdot \\ \cdot \\ \lambda \end{bmatrix} \to \begin{bmatrix} \eta_{1k} \\ \cdot \\ \cdot \\ \cdot \\ \eta_{mk} \end{bmatrix}^{n_1, ..., n_m}_{k<\gamma} ,$$

where the η_{ik} are infinite cardinals with $\eta'_{1k} > \eta_{2k}, ..., \eta_{mk}$ (for each k). Suppose κ is an infinite cardinal with $\lambda < \kappa'$ and $|\gamma|^{|\sigma|} < \kappa'$ whenever $\sigma < \lambda$. Then

$$\begin{bmatrix} \kappa \\ \cdot \\ \cdot \\ \cdot \\ \kappa \end{bmatrix} \to \begin{bmatrix} \eta_{1k} \\ \cdot \\ \cdot \\ \cdot \\ \eta_{mk} \end{bmatrix}^{n_1+1, n_2, ..., n_m}_{k<\gamma} .$$

Proof. Suppose all the conditions stated in the theorem are satisfied, and take any disjoint partition $\mathbf{\Delta} = \{\Delta_k; k < \gamma\}$ of $[\kappa]^{n_1+1} \times [\kappa]^{n_2} \times ... \times [\kappa]^{n_m}$. Define a ramification system $\mathfrak{R}$ on κ of height λ as follows. Take σ with $\sigma < \lambda$, and suppose $S(\mathbf{v})$ has already been defined for each $\mathbf{v}$ in $N \cap SEQ_\sigma$. If $S(\mathbf{v}) = \emptyset$, put $F(\mathbf{v}) = \emptyset$ and $n(\mathbf{v}) = 0$. Otherwise, choose $x(\mathbf{v})$ in $S(\mathbf{v})$ and put $F(\mathbf{v}) = \{x(\mathbf{v})\}$.

Place $G(\mathbf{v}) = \bigcup\{F(\mathbf{v}\lceil\tau); \tau \leqslant \sigma\}$. Define a partition $\mathbf{\Gamma}(\mathbf{v})$ of $S(\mathbf{v}) - F(\mathbf{v})$ by requiring:

$$y \equiv z (\mathrm{mod}\ \mathbf{\Gamma}(\mathbf{v})) \Leftrightarrow \forall a_1 \in [G(\mathbf{v})]^{n_1} \ldots \forall a_m \in [G(\mathbf{v})]^{n_m}$$

$$(\langle a_1 \cup \{y\}, a_2, \ldots, a_m\rangle \equiv \langle a_1 \cup \{z\}, a_2, \ldots, a_m\rangle (\mathrm{mod}\ \mathbf{\Delta}))\ .$$

Put $n(\mathbf{v}) = |\Gamma(\mathbf{v})|$ and let $S(\boldsymbol{\mu})$ for $\boldsymbol{\mu}$ in $N \cap SEQ_{\sigma+1}$ with $\boldsymbol{\mu}\lceil\sigma = \mathbf{v}$ range over the classes of $\Gamma(\mathbf{v})$. This defines $\mathfrak{R}$.

If $\sigma < \lambda$, put $\lambda_\sigma = |\gamma|^\theta$ where $\theta = |\sigma|^{n_1 \ldots n_m}$. Then $|n(\mathbf{v}\lceil\tau)| \leqslant \lambda_\sigma$ whenever $\mathbf{v} \in N \cap SEQ_\sigma$ and $\tau < \sigma$. Moreover, if $\sigma < \lambda$ then $\lambda_\sigma^{|\sigma|} < \kappa'$. Hence the Ramification Lemma (Theorem 2.2.3) applies to $\mathfrak{R}$, and we may choose a sequence $\mathbf{v}$ in $N \cap SEQ_\lambda$ for which $S(\mathbf{v}) \neq \emptyset$. Then always $S(\mathbf{v}\lceil\sigma) \neq \emptyset$ so $F(\mathbf{v}\lceil\sigma) = \{x(\mathbf{v}\lceil\sigma)\}$. Choose x_λ from $S(\mathbf{v})$ and for σ with $\sigma < \lambda$ put $x_\sigma = x(\mathbf{v}\lceil\sigma)$. Then $x_\tau \neq x_\sigma$ when $\tau < \sigma \leqslant \lambda$.

If $\tau < \sigma < \lambda$ then $x_\sigma \in F(\mathbf{v}\lceil\sigma) \subseteq S(\mathbf{v}\lceil\sigma) \subseteq S(\mathbf{v}\lceil\tau + 1)$, and $x_\lambda \in S(\mathbf{v}) \subseteq S(\mathbf{v}\lceil\tau + 1)$ so both $x_\sigma, x_\lambda \in S(\mathbf{v}\lceil\tau + 1)$. Thus both x_σ and x_λ are in the same class of $\Gamma(\mathbf{v}\lceil\tau)$, and so if $a_i \in [\{x_\rho; \rho \leqslant \tau\}]^{n_i}$ (where $i = 1, \ldots, m$) then

$$\langle a_1 \cup \{x_\sigma\}, a_2, \ldots, a_m\rangle \equiv \langle a_1 \cup \{x_\lambda\}, a_2, \ldots, a_m\rangle (\mathrm{mod}\ \mathbf{\Delta})\ . \tag{1}$$

Put $X = \{x_\tau; \tau < \lambda\}$, so $|X| = \lambda$, and consider the partition

$$[X]^{n_1} \times \ldots \times [X]^{n_m} = \bigcup\{\Delta'_k; k < \gamma\}$$

where

$$\langle a_1, \ldots, a_m\rangle \in \Delta'_k \Leftrightarrow \langle a_1 \cup \{x_\lambda\}, a_2, \ldots, a_m\rangle \in \Delta_k\ .$$

By the partition relation for λ, there are k with $k < \gamma$ and a sequence $H_1, \ldots, H_m$ where $H_i \in [X]^{\eta_{ik}}$ such that $[H_1]^{n_1} \times \ldots \times [H_m]^{n_m} \subseteq \Delta'_k$. Since $\eta'_{1k} > \eta_{2k}, \ldots, \eta_{ml}$, there is H in $[H_1]^{\eta_{1k}}$ such that if $x_\sigma \in H$ and $x_\tau \in H_2 \cup \ldots \cup H_m$ then $\tau < \sigma$. Hence by (1), $[H]^{n_1+1} \times [H_2]^{n_2} \times \ldots \times [H_m]^{n_m} \subseteq \Delta_k$. This completes the proof.

Theorem 4.3.2. *Let $m, n_1, \ldots, n_m \geqslant 1$. Let λ be a cardinal such that*

$$\begin{bmatrix}\lambda\\ \vdots\\ \lambda\end{bmatrix} \to \begin{bmatrix}\eta_{1k}\\ \vdots\\ \eta_{mk}\end{bmatrix}^{n_1,\ldots,n_m}_{k<\gamma}\ ,$$

where the η_{ik} are infinite cardinals with $\lambda' > \eta_{1k}, \ldots, \eta_{mk}$ (for each k). Suppose κ is an infinite cardinal with $\lambda < \kappa'$ and $|\gamma|^{|\sigma|} < \kappa'$ whenever $\sigma < \lambda$. Then

$$\begin{bmatrix} \kappa \\ \kappa \\ \cdot \\ \cdot \\ \cdot \\ \kappa \end{bmatrix} \to \begin{bmatrix} \lambda \\ \eta_{1k} \\ \cdot \\ \cdot \\ \cdot \\ \eta_{mk} \end{bmatrix}^{1, n_1, \ldots, n_m}_{k<\gamma}$$

Proof. The proof is similar to that of Theorem 4.3.1. Define a ramification system similar to that in the last proof, except that this time the partition $\mathbf{\Gamma}(\mathbf{v})$ of $S(\mathbf{v}) - F(\mathbf{v})$ should satisfy

$$y \equiv z \ (\mathrm{mod}\ \mathbf{\Gamma}(\mathbf{v})) \Leftrightarrow \forall a_1 \in [G(\mathbf{v})]^{n_1} \ldots \forall a_m \in [G(\mathbf{v})]^{n_m}$$
$$(\langle y, a_1, \ldots, a_m \rangle \equiv \langle z, a_1, \ldots, a_m \rangle \ (\mathrm{mod}\ \mathbf{\Delta}))\ .$$

An application of the Ramification Lemma yields distinct elements x_σ where $\sigma \leqslant \lambda$ such that if $a_i \in [\{x_\rho; \rho < \sigma\}]^{n_i}$ (where $i = 1, \ldots, m$) then

$$\langle x_\sigma, a_1, \ldots, a_m \rangle \equiv \langle x_\lambda, a_1, \ldots, a_m \rangle \ (\mathrm{mod}\ \mathbf{\Delta})\ .$$

Put $X = \{x_\tau; \tau < \lambda\}$ and define the partition

$$[X]^{n_1} \times \ldots \times [X]^{n_m} = \bigcup \{\Delta_k'; k < \gamma\}$$

where

$$\langle a_1, \ldots, a_m \rangle \in \Delta_k' \Leftrightarrow \langle x_\lambda, a_1, \ldots, a_m \rangle \in \Delta_k\ .$$

There are k with $k < \gamma$ and a sequence $H_1, \ldots, H_m$ where $H_i \in [X]^{\eta_{ik}}$ such that $[H_1]^{n_1} \times \ldots \times [H_m]^{n_m} \subseteq \Delta_k'$. Put

$$H = \{x_\sigma \in H; \forall \tau \in \lambda (x_\tau \in H_1 \cup \ldots \cup H_m \Rightarrow \tau < \sigma)\}\ .$$

Then $|H| = \lambda$ and $H \times [H_1]^{n_1} \times \ldots \times [H_m]^{n_m} \subseteq \Delta_k$. Thus Theorem 4.3.2 is proved.

As particular applications of the last two theorems, we mention the following (assuming GCH). Applying Theorem 4.3.2 to the relation $\kappa^+ \to (\kappa)^2_\gamma$ (where $\gamma < \kappa'$) from Theorem 2.2.4 leads to the relation

$$\begin{pmatrix} \kappa^{++} \\ \kappa^{++} \end{pmatrix} \to \begin{pmatrix} \kappa^{+} \\ \kappa \end{pmatrix}^{1,2}_{\gamma}\ .$$

Then using Theorem 4.3.1 gives

$$\begin{pmatrix} \kappa^{+++} \\ \kappa^{+++} \end{pmatrix} \to \begin{pmatrix} \kappa^{+} \\ \kappa \end{pmatrix}^{2,2}_{\gamma}\ .$$

It is not clear how close these results are to best possible. There is the following rather trivial negative relation.

Theorem 4.3.3. *Suppose κ is a strong limit cardinal (that is $2^\lambda < \kappa$ whenever $\lambda < \kappa$). Then*

$$\binom{2^\kappa}{\kappa} \nrightarrow \begin{pmatrix} \kappa & 2 \\ 1 & \kappa \end{pmatrix}^{2,1}, \quad \binom{2^\kappa}{2^\kappa} \nrightarrow \begin{pmatrix} \kappa & 2 \\ 2 & \kappa \end{pmatrix}^{2,2}.$$

Proof. We shall prove the second relation; the first is similar. For distinct elements f, g in ${}^\kappa 2$, as before let $\delta(f, g)$ be the least α where $f(\alpha) \neq g(\alpha)$. Define a disjoint partition $[{}^\kappa 2]^2 \times [{}^\kappa 2]^2 = \Delta_0 \cup \Delta_1$ by

$$\langle \{f_0, g_0\}, \{f_1, g_1\} \rangle \in \Delta_0 \Leftrightarrow \delta(f_0, g_0) \leqslant \delta(f_1, g_1).$$

Take distinct f_1, g_1 from ${}^\kappa 2$, so $\delta(f_1, g_1) < \kappa$. Then if f_0, g_0 are such that $\delta(f_0, g_0) \leqslant \delta(f_1, g_1)$ then f_0 and g_0 differ no later than at $\delta(f_1, g_1)$. Hence if H is a subset of ${}^\kappa 2$ with the property $[H]^2 \times \{\{f_1, g_1\}\} \subseteq \Delta_0$ then $|H| \leqslant 2^{|\alpha+1|}$ where $\alpha = \delta(f_1, g_1)$, so $|H| < \kappa$. Similarly, if $\{\{f_0, g_0\}\} \times [H]^2 \subseteq \Delta_1$ then $|H| < \kappa$, and the proof is complete.

From Theorem 4.1.12 it follows that

$$\binom{\kappa^+}{\kappa^+} \rightarrow \binom{\kappa}{\kappa}^{1,1}_2,$$

so applying Theorem 4.3.2 shows

$$\begin{bmatrix} \kappa^{++} \\ \kappa^{++} \\ \kappa^{++} \end{bmatrix} \rightarrow \begin{bmatrix} \kappa^+ \\ \kappa \\ \kappa \end{bmatrix}^{1,1,1}_2$$

In fact it is not known whether the better relation

$$\begin{bmatrix} \kappa^{+} \\ \kappa^{+} \\ \kappa^{+} \end{bmatrix} \overset{?}{\rightarrow} \begin{bmatrix} \kappa \\ \kappa \\ \kappa \end{bmatrix}^{1,1,1}_2$$

holds (see [24, Problem 28]). However, this particular relation, if true, cannot be improved to apply to partitions into 3 classes, as the following theorem of Sierpinksi [88] shows.

Theorem 4.3.4 (GCH). *For all infinite κ,*

$$\begin{bmatrix} \kappa^+ \\ \kappa^+ \\ \kappa^+ \end{bmatrix} \nrightarrow \begin{bmatrix} \kappa & 1 & 1 \\ 1 & \kappa & 1 \\ 1 & 1 & \kappa \end{bmatrix}^{1,1,1} .$$

Proof. (This is the second case of an inductive construction of which the first case is Theorem 4.1.4. See Kuratowski [63].) For each ordinal α where $\alpha < \kappa^+$ we have $|\alpha + 1| \leqslant \kappa$ and so we may write $\{\beta; \beta \leqslant \alpha\} = \{\alpha(\nu); \nu < \kappa\}$. Form a partition $\kappa^+ \times \kappa^+ \times \kappa^+ = \Delta_0 \cup \Delta_1 \cup \Delta_2$ as follows. Given α, β, γ if $\alpha \geqslant \beta, \gamma$ and $\beta = \alpha(\mu)$, $\gamma = \alpha(\nu)$

$$\langle \alpha, \beta, \gamma \rangle \in \Delta_1 \text{ if } \mu \leqslant \nu, \langle \alpha, \beta, \gamma \rangle \in \Delta_2 \text{ if } \nu < \mu ;$$

if $\alpha < \beta$, $\beta \geqslant \gamma$ and $\alpha = \beta(\mu)$, $\gamma = \beta(\nu)$

$$\langle \alpha, \beta, \gamma \rangle \in \Delta_0 \text{ if } \mu \leqslant \nu, \langle \alpha, \beta, \gamma \rangle \in \Delta_2 \text{ if } \nu < \mu ;$$

if $\alpha, \beta < \gamma$ and $\alpha = \gamma(\mu)$, $\beta = \gamma(\nu)$

$$\langle \alpha, \beta, \gamma \rangle \in \Delta_0 \text{ if } \mu \leqslant \nu, \langle \alpha, \beta, \gamma \rangle \in \Delta_1 \text{ if } \nu < \mu .$$

This partition has the right properties. For suppose there are a subset H_1 of κ^+ and members β, γ of κ^+ with $H_1 \times \{\beta\} \times \{\gamma\} \subseteq \Delta_0$. Then $\alpha < \beta, \gamma$ for all α in H_1. Suppose $\beta \geqslant \gamma$, and $\gamma = \beta(\nu)$. Then for any α in H_1, we must have $\alpha = \beta(\mu)$ for some μ with $\mu \leqslant \nu$, so that $|H_1| \leqslant |\nu| < \kappa$. Likewise, if $\beta < \gamma$ and $\beta = \gamma(\nu)$ then for any α in H_1, we have $\alpha = \gamma(\mu)$ where $\mu \leqslant \nu$, so again $|H_1| < \kappa$. Similarly, if $\{\alpha\} \times H_2 \times \{\gamma\} \subseteq \Delta_1$ or $\{\alpha\} \times \{\beta\} \times H_3 \subseteq \Delta_2$, then $|H_2|, |H_3| < \kappa$. Thus the theorem is proved.

As a final remark, we comment that Hajnal [55] has shown that κ is an uncountable two-valued measurable cardinal, then (in the obvious notation)

$$\begin{pmatrix} \kappa^+ \\ \kappa \end{pmatrix} \to \begin{pmatrix} \kappa \\ \kappa \end{pmatrix}_\gamma^{1,<\aleph_0}$$

for each γ where $\gamma < \kappa$. A similar argument shows

$$\begin{pmatrix} \aleph_1 \\ \aleph_0 \end{pmatrix} \to \begin{pmatrix} \aleph_0 \\ \aleph_0 \end{pmatrix}_\gamma^{1,n}$$

for all finite n and γ. (This result was first proved by Galvin.)

CHAPTER 5

THEORY OF INFINITE GRAPHS

§1. The chromatic number

A *graph* $\boldsymbol{G}$ is a pair $\langle G, E\rangle$ where $E \subseteq [G]^2$. The elements of G are the *vertices* of $\boldsymbol{G}$, the elements of E the *edges* of $\boldsymbol{G}$. Two vertices x, y are *adjacent*, or *joined by an edge*, if $\{x, y\} \in E$. A subset S of G no two elements of which are joined by an edge is called an *independent* set. A *subgraph* of the graph $\boldsymbol{G}$ is a graph $\langle H, D\rangle$ where $H \subseteq G$ and $D \subseteq E$. The subgraph of $\boldsymbol{G}$ *spanned* by a subset H of G is the graph $\langle H, E \cap [H]^2\rangle$. A *circuit* in the graph $\boldsymbol{G}$ is a finite set of vertices $x_1, ..., x_n$ such that $\{x_1, x_2\}, ..., \{x_{n-1}, x_n\}, \{x_n, x_1\} \in E$.

A graph of the form $\langle G, [G]^2\rangle$ in which every two vertices are joined is called a *complete* graph, and will be denoted by $\boldsymbol{K}_\eta$ where $\eta = |G|$. A graph of the form $\langle G \cup H, E\rangle$ where G and H are disjoint and $E = \{\{x, y\}; x \in G$ and $y \in H\}$ is called a *complete bipartite* graph, and will be denoted by $\boldsymbol{K}_{\eta,\theta}$ where $\eta = |G|, \theta = |H|$.

With every graph $\boldsymbol{G} = \langle G, E\rangle$ there is associated a partition of $[G]^2$ into the two classes E and $[G]^2 - E$, so the study of graphs amounts to the investigation of partitions of pairs into two classes. However, the language of graph theory often provides a more natural way of expressing the results. In particular, the partition symbol $\kappa \to (\eta_0, \eta_1)^2$ asserts that for any graph $\boldsymbol{G}$ on κ vertices, if $\boldsymbol{G}$ contains no independent set of power η_0, then $\boldsymbol{G}$ must contain a $\boldsymbol{K}_{\eta_1}$ subgraph. In this section, we shall replace the condition "$\boldsymbol{G}$ contains no independent set of power η_0" by weaker conditions of the form "$\boldsymbol{G}$ has large colouring number" or "$\boldsymbol{G}$ has large chromatic number", and ask what effect this has on the subgraphs of $\boldsymbol{G}$.

Before giving the definitions of colouring number and chromatic number, it is convenient to introduce the following notation. Given a graph $\boldsymbol{G} = \langle G, E\rangle$, for a vertex x of $\boldsymbol{G}$ and subset A of G, let $\boldsymbol{G}(x, A)$ be the set of vertices in A adjacent in $\boldsymbol{G}$ to x, that is

$$\boldsymbol{G}(x, A) = \{y \in A; \{x, y\} \in E\}.$$

Definition 5.1.1. An *η-colouring* of a graph $\boldsymbol{G} = \langle G, E\rangle$ is a well ordering $\prec$ of G such that $|\boldsymbol{G}(x, \{y \in G; y \prec x\})| < \eta$ for each vertex x of $\boldsymbol{G}$. The *colouring number* $\mathrm{Col}(\boldsymbol{G})$ of $\boldsymbol{G}$ is the least cardinal η such that $\boldsymbol{G}$ has an η-colouring.

Definition 5.1.2. The *chromatic number* $\mathrm{Chr}(\boldsymbol{G})$ of the graph $\boldsymbol{G}$ is the least cardinal number η such that G can be decomposed into η independent sets.

Thus the chromatic number gives the smallest number of colours needed to paint the vertices of $\boldsymbol{G}$ so that no two vertices of the same colour are joined by an edge. The importance of the chromatic number in connection with the four colour theorem for finite graphs is well known. However, we shall be concerned only with infinite graphs.

We show in Lemma 5.1.3 that the colouring number and the chromatic number are related by the inequality $\mathrm{Chr}(\boldsymbol{G}) \leqslant \mathrm{Col}(\boldsymbol{G})$. That there can be no reverse inequality is shown by the example of the complete bipartite graph $\boldsymbol{K}_{\eta,\eta}$ which has chromatic number 2, yet colouring number $\eta \dotplus 1$.

Lemma 5.1.3. *For every graph* $\boldsymbol{G}$, $\mathrm{Chr}(\boldsymbol{G}) \leqslant \mathrm{Col}(\boldsymbol{G})$.

Proof. Suppose $\prec$ is an η-colouring of $\boldsymbol{G} = \langle G, E\rangle$. Define subsets S_ν of G for ν with $\nu < \eta$ by induction on the ordering $\prec$ as follows:

$$x \in S_\nu \Leftrightarrow \nu \text{ is least such that } \forall y \prec x(\{x, y\} \in E \Rightarrow y \notin S_\nu)\,.$$

Since $\prec$ is an η-colouring this is a good definition. Then $G = \bigcup\{S_\nu; \nu < \eta\}$ and each S_ν is independent, so $\mathrm{Chr}(\boldsymbol{G}) \leqslant \eta$.

As a final introductory remark, let us note that if the graph $\boldsymbol{G} = \langle G, E\rangle$ has colouring number η, then in fact $\boldsymbol{G}$ has an η-colouring of the smallest possible order type, namely $|G|$.

Theorem 5.1.4. *Let* $\boldsymbol{G}$ *be a graph on* κ *vertices, and suppose* $\mathrm{Col}(\boldsymbol{G}) = \eta$. *Then* $\boldsymbol{G}$ *has an* η*-colouring* $\prec$ *such that* $\mathrm{tp}(G, \prec) = \kappa$.

The theorem is trivial if $\eta \geqslant \kappa$, and when $\eta < \kappa$ it is a consequence of the following lemma.

Lemma 5.1.5. *Let* $\boldsymbol{G}$ *be a graph on* κ *vertices and suppose* $\boldsymbol{G}$ *has an* η*-colouring where* $\eta < \kappa$. *Then* $\boldsymbol{G}$ *has an* η*-colouring* $\prec$ *such that* $\mathrm{tp}(G, \prec) = \kappa$.

Proof. Let $\prec$ be an η-colouring of $\boldsymbol{G}$, and suppose $\mathrm{tp}(G, \prec) = \xi$ (so $|\xi| = \kappa$).

Let $\langle x_\gamma; \gamma < \xi\rangle$ be the sequence of elements of G, in their $\mathrel{-\!\!<}$-ordering. We shall in fact construct a colouring $\prec$ of $\boldsymbol{G}$, with $\mathrm{tp}(G, \prec) = \kappa$, such that for each y in G

$$\boldsymbol{G}(y,\ \{x \in G; x \prec y\}) = \boldsymbol{G}(y,\ \{x \in G; x \mathrel{-\!\!<} y\})\,, \tag{1}$$

so certainly $\prec$ will be an η-colouring.

For each y in G, define the set $\mathrm{PR}(y)$ as follows: inductively put

$$\mathrm{PR}_0(y) = \{y\};\ \mathrm{PR}_{n+1}(y) = \bigcup\{\boldsymbol{G}(x,\ \{w; w \mathrel{-\!\!<} x\}); x \in \mathrm{PR}_n(y)\}\,,$$

and then

$$\mathrm{PR}(y) = \bigcup\{\mathrm{PR}_n(y); n < \omega\}\,.$$

Take any map $f : \kappa \to \xi$ from κ onto ξ. Define sets G_α for α with $\alpha < \kappa$ by

$$G_\alpha = \mathrm{PR}(x_{f(\alpha)}) - \bigcup\{\mathrm{PR}(x_{f(\beta)}); \beta < \alpha\}\,.$$

Then G is a disjoint union, $G = \bigcup\{G_\alpha; \alpha < \kappa\}$. Define a well ordering $\prec$ on G as follows: for x, y in G, if $x \in G_\alpha$ and $y \in G_\beta$ then

$$x \prec y \Leftrightarrow \alpha < \beta \text{ or } (\alpha = \beta \text{ and } x \mathrel{-\!\!<} y)\,.$$

If $x \mathrel{-\!\!<} y$, $y \in G_\beta$ and x is adjacent to y in $\boldsymbol{G}$ then both x, y are in $\mathrm{PR}(x_{f(\beta)})$ so either $x \in G_\beta$ or $x \in G_\alpha$ where $\alpha < \beta$; either way $x \prec y$. Suppose $x \prec y$ and that x, y are adjacent in $\boldsymbol{G}$. If both x, y are in the same G_α, surely $x \mathrel{-\!\!<} y$. So suppose $x \in G_\alpha$, $y \in G_\beta$ with $\alpha \neq \beta$; thus $\alpha < \beta$. Hence $x \in \mathrm{PR}_n(x_{f(\alpha)})$ for some n. If $y \mathrel{-\!\!<} x$ then we would have $y \in \mathrm{PR}_{n+1}(x_{f(\alpha)})$ so $y \in \mathrm{PR}(x_{f(\alpha)})$, which is incompatible with $y \in G_\beta$ where $\beta > \alpha$; thus $x \mathrel{-\!\!<} y$. This establishes (1).

To prove the lemma, we need only show $\mathrm{tp}(G, \prec) = \kappa$. With this aim, note that always $|\mathrm{PR}(y)| \leqslant \eta \dotplus \aleph_0$, for a trivial induction shows that $|\mathrm{PR}_n(y)| \leqslant \eta \dotplus \aleph_0$ for each n. Since $G_\beta \subseteq \mathrm{PR}(x_{f(\beta)})$, also $|G_\beta| \leqslant \eta \dotplus \aleph_0$. Take any y in G, say $y \in G_\beta$. Then $\{x \in G; x \prec y\} \subseteq \bigcup\{G_\alpha; \alpha \leqslant \beta\}$, so

$$|\{x \in G; x \prec y\}| \leqslant (\eta \dotplus \aleph_0) \cdot |\beta| < \kappa\,.$$

Hence $\mathrm{tp}(G, \prec) \leqslant \kappa$.

In about 1949 Tutte [9] and Zykov [96] showed that for every integer n there is a (finite) graph which has chromatic number at least n yet contains no triangle (that is, a circuit of length 3). Erdös [15] improved this by showing that for every pair of integers n, k there is a (finite) graph which has chromatic number at least n and contains no circuits of length shorter than k. One can ask if this result will hold for infinite chromatic number n as well. The theorem of Erdös [15] shows easily that this is so for $n = \aleph_0$. Otherwise, the first result in this direction was by Erdös and Rado [30], where they proved that for every κ and every finite k, there is a graph with chromatic number $\geqslant\kappa$ and no *odd* circuits of length shorter than k. In [21], Erdös and Hajnal constructed such an example with just κ vertices. This construction is repeated in Theorem 5.1.9 below. However, also in [21], Erdös and Hajnal show that the general answer to the problem above is in the negative, for they show that a graph which contains no quadrilateral has chromatic number at most $\aleph_0$. This result follows from Corollary 5.1.11.

We start with a couple of lemmas, which will enable us to give the Erdös-Hajnal example of the graph without odd circuits. The construction is an extension of one used by Specker [93]. We shall use the partition symbol $\alpha \to (\beta)^1_\gamma$ for ordinal numbers α and β, which by definition means that whenever a set well ordered with order type α is partitioned into γ classes, at least one of the classes has order type at least β.

Lemma 5.1.6. *Let α, β be ordinals such that $\alpha \to (\alpha)^1_\gamma$ and $\beta \to (\beta)^1_\gamma$. Then $\alpha\beta \to (\alpha\beta)^1_\gamma$.*

Proof. Let S be a set well ordered with order type $\alpha\beta$, so we may suppose that $S = \beta \times \alpha$, with the lexicographic ordering. Suppose S is partitioned, $S = \bigcup\{\Delta_k; k<\gamma\}$. For x in β, put $\Delta_k(x) = \{y<\alpha; \langle x, y\rangle \in \Delta_k\}$. Then for each x, we have $\alpha = \bigcup\{\Delta_k(x); k<\gamma\}$ and by the relation $\alpha\to(\alpha)^1_\gamma$ there is $k(x)$ such that $\mathrm{tp}(\Delta_{k(x)}(x)) = \alpha$. Put $\Gamma_k = \{x<\beta; k(x)=k\}$, so $\beta = \bigcup\{\Gamma_k; k<\gamma\}$. By the relation $\beta\to(\beta)^1_\gamma$ there is k_0 such that $\mathrm{tp}(\Gamma_{k_0})=\beta$. Put $R = \{\langle x,y\rangle\in S; x\in\Gamma_{k_0} \text{ and } y\in\Delta_{k_0}(x)\}$, then $R\subseteq\Delta_{k_0}$ and $\mathrm{tp}(R) = \alpha\beta$, so $\mathrm{tp}(\Delta_{k_0}) = \alpha\beta$.

Corollary 5.1.7. *If κ is a regular cardinal, $\gamma<\kappa$ and $m<\omega$ then $\kappa^m \to(\kappa^m)^1_\gamma$, where κ^m is the ordinal power.*

Proof. By induction on m, using Lemma 5.1.6 and noting $\kappa \to (\kappa)^1_\gamma$.

Lemma 5.1.8. *Let κ be regular, let m be finite. Take any subset X of ${}^m\kappa$ of order type κ^m (in the lexicographic ordering of the sequences in ${}^m\kappa$). For each l with $l<m$ and each sequence $\langle\alpha_0, ..., \alpha_{l-1}\rangle$ from ${}^l\kappa$ there is a set $T(\alpha_0, ..., \alpha_{l-1})$ in $[\kappa]^\kappa$ such that*

$$\forall l<m(\alpha_l\in T(\alpha_0, ..., \alpha_{l-1}))\Rightarrow\langle\alpha_0, ..., \alpha_{m-1}\rangle\in X. \tag{1}$$

Proof. By induction on m. Trivially for $m = 1$, put $T(\emptyset) = X$. So suppose $m\geqslant 2$. For each α, put

$$X(\alpha) = \{\langle\alpha_1, ..., \alpha_{m-1}\rangle; \langle\alpha, \alpha_1, ..., \alpha_{m-1}\rangle\in X\},$$

so $\text{tp}(X(\alpha)) \leqslant \kappa^{m-1}$. Write

$$T = \{\alpha < \kappa ; \text{tp}(X(\alpha)) = \kappa^{m-1}\} .$$

We show that $|T| = \kappa$. For otherwise, $|T| < \kappa$ and so T is not cofinal in κ, by the regularity of κ. Hence $T \subseteq \beta$ for some β where $\beta < \kappa$. Thus for each α with $\alpha \geqslant \beta$ there is an ordinal $\delta(\alpha)$ with $\delta(\alpha) < \kappa$ such that $\text{tp}(X(\alpha)) \leqslant \kappa^{m-2}\delta(\alpha)$. Now for γ with $\gamma < \alpha$, we know $|\Sigma_0(\delta(\alpha); \beta \leqslant \alpha < \gamma)| < \kappa$ so that $\Sigma_0(\delta(\alpha); \beta \leqslant \alpha < \gamma) < \kappa$ (where here Σ_0 stands for ordinal sum). Hence if $\sigma = \Sigma_0(\delta(\alpha); \beta \leqslant \alpha < \nu)$ then $\sigma \leqslant \kappa$. Since

$$\text{tp}(X) \leqslant \Sigma_0(\kappa^{m-1}; \alpha < \beta) + \Sigma_0(\kappa^{m-2}\delta(\alpha); \beta \leqslant \alpha < \kappa) ,$$

we would have $\text{tp}(X) \leqslant \kappa^{m-1}\,\beta + \kappa^{m-2}\,\sigma \leqslant \kappa^{m-1}(\beta + 1) < \kappa^m$. This contradiction establishes that indeed $|T| = \kappa$.

We are now ready to define the sets $T(\alpha_0, ..., \alpha_{l-1})$. Put $T(\emptyset) = T$. For each α in $T(\emptyset)$, apply the inductive hypothesis to the set $X(\alpha)$, and so obtain sets $T(\alpha, \alpha_1, ..., \alpha_{l-1})$ (for l with $1 \leqslant l < m$) of power κ such that

$$\text{if } \alpha_l \in T(\alpha, \alpha_1, ..., \alpha_{l-1})(1 \leqslant l < m) \text{ then } \langle \alpha, \alpha_1, ..., \alpha_{m-1} \rangle \in X(\alpha) .$$

If $\alpha \notin T(\emptyset)$, put $T(\alpha, \alpha_1, ..., \alpha_{l-1}) = \kappa$. Then (1) holds, and the lemma is proved.

We are now ready to give the example of a graph of large chromatic number but without small odd circuits.

Theorem 5.1.9. *For each infinite cardinal κ and each finite j, there is a graph on κ vertices with chromatic number κ, which does not contain circuits of length $2i + 1$ for any i with $1 \leqslant i \leqslant j$.*

Proof. Suppose first that κ is regular. Put $G = {}^{2j^2+1}\kappa$, the set of $(2j^2 + 1)$-place sequences with entries in κ. Let $\prec$ be the lexicographic ordering on G. For each sequence $\boldsymbol{a}$ in G, let a_k denote the k-th component of $\boldsymbol{a}$. Let E be the subset of $[G]^2$ such that, for $\boldsymbol{a}, \boldsymbol{b}$ in G with $\boldsymbol{a} \prec \boldsymbol{b}$,

$$\{\boldsymbol{a}, \boldsymbol{b}\} \in E \Leftrightarrow a_j < b_0 < a_{j+1} < b_1 < ... < a_{2j^2} < b_{2j^2-j} ,$$

and let $\boldsymbol{G}$ be the graph $\langle G, E \rangle$. We shall show $\boldsymbol{G}$ has the required properties. Clearly $|G| = \kappa$.

We note first that any subset of G of order type κ^{2j^2+1} (in the ordering $\prec$) is not independent in $\boldsymbol{G}$. For suppose $X \subseteq G$ and $\text{tp}(X) = \kappa^{2j^2+1}$. Apply Lemma 5.1.8 with this choice of X, to obtain sets $T(\alpha_0, ..., \alpha_{l-1})$ in $[\kappa]^\kappa$ so that

$$\forall l < 2j^2 + 1(\alpha_l \in T(\alpha_0, ..., \alpha_{l-1})) \Rightarrow \langle \alpha_0, ..., \alpha_{2j^2} \rangle \in X .$$

Since κ is infinite, we may choose ordinals $a_0, a_1, ..., a_{2j^2}$ and $b_0, b_1, ..., b_{2j^2}$ from κ so that

$$a_0 \in T(\emptyset)\ ,$$

$$a_1 \in T(a_0) \text{ and } a_1 > a_0\ ,$$

$$\vdots$$

$$a_j \in T(a_0, ..., a_{j-1}) \text{ and } a_j > a_{j-1},$$

$$b_0 \in T(\emptyset) \text{ and } b_0 > a_j\ ,$$

$$a_{j+1} \in T(a_0, ..., a_j) \text{ and } a_{j+1} > b_0\ ,$$

$$b_1 \in T(b_0) \text{ and } b_1 > a_{j+1}\ ,$$

$$\vdots$$

$$a_{2j^2} \in T(a_0, ..., a_{2j^2-1}) \text{ and } a_{2j^2} > b_{2j^2-j-1}\ ,$$

$$b_{2j^2-j} \in T(b_0, ..., b_{2j^2-j-1}) \text{ and } b_{2j^2-j} > a_{2j^2}\ ,$$

$$b_{2j^2-j+1} \in T(b_0, ..., b_{2j^2-j}) \text{ and } b_{2j^2-j+1} > b_{2j^2-j}\ ,$$

$$\vdots$$

$$b_{2j^2} \in T(b_0, ..., b_{2j^2-1}) \text{ and } b_{2j^2} > b_{2j^2-1}\ .$$

Put $\boldsymbol{a} = \langle a_0, ..., a_{2j^2}\rangle$, $\boldsymbol{b} = \langle b_0, ..., b_{2j^2}\rangle$. Then $\boldsymbol{a}, \boldsymbol{b} \in X$, $\boldsymbol{a} \prec \boldsymbol{b}$ and $\{\boldsymbol{a}, \boldsymbol{b}\} \in E$, so that X is not independent.

We can now show that $\boldsymbol{G}$ has chromatic number κ. For suppose, on the contrary, that $\mathrm{Chr}(\boldsymbol{G}) = \lambda$ where $\lambda < \kappa$. Thus $G = \bigcup\{G_\nu; \nu < \lambda\}$ where each G_ν is independent. By Corollary 5.1.7, at least one of the G_ν must have order type κ^{2j^2+1}, and this is impossible with G_ν independent. Hence $\mathrm{Chr}(\boldsymbol{G}) = \kappa$.

It remains to check that $\boldsymbol{G}$ has no odd circuits of lengths between 3 and $2j + 1$. This is a tedious, but trivial, verification. For example, suppose in fact $\boldsymbol{a}, \boldsymbol{b}, \boldsymbol{c}$ gave a triangle in $\boldsymbol{G}$. Clearly we may suppose $\boldsymbol{a} \prec \boldsymbol{b} \prec \boldsymbol{c}$. Since $\{\boldsymbol{a}, \boldsymbol{b}\} \in E$ we know $a_{2j} < b_j$, from $\{\boldsymbol{b}, \boldsymbol{c}\} \in E$ we have $b_j < c_0$, so that $a_{2j} < c_0$. Yet with $\{\boldsymbol{a}, \boldsymbol{c}\} \in E$ we have $c_0 < a_{j+1} < a_{2j}$, which is impossible. The verification for odd circuits of length 5 or more is similar. This completes the proof when κ is regular.

If κ is singular, take an increasing sequence $\langle \kappa_\sigma; \sigma < \kappa' \rangle$ of regular cardinals below κ with $\kappa = \Sigma(\kappa_\sigma; \sigma < \kappa')$. For each σ, there is a graph $\boldsymbol{G}_\sigma = \langle G_\sigma, E_\sigma\rangle$ with $|G_\sigma| = \kappa_\sigma$ and $\mathrm{Chr}(\boldsymbol{G}_\sigma) = \kappa_\sigma$ which has no circuits of lengths $2i + 1$ for i with $1 \leqslant i \leqslant j$. We may suppose the G_σ are pairwise disjoint. Then clearly the graph $G = \langle \bigcup\{G_\sigma; \sigma < \kappa'\}, \bigcup\{E_\sigma; \sigma < \kappa'\}\rangle$ has no circuits of odd length at most $2j + 1$, and $|G| = \kappa$ with $\mathrm{Chr}(G) = \kappa$.

The following theorem will show, as opposed to Theorem 5.1.9, that any graph with uncountable chromatic number contains *even* circuits of any length. In fact more holds, for the graph must contain large complete bipartite subgraphs. The next several theorems first appeared in Erdös and Hajnal [21].

Theorem 5.1.10 (GCH). *Let λ be infinite and suppose $\theta^+ < \lambda$. Let $\boldsymbol{G}$ be any graph which has no $\boldsymbol{K}_{\lambda^+,\theta}$ complete bipartite subgraph. Then* $\mathrm{Col}(\boldsymbol{G}) \leqslant \lambda$. *(If θ is finite, GCH is not needed.)*

Corollary 5.1.11. *If $\boldsymbol{G}$ is any graph with* $\mathrm{Col}(\boldsymbol{G}) > \aleph_0$ *then $\boldsymbol{G}$ contains complete bipartite subgraphs $\boldsymbol{K}_{\aleph_1,i}$ for each finite i.*

Proof (of Theorem 5.1.10) Take the graph $\boldsymbol{G} = \langle G, E\rangle$ which has no $\boldsymbol{K}_{\lambda^+,\theta}$ subgraph. Put $|G| = \kappa$, and use induction on κ. If $\kappa \leqslant \lambda$ then the theorem is trivial. So suppose that $\kappa > \lambda$ and that the statement of the theorem holds for any graph on fewer than κ vertices. We use an idea similar to that used in the proof of Theorem 1.3.12 to construct a decomposition $G = \bigcup\{H_\gamma; \gamma < \kappa\}$ of G such that the inductive hypothesis can be used to show that $\mathrm{Col}(\boldsymbol{H}_\gamma) \leqslant \lambda$, where $\boldsymbol{H}_\gamma$ is the subgraph of $\boldsymbol{G}$ spanned by H_γ, and moreover the orderings that give the colourings of the $\boldsymbol{H}_\gamma$ can be pieced together to give a λ-colouring of $\boldsymbol{G}$.

Let $\langle \omega(\alpha); \alpha < \kappa\rangle$ be the sequence of all limit ordinals less than κ, in their natural order. Write G as a disjoint union, $G = \bigcup\{G_{\omega(\alpha)}; \alpha < \kappa\}$, where always $|G_{\omega(\alpha)}| \leqslant \lambda$. First define inductively subsets X_α of G where $\alpha < \kappa$ as follows. Suppose X_β already defined for all β with $\beta < \alpha$, and put

$$X_\alpha^* = \begin{cases} G_{\omega(\gamma)} \cup \bigcup\{X_\beta; \beta < \alpha\} \text{ if } \alpha = \omega(\gamma) + 1 \text{ for some } \gamma\,, \\ \bigcup\{X_\beta; \beta < \alpha\} \text{ otherwise.} \end{cases}$$

Then define X_α by

$$X_\alpha = X_\alpha^* \cup \{x \in G; |\boldsymbol{G}(x, X_\alpha^*)| \geqslant \theta^+\}\,.$$

We show by induction on α that $|X_\alpha| \leqslant \lambda \dotplus |\alpha|$. So suppose that whenever $\beta < \alpha$ then $|X_\beta| \leqslant \lambda \dotplus |\beta|$. Then

$$\begin{aligned} |X_\alpha^*| &\leqslant \lambda \dotplus \Sigma(|X_\beta|; \beta < \alpha) \\ &\leqslant \lambda \dotplus \Sigma(\lambda \dotplus |\beta|; \beta < \alpha) \\ &\leqslant \lambda \dotplus (\lambda \dotplus |\alpha|)\cdot|\alpha| = \lambda \dotplus |\alpha|\,. \end{aligned}$$

Consider the family $\mathcal{A}_\alpha$ where $\mathcal{A}_\alpha = \{G(x, X_\alpha^*); x \in X_\alpha$ and $|G(x, X_\alpha^*)| \geq \theta^+\}$, so $\mathcal{A}_\alpha$ is a family of subsets of X_α^* all of power at least θ^+. Since $\boldsymbol{G}$ has no $\boldsymbol{K}_{\lambda^+,\theta}$ subgraph, certainly

$$\text{if } \mathcal{A} \subseteq \mathcal{A}_\alpha \text{ with } |\mathcal{A}| \geq (\lambda \dotplus |\alpha|)^+, \text{ then } |\bigcap \mathcal{A}| < \theta .$$

By Corollary 3.2.4, it follows that $|\mathcal{A}_\alpha| \leq \lambda \dotplus |\alpha|$. Considering again that $\boldsymbol{G}$ has no $\boldsymbol{K}_{\lambda^+,\theta}$ subgraph, for any A in $\mathcal{A}_\alpha$ surely

$$|\{x \in X_\alpha; G(x, X_\alpha^*) = A\}| < \lambda^+ .$$

Consequently

$$|\{x \in G; |G(x, X_\alpha^*)| \geq \theta^+\}| \leq \lambda \cdot |\mathcal{A}_\alpha| \leq \lambda \cdot (\lambda \dotplus |\alpha|) = \lambda \dotplus |\alpha| ,$$

and so

$$|X_\alpha| \leq |X_\alpha^*| \dotplus (\lambda \dotplus |\alpha|) = \lambda \dotplus |\alpha| ,$$

as claimed.

Clearly $X_\alpha^* \subseteq X_\alpha \subseteq X_{\alpha+1}^*$, and since $\omega(\gamma)$ is a limit ordinal, it follows from this that

$$\bigcup\{X_\alpha^*; \alpha < \omega(\gamma)\} = \bigcup\{X_\alpha; \alpha < \omega(\gamma)\} = X_{\omega(\gamma)}^* . \tag{1}$$

We are now ready to define the sets H_γ where $\gamma < \kappa$. Put

$$H_\gamma = X_{\omega(\gamma+1)}^* - X_{\omega(\gamma)}^* .$$

It follows that

$$X_{\omega(\gamma)}^* = \bigcup\{H_\beta; \beta < \gamma\} . \tag{2}$$

For if $\beta < \gamma$ then

$$H_\beta \subseteq X_{\omega(\beta+1)}^* = \bigcup\{X_\alpha^*; \alpha < \omega(\beta + 1)\} \subseteq \bigcup\{X_\alpha^*; \alpha < \omega(\gamma)\} = X_{\omega(\gamma)}^* ,$$

so certainly $\bigcup\{H_\beta; \beta < \gamma\} \subseteq X_{\omega(\gamma)}^*$. We show by induction on γ that also $X_{\omega(\gamma)}^* \subseteq \bigcup\{H_\beta; \beta < \gamma\}$. Note that if $\alpha < \omega(\gamma)$ there is β with $\beta < \gamma$ such that $\omega(\beta) \leq \alpha < \omega(\beta + 1)$, so that

$$X_\alpha^* \subseteq X_{\omega(\beta+1)}^* \subseteq H_\beta \cup X_{\omega(\beta)}^* \subseteq H_\beta \cup \bigcup\{H_\delta ; \delta < \beta\}$$

(making use of the inductive hypothesis). Hence

$$X_{\omega(\gamma)}^* = \bigcup\{X_\alpha^*; \alpha < \omega(\gamma)\} \subseteq \bigcup\{H_\beta; \beta < \gamma\} ,$$

and the claim is established.

Since $G = \bigcup\{G_{\omega(\gamma)}; \gamma < \kappa\}$ and $G_{\omega(\gamma)} \subseteq X_{\omega(\gamma)+1}^*$, it follows that $G = \bigcup\{H_\gamma; \gamma < \kappa\}$. Furthermore, the H_γ are pairwise disjoint.

Take any y in G, and suppose in fact $y \in H_\gamma$. Put

$$H(y) = G(y, \bigcup\{H_\delta; \delta < \gamma\}) .$$

The construction of the sets H_γ ensures that $|H(y)| < \lambda$. To see this, note that by (1) and (2),

$$H(y) = G(y, \bigcup\{X_\alpha^*; \alpha < \omega(\gamma)\}) = \bigcup\{G(y, X_\alpha^*); \alpha < \omega(\gamma)\} . \tag{3}$$

Since $y \in H_\gamma$, by definition of H_γ we know $y \notin X^*_{\omega(\gamma)}$ so from (1), $y \notin X^*_{\alpha+1}$ for any α with $\alpha < \omega(\gamma)$. Hence $y \notin X_\alpha$, so that $|G(y, X_\alpha^*)| \leqslant \theta$. Also $G(y, X_\alpha^*) \subseteq G(y, X_\beta^*)$ if $\alpha < \beta$, since then $X_\alpha^* \subseteq X_\beta^*$. Thus (3) expresses $H(y)$ as an increasing union of sets all of power at most θ. Hence $|H(y)| \leqslant \theta^+$. By assumption, $\theta^+ < \lambda$, and so indeed always $|H(y)| < \lambda$.

We can now give a λ-colouring of G, and so show $\mathrm{Col}(\boldsymbol{G}) \leqslant \lambda$. Let $\boldsymbol{H}_\gamma$ be the subgraph of $\boldsymbol{G}$ spanned by H_γ. Since $|H_\gamma| \leqslant |X^*_{\omega(\gamma+1)}| \leqslant \lambda \dotplus |\omega(\gamma+1)| < \kappa$, and necessarily each $\boldsymbol{H}_\gamma$ has no $\boldsymbol{K}_{\lambda^+,\theta}$ subgraph, the original inductive assumption provides a λ-colouring $\prec_\gamma$ of $\boldsymbol{H}_\gamma$. Define a well ordering $\prec$ on G as follows: for $x, y \in G$, if $x \in H_\beta$ and $y \in H_\gamma$,

$$x \prec y \Leftrightarrow \beta < \gamma \text{ or } (\beta = \gamma \text{ and } x \prec_\gamma y) .$$

This gives a λ-colouring of $\boldsymbol{G}$, since for any y in G, say $y \in H_\gamma$,

$$\{x \in \boldsymbol{G}(y, G); x \prec y\} \subseteq H(y) \cup \{x \in \boldsymbol{H}_\gamma(y, H_\gamma); x \prec_\gamma y\} ,$$

and both sets in the union have power less than λ.

One can weaken the condition imposed on the graph $\boldsymbol{G}$ in Theorem 5.1.10 to allow the possibility $\theta^+ = \lambda$, and ask if still $\mathrm{Col}(\boldsymbol{G}) \leqslant \lambda$. The next theorem gives a positive answer if $|G|$ is sufficiently small. When $|G|$ is larger than the bound given, the answer is not known (see Problem 50 in [24]).

Theorem 5.1.12 (GCH). *Let θ be infinite and let θ_0 be the least cardinal greater than θ such that $\theta_0' = \theta'$.* Suppose $\lambda = \theta^+$ *and let* $\boldsymbol{G}$ *be any graph on at most θ_0 vertices which has no $\boldsymbol{K}_{\lambda^+,\theta}$ subgraph. Then* $\mathrm{Col}(\boldsymbol{G}) \leqslant \lambda$.

Proof. Proceed as in the proof of Theorem 5.1.10, except define X_α as

$$X_\alpha = X_\alpha^* \cup \{x \in G; |\boldsymbol{G}(x, X_\alpha^*)| \geqslant \theta\} .$$

Lemma 3.2.3 shows that still $|\mathcal{A}_\alpha| \leqslant \lambda \dotplus |\alpha|$; the restriction $|G| \leqslant \theta_0$ ensures $(\lambda \dotplus |\alpha|)' \neq \theta'$ so that the conditions of the lemma are met. With this modification, it follows that $|H(y)| \leqslant \theta$, so still $|H(y)| < \lambda$.

Attempting to weaken the condition still further, to the case $\theta = \lambda$, meets with failure however. There is a relatively simple example of a graph $\boldsymbol{G}$ with $\mathrm{Col}(\boldsymbol{G}) = \lambda^+$ which contains no $\boldsymbol{K}_{\lambda^+,\lambda}$ subgraph (in fact, no $\boldsymbol{K}_{\lambda,\lambda}$ subgraph). A rather more complicated example is possible with $\mathrm{Chr}(\boldsymbol{G}) = \lambda^+$.

Theorem 5.1.13. *There is a graph* $\boldsymbol{G}$ *on* λ^+ *vertices which has no* $\boldsymbol{K}_{\lambda,\lambda}$ *subgraph but yet* $\mathrm{Col}(\boldsymbol{G}) = \lambda^+$.

Proof. By Theorem 1.1.4, there is a pairwise almost disjoint family $\mathscr{A}$ of λ^+ subsets of the set λ each of power λ. Put $G = \lambda \cup \mathscr{A}$ and $E = \{\{x, y\} \in [G]^2;\ y \in \mathscr{A} \text{ and } x \in y\}$; consider the graph $\boldsymbol{G} = \langle G, E\rangle$. Thus $\boldsymbol{G}$ is a graph with $|G| = \lambda^+$, and the almost disjointedness of $\mathscr{A}$ ensures that $\boldsymbol{G}$ has no $\boldsymbol{K}_{\lambda,\lambda}$ subgraph. We show $\mathrm{Col}(\boldsymbol{G}) = \lambda^+$. Suppose for a contradiction that $\mathrm{Col}(\boldsymbol{G}) \leqslant \lambda$. By Lemma 5.1.5, $\boldsymbol{G}$ has a λ-colouring $\prec$ with $\mathrm{tp}(G, \prec) = \lambda^+$. Since λ is not cofinal in λ^+, there is x in G with $\lambda \subseteq \{y \in G; y \prec x\}$, so in fact $x \in \mathscr{A}$. Now $\boldsymbol{G}(x, \{y \in G; y \prec x\}) = x$, so $|\boldsymbol{G}(x, \{y \in G; y \prec x\})| = \lambda$, which is contrary to $\prec$ being a λ-colouring of $\boldsymbol{G}$. Thus indeed $\mathrm{Col}(\boldsymbol{G}) = \lambda^+$, and this example proves the theorem.

The graph just constructed clearly has chromatic number 2. To produce a similar example with chromatic number λ^+ is more difficult.

Theorem 5.1.14 (GCH). *There is a graph* $\boldsymbol{G}$ *on* λ^+ *vertices which has no* $\boldsymbol{K}_{\lambda,\lambda}$ *subgraph and for which* $\mathrm{Chr}(\boldsymbol{G}) = \lambda^+$.

Proof. We shall construct the graph on the vertex set $G = \lambda^+ \times \lambda^+$. To define the edges, we shall use an almost disjoint family $\mathscr{A}$ of λ^+ subsets of G each of power λ. First we define the family $\mathscr{A}$. Let

$$\mathscr{X} = \{X \subseteq G;\ |\mathrm{dom}\, X| = \lambda \text{ and } \forall_\alpha \in \mathrm{dom}\, X(|\{\beta; \langle\alpha, \beta\rangle \in X\}| = \lambda)\}\,.$$

Then $\mathscr{X} \subseteq [G]^\lambda$ and $|\mathscr{X}| = \lambda^+$. Write $\mathscr{X} = \{X_\mu; \mu < \lambda^+\}$. We want to define the members A_ν (where $\nu < \lambda^+$) of $\mathscr{A}$ so that

$$X_\mu \cap A_\nu \neq \emptyset \text{ if } \mu < \nu\,. \tag{1}$$

Inductively suppose that for some ν with $\nu < \lambda^+$ the A_μ for μ with $\mu < \nu$ have already been defined so that $A_\mu \in [G]^\lambda$, the A_μ are pairwise almost disjoint and

$$|\{\beta; \langle\alpha, \beta\rangle \in A_\mu\}| \leqslant 1 \text{ for each } \alpha\,. \tag{2}$$

Write $\{A_\mu; \mu < \nu\} = \{A_{\nu\rho}; \rho < \lambda\}$ and $\{X_\mu; \mu < \nu\} = \{X_{\nu\rho}; \rho < \lambda\}$. Since always $|\mathrm{dom}\, X_{\nu\rho}| = \lambda$, inductively we can choose α_ρ (where $\rho < \lambda$) so that

$$\alpha_\rho \in \mathrm{dom}\, X_{\nu\rho} - \{\alpha_\pi; \pi < \rho\} .$$

Now choose β_ρ such that

$$\langle \alpha_\rho, \beta_\rho \rangle \in X_{\nu\rho} - \bigcup\{A_{\nu\pi}; \pi < \rho\} ;$$

since $|\{\beta; \langle \alpha_\rho, \beta \rangle \in X_{\nu\rho}\}| = \lambda$ it follows from (2) that such a choice of β_ρ is possible. Finally put

$$A_\nu = \{\langle \alpha_\rho, \beta_\rho \rangle; \rho < \lambda\} .$$

Then $A_\nu \in [G]^\lambda$, $|A_\mu \cap A_\nu| < \lambda$ whenever $\mu < \nu$, $|\{\beta; \langle \alpha, \beta \rangle \in A_\nu\}| \leqslant 1$ for each α, and (1) holds. This completes the definition.

We can now specify the edges of $\boldsymbol{G}$. Let $f : G \to \lambda^+$ be a one-to-one map. For $\langle \alpha, \beta \rangle, \langle \gamma, \delta \rangle$ from G, put

$$\{\langle \alpha, \beta \rangle, \langle \gamma, \delta \rangle\} \in E \Leftrightarrow \alpha < \gamma \text{ and } \beta > \delta \text{ and } \langle \gamma, \delta \rangle \in A_{f(\alpha,\beta)} \cdot \tag{3}$$

Let us check that the graph $\boldsymbol{G} = \langle G, E \rangle$ has no $\boldsymbol{K}_{\lambda,\lambda}$ subgraph. Suppose, on the contrary, that there are disjoint sets H_0, H_1 in $[G]^\lambda$ with $\{x, y\} \in E$ whenever $x \in H_0$, $y \in H_1$. We may suppose that the smallest first component of any pair from $H_0 \cup H_1$ is from a pair in H_0. Inductively define subsets H_{0n}, H_{1n} (where $n < \omega$) as follows: H_{0n} is that subset of $H_0 - \bigcup\{H_{0m}; m < n\}$ maximal with the property that the first component of each pair in H_{0n} is less than the first component of each pair in $H_1 - \bigcup\{H_{1m}; m < n\}$, and H_{1n} is that subset of $H_1 - \bigcup\{H_{1m}; m < n\}$ maximal with the property that the first component of each pair in H_{1n} is less than that of each pair in $H_0 - \bigcup\{H_{0m}; m \leqslant n\}$. Only finitely many of the H_{0n} and H_{1n} can be non-empty. For otherwise, all are non-empty so we can choose $\langle \alpha_{2n}, \beta_{2n} \rangle$ from H_{0n} and $\langle \alpha_{2n+1}, \beta_{2n+1} \rangle$ from H_{1n}. By (3), the sequence $\langle \beta_n; n < \omega \rangle$ gives an infinite descending chain of ordinals, which is impossible. Thus there are k, l such that $|H_{0k}| = \lambda$ and $|H_{1l}| = \lambda$. Suppose say $k \leqslant l$. Again by (3), for each $\langle \alpha, \beta \rangle$ in H_{0k} we have $H_{1l} \subseteq A_{f(\alpha,\beta)}$. This contradicts the almost disjoint property of the $A_{f(\alpha,\beta)}$.

Finally, we show that $\boldsymbol{G}$ has chromatic number λ^+. For a contradiction, suppose in fact $\mathrm{Chr}(\boldsymbol{G}) \leqslant \lambda$, so there is a decomposition $G = \bigcup\{G_\sigma; \sigma < \lambda\}$ where each G_σ is independent in $\boldsymbol{G}$. In particular, since λ^+ is regular, there is σ (with $\sigma < \lambda$) for which, if

$$D_\sigma = \{\alpha; |\{\beta; \langle \alpha, \beta \rangle \in G_\sigma\}| = \lambda^+\}$$

then $|D_\sigma| = \lambda^+$. By the choice of D_σ, there are X in $\mathfrak{X}$ and α in D_σ such that $X \subseteq G_\sigma$ and $\alpha < \gamma$ for all γ in dom X. Suppose in fact $X = X_\mu$. Since $|\{\beta;$

$\langle\alpha, \beta\rangle \in G_\sigma\}| = \lambda^+$, there is β, with $\langle\alpha, \beta\rangle \in G_\sigma$ and $f(\alpha, \beta) > \mu$, which is greater than the second component of every pair in X. By (1), there is $\langle\gamma, \delta\rangle$ in G with $\langle\gamma, \delta\rangle \in X_\mu \cap A_{f(\alpha,\beta)}$. Then by (3), $\langle\alpha, \beta\rangle$ and $\langle\gamma, \delta\rangle$ are joined by an edge in $\boldsymbol{G}$. Since both $\langle\alpha, \beta\rangle$ and $\langle\gamma, \delta\rangle$ are in G_σ, this contradicts the independence of G_σ. Thus $\mathrm{Chr}(\boldsymbol{G}) = \lambda^+$, and the theorem is proved.

The example just given to prove Theorem 5.1.14 clearly contains no complete $\boldsymbol{K}_{\aleph_0}$ subgraph. In fact Hajnal [56] has constructed a more involved example which contains no triangle.

§2. A colouring problem

In this section we shall discuss questions of the following form. Given a graph $\boldsymbol{G}$ on κ vertices without too many edges (specifically, without complete $\boldsymbol{K}_\lambda$ subgraphs for large values of λ) when is it possible to paint the vertices of $\boldsymbol{G}$ with few colours so that there are no $\boldsymbol{K}_\theta$ subgraphs of $\boldsymbol{G}$ all the vertices of which are the one colour? Problems of this type have been discussed by Erdös and Hajnal [22]. We introduce the following notation.

Definition 5.2.1. The symbol $[\kappa, \lambda] \to [\eta, \theta]$ means the following. Whenever a graph $\boldsymbol{G} = \langle G, E\rangle$ is given for which $|G| = \kappa$ and $\boldsymbol{G}$ contains no $\boldsymbol{K}_\lambda$ subgraph, then there is a pairwise disjoint decomposition $G = \bigcup\{G_\mu; \mu < \eta\}$ of the vertex set such that none of the subgraphs of $\boldsymbol{G}$ spanned by G_μ (where $\mu < \eta$) contain a $\boldsymbol{K}_\theta$ subgraph.

We shall refer to the decomposition $G = \bigcup\{G_\mu; \mu < \eta\}$ of the vertices of $\boldsymbol{G}$ as a *colouring of* **G** *by* η *colours* (not to be confused with an η-colouring $\prec$ of $\boldsymbol{G}$ as in the previous section).

A few trivial remarks are in order. If the relation $[\kappa, \lambda] \to [\eta, \theta]$ holds then if κ or λ are decreased, or if η or θ are increased, the relation will remain in force. We shall always assume $\lambda, \theta > 1$ and $\eta \leqslant \kappa$. The relation $[\kappa, \lambda] \to [\kappa, \theta]$ is always true. The relation $[\kappa, \lambda] \to [1, \theta]$ is true if and only if $\theta \geqslant \lambda$. The case when $\lambda > \kappa$ is without interest, since no restriction is then placed on the graph $\boldsymbol{G}$. If $\lambda > \kappa$, obvious colourings of the complete graph on κ vertices show that then $[\kappa, \lambda] \nrightarrow [\eta, \theta]$ if $\eta, \theta < \kappa$ and also $[\kappa, \lambda] \nrightarrow [\eta, \kappa]$ if $\eta < \kappa'$; yet clearly $[\kappa, \lambda] \to [\kappa', \kappa]$ does hold. In view of these remarks, we may restrict the discussion of the symbol $[\kappa, \lambda] \to [\eta, \theta]$ by supposing that always $1 < \theta < \lambda \leqslant \kappa$ and $1 < \eta < \kappa$.

The case $\theta = 2$ is of particular interest, for a subgraph of $\boldsymbol{G}$ withput a $\boldsymbol{K}_2$

subgraph is an independent set in G. Thus the relation $[\kappa, \lambda] \to [\eta, 2]$ is true if and only if every graph on κ vertices without a K_λ subgraph has chromatic number at most η. The example provided by Theorem 5.1.9 of a graph on κ vertices with chromatic number κ and yet containing no triangle shows that $[\kappa, 3] \nrightarrow [\eta, 2]$ for all η with $\eta < \kappa$. In the first theorem below, we shall prove the following generalization of Theorem 5.1.9: for all finite n with $n \geqslant 2$, the relation $[\kappa, n+1] \nrightarrow [\eta, n]$ holds whenever $\eta < \kappa$. The example which establishes this negative relation is similar to that given in the proof of Theorem 5.1.9. For infinite values of θ, in Theorem 5.2.4 we shall show by a different method that $[\kappa^\theta, \theta^+] \nrightarrow [\eta, \theta]$ whenever $\eta < \kappa$.

Theorem 5.2.2. *Let κ be infinite and let n be finite with $n \geqslant 2$. Then for every η with $\eta < \kappa$, the relation $[\kappa, n+1] \nrightarrow [\eta, n]$ holds.*

Proof. Let η be given with $\eta < \kappa$. There is a regular cardinal κ_0 with $\eta < \kappa_0 \leqslant \kappa$. In view of the earlier remarks, it suffices to prove the negative relation with κ replaced by κ_0. So in fact we may suppose that κ is regular.

Put $G = \{\boldsymbol{a} \in {}^{n+1}\kappa; a_0 < a_1 < \dots < a_n\}$, where for a sequence $\boldsymbol{a}$ from ${}^{n+1}\kappa$, we are again denoting by a_k the k-th component of $\boldsymbol{a}$. Let E be the subset of $[G]^2$ such that, for distinct $\boldsymbol{a}, \boldsymbol{b}$ in G,

$$\{\boldsymbol{a}, \boldsymbol{b}\} \in E \Leftrightarrow a_k < b_0 < a_{k+1} < b_1 \text{ for some } k \text{ with } 1 \leqslant k < n.$$

Let $\boldsymbol{G}$ be the graph $\langle G, E\rangle$; we show $\boldsymbol{G}$ has no $\boldsymbol{K}_{n+1}$ subgraph and that in any colouring of $\boldsymbol{G}$ by η colours there is a monochromatic $\boldsymbol{K}_n$ subgraph.

To see that $\boldsymbol{G}$ has no $\boldsymbol{K}_{n+1}$ subgraph, take distinct sequences $\boldsymbol{a}^0, \boldsymbol{a}^1, \dots, \boldsymbol{a}^n$ from G, where we may suppose $a_0^0 \leqslant a_0^1 \leqslant \dots \leqslant a_0^n$. Suppose $\boldsymbol{a}^0$ is joined in $\boldsymbol{G}$ to all of $\boldsymbol{a}^1, \dots, \boldsymbol{a}^n$. For each i with $1 \leqslant i \leqslant n$ there is then l with $1 \leqslant l < n$ for which $a_l^0 < a_0^i < a_{l+1}^0 < a_1^i$. Since there are n values of i and only $n-1$ values of l, there must be distinct i, j with the same l value. So $a_l^0 < a_0^i, a_0^j < a_{l+1}^0 < a_1^i, a_1^j$. Thus in particular $a_0^i < a_1^j$ and so there is no k (with $1 \leqslant k < n$) for which $a_k^j < a_0^i$, and also $a_0^j < a_1^i$ so there is no k for which $a_k^i < a_0^j$. Hence $\{\boldsymbol{a}^i, \boldsymbol{a}^j\} \notin E$, and so $\boldsymbol{G}$ contains no $\boldsymbol{K}_{n+1}$.

For any finite m, let $\prec$ be the lexicographic ordering of the sequences in ${}^m\kappa$. It is easy to show by induction on m that for every α with $\alpha < \kappa$, the set $\{\boldsymbol{a} \in {}^m\kappa; \alpha < a_0 < a_1 < \dots < a_{m-1}\}$ has order type κ^m (ordinal exponentiation) under $\prec$. In particular then, $\text{tp}(G) = \kappa^{n+1}$. Now suppose $\boldsymbol{G}$ is coloured by η colours, so there is a decomposition $G = \bigcup\{G_\mu; \mu < \eta\}$. Since κ is regular, by Corollary 5.1.7 there is some μ for which $\text{tp}(G_\mu) = \kappa^{n+1}$. We show that for such a μ, the graph $\boldsymbol{G}_\mu$ contains a $\boldsymbol{K}_n$ subgraph. This will suffice to prove the theorem.

We shall use Lemma 5.1.8 to choose ordinals a_k^i where $0 \leqslant i < n$ and $0 \leqslant k \leqslant n$ such that if $\boldsymbol{a}^i = \langle a_0^i, ..., a_n^i \rangle$ then each $\boldsymbol{a}^i$ is in G_μ and all the $\boldsymbol{a}^i$ are adjacent in $\boldsymbol{G}$. For pairs $\langle i, k \rangle, \langle i', k' \rangle$ where $0 \leqslant i, i' < n$ and $0 \leqslant k, k' \leqslant n$ define an order $\prec$ by

$$\langle i, k \rangle \prec \langle i', k' \rangle \Leftrightarrow i + k < i' + k' \text{ or } (i + k = i' + k' \text{ and } i < i') .$$

We shall choose the ordinals a_k^i according to the $\prec$ ordering of the pairs $\langle i, k \rangle$. Apply Lemma 5.1.8 with $X = G_\mu$ to obtain sets $T(\alpha_0, ..., \alpha_l)$ in $[\kappa]^\kappa$ such that

$$\forall l < n + 1(\alpha_l \in T(\alpha_0, ..., \alpha_{l-1})) \Rightarrow \langle \alpha_0, ..., \alpha_n \rangle \in G_\mu . \tag{1}$$

Since each $T(\alpha_0, ..., \alpha_l)$ has power κ, we can choose a_k^i so that

$$a_k^i > a_l^j \text{ whenever } \langle j, l \rangle \prec \langle i, k \rangle$$

and

$$a_k^i \in T(\emptyset) \text{ if } k = 0; a_k^i \in T(a_0^i, ..., a_{k-1}^i) \text{ if } k > 0 .$$

Then if $\boldsymbol{a}^i = \langle a_0^i, ..., a_n^i \rangle$, by (1) always $\boldsymbol{a}^i \in G_\mu$. Consider $\boldsymbol{a}^i, \boldsymbol{a}^j$ where $i < j$. Note that

$$\langle i, j - i \rangle \prec \langle j, 0 \rangle \prec \langle i, j - i + 1 \rangle \prec \langle j, j - i + 1 \rangle ,$$

so $a_{j-i}^i < a_0^j < a_{j-i+1}^i < a_1^j$; also $1 \leqslant j - i < n$, so that $\{\boldsymbol{a}^i, \boldsymbol{a}^j\} \in E$. Hence the vertices $\boldsymbol{a}^0, ..., \boldsymbol{a}^{n-1}$ give a $\boldsymbol{K}_n$ subgraph in G_μ, and the proof is complete.

We shall turn now to infinite values of θ in our discussion of the symbol $[\kappa, \lambda] \to [\eta, \theta]$. The principal result is again a negative result, namely $[\kappa^\theta, \theta^+] \not\to [\eta, \theta]$ (where $\eta < \kappa$). An example of a graph exhibiting this property is based on the vertex set ${}^\theta\kappa$ of all functions from θ into κ. As before, let $\prec$ be the lexicographic ordering of ${}^\theta\kappa$. We need first the following lemma.

Lemma 5.2.3. *Suppose ${}^\theta\kappa$ is decomposed into fewer than κ classes, say ${}^\theta\kappa = \bigcup\{X_\mu; \mu < \eta\}$ where $\eta < \kappa$. Then one of the classes X_μ has a subset of power θ which is well ordered by $\succ$ (the converse of the lexicographic ordering).*

Proof. Consider a subset X of ${}^\theta\kappa$ which has no θ-size subset well ordered by $\succ$. Then for each f in X there is α with $\alpha < \theta$ such that

$$\text{if } g \in X \text{ and } g \restriction \alpha = f \restriction \alpha, \text{ then } g(\alpha) \leqslant f(\alpha) . \tag{1}$$

For if this failed for some f in X, for all α there would be g_α in X such that

$g_\alpha \upharpoonright \alpha = f \upharpoonright \alpha$ and $f(\alpha) < g_\alpha(\alpha)$. Then whenever $\alpha < \beta < \theta$ it would follow that $g_\alpha \upharpoonright \alpha = f \upharpoonright \alpha = g_\beta \upharpoonright \alpha$ and $g_\beta(\alpha) = f(\alpha) < g_\alpha(\alpha)$; hence $g_\alpha \succ g_\beta$. Thus $\{g_\alpha;\ \alpha < \theta\}$ would be a θ-size subset of X well ordered by $\succ$, contrary to the choice of X.

Take a pairwise disjoint family $\{X_\mu;\ \mu < \eta\}$ of subsets of ${}^\theta\kappa$ where each X_μ has no θ-size subset well ordered by $\succ$. Put $A = \bigcup\{X_\mu;\ \mu < \eta\}$; we show $A \neq {}^\theta\kappa$. This will prove the lemma. For each f in A, let α_f be the least α with the property (1) taken with $X = X_\mu$ for that X_μ such that $f \in X_\mu$. Hence for f, g in X_μ,

$$\text{if } \alpha_f = \alpha_g \text{ and } f \upharpoonright \alpha_f = g \upharpoonright \alpha_g \text{ then } f(\alpha_f) = g(\alpha_g)\,. \tag{2}$$

For each α, put $A_\alpha = \{f \in A;\ \alpha_f = \alpha\}$ so then $A = \bigcup\{A_\alpha;\ \alpha < \theta\}$.

Now define a function h in ${}^\theta\kappa$ by using transfinite recursion to define the values $h(\alpha)$ for α with $\alpha < \theta$ as follows. Suppose for some α that all the values $h(\beta)$ where $\beta < \alpha$ are known (so $h \upharpoonright \alpha$ is known). By (2), for each μ with $\mu < \eta$ there is an ordinal $\xi_{\alpha\mu}$ such that for all g in $A_\alpha \cap X_\mu$,

$$g \upharpoonright \alpha = h \upharpoonright \alpha \Rightarrow g(\alpha) = \xi_{\alpha\mu}\,.$$

Choose $h(\alpha)$ so that

$$h(\alpha) \in \kappa - \{\xi_{\alpha\mu};\ \mu < \eta\}\,.$$

This completes the definition of h. Then $h \notin A_\alpha \cap X_\mu$ whenever $\alpha < \theta$ and $\mu < \eta$. Since $A = \bigcup\{A_\alpha \cap X_\mu;\ \alpha < \theta \text{ and } \mu < \eta\}$, this means $h \notin A$. Thus $A \neq {}^\theta\kappa$, and the lemma is proved.

Theorem 5.2.4. *Let θ be infinite, suppose $\eta < \kappa$. Then* $[\kappa^\theta, \theta^+] \nrightarrow [\eta, \theta]$.

Proof. Take any well ordering $\preccurlyeq$ of ${}^\theta\kappa$, and put

$$E = \{\{f, g\} \in [{}^\theta\kappa]^2;\ f \preccurlyeq g \text{ and } f \succ g\}\,.$$

Let $\boldsymbol{G}$ be the graph $\langle {}^\theta\kappa, E\rangle$. Thus $\boldsymbol{G}$ is a graph induced by the partition of $[{}^\theta\kappa]^2$ used in proving Theorem 2.5.5. As noted in equation (2) of that proof, there is no subset X of ${}^\theta\kappa$ with $|X| = \theta^+$ which is well ordered by $\succ$. Thus certainly $\boldsymbol{G}$ can contain no $\boldsymbol{K}_{\theta^+}$ subgraph. We shall show that in any colouring of $\boldsymbol{G}$ with η colours there is a monochromatic $\boldsymbol{K}_\theta$ subgraph. Thus $\boldsymbol{G}$ will provide the example we need.

So take any colouring of $\boldsymbol{G}$ by η colours, $G = \bigcup\{G_\mu;\ \mu < \eta\}$. By Lemma 5.2.3, there is some G_μ which has a subset X well ordered by $\succ$, where $|X| = \theta$. Let $\boldsymbol{X}$ be the subgraph of $\boldsymbol{G}$ spanned by X. We shall apply to $\boldsymbol{X}$ the partition relation

$$\theta \to (\aleph_0, \theta)^2 . \tag{1}$$

This relation holds by Ramsey's theorem if $\theta = \aleph_0$, by Theorem 2.2.6 if θ is regular and by Theorem 2.4.3 if θ is singular. Note that Theorem 2.4.3 depends on GCH. In fact the relation (1) can be established without appeal to GCH – see Erdös and Rado [29, Theorem 44]. Applied to $\boldsymbol{X}$, the relation (1) implies that either there is H in $[X]^{\aleph_0}$ which is independent in $\boldsymbol{X}$, or else $\boldsymbol{X}$ contains a $\boldsymbol{K}_\theta$ subgraph. The first possibility cannot occur, however. For if $\langle f_n; n < \omega \rangle$ is an enumeration of H in increasing $\preccurlyeq$-order, we would have an infinite descending $\succ$-chain, $\ldots \succ f_2 \succ f_1 \succ f_0$, contrary to $\succ$ being a well ordering on X. Thus $\boldsymbol{X}$ contains a $\boldsymbol{K}_\theta$ subgraph, and this gives a $\boldsymbol{K}_\theta$ subgraph of $\boldsymbol{G}$ in G_μ.

If we assume GCH, then the preceeding two theorems show that all the non-trivial cases of the relation $[\kappa, \lambda] \to [\eta, \theta]$ are false.

Theorem 5.2.5 (GCH). *Suppose that* $1 < \theta < \lambda \leqslant \kappa$ *and* $1 < \eta < \kappa$. *Then* $[\kappa, \lambda] \not\to [\eta, \theta]$.

Proof. It suffices to show that $[\kappa, \theta^+] \not\to [\eta, \theta]$. This follows from Theorem 5.2.2 if θ is finite, and from Theorem 5.2.4 if θ is infinite and κ regular (for then $\kappa^\theta = \kappa$ by GCH). If κ is singular, there is a regular cardinal κ_0 with $\theta^+, \eta < \kappa_0 < \kappa$. Then $[\kappa_0, \theta^+] \not\to [\eta, \theta]$ and hence $[\kappa, \theta^+] \not\to [\eta, \theta]$.

§3. Chromatic number of subgraphs

A well-known theorem of de Bruijn and Erdös [6] is the following: for any finite k, a graph has chromatic number at most k if and only if every finite subgraph of it has chromatic number at most k. In this section we consider generalizations of this theorem. We ask questions of the form: if every subgraph of $\boldsymbol{G}$ on λ vertices has chromatic number at most η, then does $\boldsymbol{G}$ have chromatic number at most θ? The results are mostly by way of counter-examples, and many problems remain unsolved.

We start by proving the theorem of de Bruijn and Erdös mentioned above. In fact the theorem is a standard consequence of the Compactness Theorem of mathematical logic, but we shall give a combinatorial proof.

Theorem 5.3.1. *Let* $\boldsymbol{G}$ *be any graph, let* k *be finite. Then* $\mathrm{Chr}(\boldsymbol{G}) \leqslant k$ *if and only if* $\mathrm{Chr}(\boldsymbol{H}) \leqslant k$ *for every finite subgraph* $\boldsymbol{H}$ *of* $\boldsymbol{G}$.

Proof. The "only if" is trivial, so suppose every finite subgraph of $\boldsymbol{G} = \langle G, E\rangle$ has chromatic number at most k. We must find a colouring of $\boldsymbol{G}$ with k colours in which there are no monochromatic edges. Take a well ordering of G, say $G = \{x_\alpha; \alpha < \kappa\}$. We shall define a function $f : G \to \{0, 1, ..., k-1\}$ by determining the values $f(x_\alpha)$ using induction on α. The function f will be such that colouring vertex x_α with colour i if and only $f(x_\alpha) = i$ will give a suitable colouring of G with k colours.

We shall define $f(x_\alpha)$ so that for each α the following holds:
(1.α) For all H in $[G]^{<\aleph_0}$ there is $h : H \to \{0, 1, ..., k-1\}$ such that

(i.α) if $\gamma \leqslant \alpha$ and $x_\gamma \in H$ then $h(x_\gamma) = f(x_\gamma)$,

(ii) if $\{x, y\} \in E \cap [H]^2$ then $h(x) \neq h(y)$.

Suppose for some α with $\alpha < \kappa$ that $f(x_\beta)$ has already been defined so that (1.β) holds, for all β with $\beta < \alpha$. (The inductive assumption for $\alpha = 0$ is interpreted to mean that for all H in $[G]^{<\aleph_0}$ there is $h : H \to \{0, 1, ..., k-1\}$ such that (ii) holds. This is the hypothesis of the theorem.) Suppose that none of the possibilities $f(x_\alpha) = 0, 1, ..., k-2$ are compatible with (1.α). Thus for each l where $l \leqslant k-2$ there is a finite subset H_l of G such that $x_\alpha \in H_l$ and for every function $h : H \to \{0, 1, ..., k-1\}$ with $h(x_\alpha) = l$ and $h(x_\gamma) = f(x_\gamma)$ whenever $x_\gamma \in H$ with $\gamma < \alpha$, there are vertices x, y in H, adjacent in $\boldsymbol{G}$, for which $h(x) = h(y)$. I claim that then defining $f(x_\alpha)$ by $f(x_\alpha) = k-1$ will make (1.α) hold. For take any finite subset H of G. Put $H^* = H \cup H_0 \cup ... \cup H_{k-2}$, so H^* is also a finite subset of G. There is β with $\beta < \alpha$ such that if $x_\gamma \in H$ with $\gamma < \alpha$ then $\gamma \leqslant \beta$. By (1.β) applied to H^*, there is $h^* : H^* \to \{0, 1, ..., k-1\}$ such that (i.β) and (ii) hold for H^* and h^*. In fact, it must be that $h^*(x_\alpha) = k-1$. For if not, say $h^*(x_\alpha) = l$ where $l \leqslant k-2$, since $H_l \subseteq H^*$ it follows that $h^* \upharpoonright H_l$ gives a colouring of H_l such that (ii) holds, any x_γ with $\gamma < \alpha$ is coloured $f(x_\gamma)$, and x_l is coloured l, contrary to the choice of H_l. So indeed $h^*(x_\alpha) = k-1$. Thus $h^* \upharpoonright H$ gives a suitable colouring of H under which x_α is coloured $k-1$, and $k-1 = f(x_\alpha)$. Since H was arbitrary, in fact (1.α) holds.

This completes the definition of f, and clearly f gives a colouring of $\boldsymbol{G}$ with k colours such that no two adjacent vertices are coloured the same.

Corollary 5.3.2. *Every graph with chromatic number $\aleph_0$ has a countable subgraph with chromatic number $\aleph_0$.*

Proof. Take a graph $\boldsymbol{G}$ with $\mathrm{Chr}(\boldsymbol{G}) = \aleph_0$. For every finite k then $\mathrm{Chr}(\boldsymbol{G}) > k$ so by Theorem 5.3.1 there is a finite subgraph $\boldsymbol{H}_k$ of $\boldsymbol{G}$ with $\mathrm{Chr}(\boldsymbol{H}_k) > k$.

Let H be the union of the H_k, so $\mathrm{Chr}(H) \geqslant \aleph_0$. Thus H is a countable subgraph of G with $\mathrm{Chr}(H) = \aleph_0$.

One can ask if Corollary 5.3.2 remains true when $\aleph_0$ is replaced by $\aleph_1$. The answer is in the negative. Assuming GCH, as a corollary to the next theorem we shall show that there is a graph on $\aleph_2$ vertices with chromatic number $\aleph_1$, every subgraph of which on $\aleph_1$ vertices has chromatic number at most $\aleph_0$. We first prove the following lemma.

Lemma 5.3.3. *Let n be finite with $n \geqslant 2$. Take any set S with $|S| = \beth_{n-1}(\kappa)$ for some infinite cardinal κ. Then there is a function $f : {}^nS \to \kappa$ such that for $x_0, ..., x_n \in S$,*

$$\forall i < n (x_i \neq x_{i+1}) \Rightarrow f(x_0, ..., x_{n-1}) \neq f(x_1, ..., x_n)\,. \tag{1}$$

Proof. Identify S with $\beth_{n-1}(\kappa)$. Let us say that a function f has property $P(n, \kappa)$ if f has the property in the lemma. We shall prove by induction on n that for all κ there is a function f with property $P(n, \kappa)$.

Consider first the case $n = 2$. Let κ be given, and identify $\beth_1(\kappa)$ with ${}^\kappa 2$. As before, let $\prec$ be the lexicographic ordering of ${}^\kappa 2$ and for distinct functions x, y in ${}^\kappa 2$ let $\delta(x, y)$ be the least α for which $x(\alpha) \neq y(\alpha)$. Take any one-to-one map $p : \kappa \times 2 \to \kappa$, and define f as follows:

$$f(x, y) = \begin{cases} 0 & \text{if } x = y\,, \\ p(\delta(x, y), 0) & \text{if } x \prec y\,, \\ p(\delta(x, y), 1) & \text{if } x \succ y\,. \end{cases}$$

Since $\delta(x, y) \neq \delta(y, z)$ whenever $x \prec y \prec z$, it follows that f has property $P(2, \kappa)$.

Now suppose $n > 2$, and make the inductive assumption that for all m with $2 \leqslant m < n$ and for every infinite λ there is a function with the property $P(m, \lambda)$. Let κ be given. There is thus a function g with property $P(2, \beth_{n-2}(\kappa))$ and a function h with property $P(n-1, \kappa)$. Now $\mathrm{dom}(g) = {}^2(\beth_1(\beth_{n-2}(\kappa))) = {}^2(\beth_{n-1}(\kappa))$. Thus we may define $f : {}^n(\beth_{n-1}(\kappa)) \to \kappa$ as follows: if $\alpha_0, ..., \alpha_{n-1} < \beth_{n-1}(\kappa)$,

$$f(\alpha_0, ..., \alpha_{n-1}) = h(g(x_0, x_1), ..., g(x_{n-2}, x_{n-1}))\,.$$

Then f satisfies (1). For given $\alpha_0, ..., \alpha_n$ from $\beth_{n-1}(\kappa)$ with $\alpha_0 \neq \alpha_1, ..., \alpha_{n-1} \neq \alpha_n$, since g has property $P(2, \beth_{n-2}(\kappa))$ if $\beta_0 = g(\alpha_0, \alpha_1), ..., \beta_{n-1} = g(\alpha_{n-1}, \alpha_n)$ then $\beta_0 \neq \beta_1, ..., \beta_{n-2} \neq \beta_{n-1}$. Since h has property $P(n-1, \kappa)$, further $h(\beta_0, ..., \beta_{n-2}) \neq h(\beta_1, ..., \beta_{n-1})$, that is, $f(\alpha_0, ..., \alpha_{n-1}) \neq f(\alpha_1, ..., \alpha_n)$. Consequently f has property $P(n, \kappa)$, and the induction step is complete.

Theorem 5.3.4. *Let n be finite with $n \geqslant 2$. For every κ there is a graph $\mathbf{G}$ on* $(\beth_{n-1}(\kappa))^+$ vertices with $\mathrm{Chr}(\mathbf{G}) > \kappa$ *that that every subgraph of $\mathbf{G}$ spanned by at most $\beth_{n-1}(\kappa)$ vertices has chromatic number at most κ.*

Proof. Write $\lambda = (\beth_{n-1}(\kappa))^+$, and let $G = [\lambda]^n$. Define a subset E of $[G]^2$ as follows: for a, b in G, say $a = \{\alpha_0, ..., \alpha_{n-1}\}_<$ and $b = \{\beta_0, ..., \beta_{n-1}\}_<$ define

$$\{a, b\} \in E \Leftrightarrow \forall i < n-1\ (\alpha_i = \beta_{i+1})\,.$$

Then the graph $\mathbf{G} = \langle G, E\rangle$ has the required properties.

To see that $\mathrm{Chr}(\mathbf{G}) > \kappa$, take any decomposition $G = \bigcup\{G_k; k < \kappa\}$ of G into κ classes. Since $G = [\lambda]^n$, the partition relation $(\beth_{n-1}(\kappa))^+ \to (\kappa^+)^n_\kappa$ from Theorem 2.3.3 applies. There must be H in $[\lambda]^{\kappa^+}$ which is homogeneous for the partition. For $\alpha_0, ..., \alpha_n$ from H with $\alpha_0 < ... < \alpha_n$ if $a = \{\alpha_0, ..., \alpha_{n-1}\}$ and $b = \{\alpha_1, ..., \alpha_n\}$ then $\{a, b\} \in E$. Since $a, b \in [H]^n$ both a, b are in the same class G_k, which is thus not independent in $\mathbf{G}$. Hence $\mathrm{Chr}(\mathbf{G}) > \kappa$.

Now take any subset H of G with $|H| \leqslant \beth_{n-1}(\kappa)$ and consider the subgraph $\mathbf{H}$ of $\mathbf{G}$ spanned by H. Since λ is regular there is α with $\alpha < \lambda$ such that $H \subseteq [\alpha]^n$, and we may suppose $|\alpha| = \beth_{n-1}(\kappa)$. A function f as in Lemma 5.3.3 for the set α induces a colouring of $[\alpha]^n$, and so also of $\mathbf{H}$, such that no two vertices joined by an edge get the same colour. Hence $\mathrm{Chr}(\mathbf{H}) \leqslant \kappa$.

The graph used in the proof above first appeared in Erdös and Hajnal [20] as an example of a graph with large chromatic number but without circuits of small odd lengths. In [23], the same authors noted that it gave an example with the properties in Theorem 5.3.4.

Corollary 5.3.5 (GCH). *Let n be finite, $n \geqslant 1$. For every ordinal α there is a graph $\mathbf{G}$ on $\aleph_{\alpha+n}$ vertices with* $\mathrm{Chr}(\mathbf{G}) = \aleph_{\alpha+1}$ *such that every subgraph of $\mathbf{G}$ spanned by at most $\aleph_{\alpha+n-1}$ vertices has chromatic number at most $\aleph_\alpha$.*

Proof. The case $n = 1$ is trivial. For n with $n \geqslant 2$, use the graph constructed in the proof of Theorem 5.3.4, with $\kappa = \aleph_\alpha$. To see that $\mathrm{Chr}(\mathbf{G}) = \aleph_{\alpha+1}$, note that by GCH, $\aleph_{\alpha+n} = \beth_{n-1}(\aleph_{\alpha+1})$ and use Lemma 5.3.3.

In particular, Corollary 5.3.5 in the special case $\alpha = 0$ and $n = 2$ gives a graph $\mathbf{G}$ on $\aleph_2$ vertices with $\mathrm{Chr}(\mathbf{G}) = \aleph_1$ so that every subgraph $\mathbf{H}$ of $\mathbf{G}$ on $\aleph_1$ vertices has $\mathrm{Chr}(\mathbf{H}) \leqslant \aleph_0$. It is not known if there is such a graph $\mathbf{G}$ with $\mathrm{Chr}(\mathbf{G}) = \aleph_0$. It is also not known if Corollary 5.3.5 can be extended to infinite values of n. It is clear that if $\mathbf{G}$ is a graph on $\aleph_\omega$ vertices such that every subgraph on fewer than $\aleph_\omega$ vertices has chromatic number at most $\aleph_0$, then

$\mathrm{Chr}(G) \leqslant \aleph_0$. The simplest unsolved problem is the following (GCH): is there a graph G on $\aleph_{\omega+1}$ vertices with $\mathrm{Chr}(G) \geqslant \aleph_1$, yet $\mathrm{Chr}(H) \leqslant \aleph_0$ for every subgraph H of G on at most $\aleph_\omega$ vertices?

To conclude this section, we note that one can ask questions similar to those above but replacing "chromatic number" by "colouring number" throughout. The direct analogue of Theorem 5.3.1 does not hold if $k > 2$. Erdös and Hajnal [21] have shown that if G is a graph such that every finite subgraph of G has colouring number at most k, then G has colouring number at most $2k - 2$. They provide an example to show that this is the best possible result. Neither the proof of their theorem nor the construction of the example is particularly simple. In the case of infinite colouring number however, Theorems 5.1.10 and 5.1.12 lead to the following positive results.

Theorem 5.3.6 (GCH). *Let G be a graph such that every subgraph of G on at most κ^+ vertices has colouring number at most κ. Then* $\mathrm{Col}(G) \leqslant \kappa^{++}$.

Proof. Let G be a graph with $\mathrm{Col}(H) < \kappa$ for every subgraph H of G on κ^+ vertices. Note that $\mathrm{Col}(K_{\kappa^+,\kappa}) = \kappa^+$, and so G can contain no $K_{\kappa^+,\kappa}$ subgraph. Thus by Theorem 5.1.10, $\mathrm{Col}(G) \leqslant \kappa^{++}$.

If Theorem 5.1.12 is used in place of Theorem 5.1.10, the following sometimes better result is obtained.

Theorem 5.3.7 (GCH). *Let κ_0 be the least cardinal such that $\kappa_0 > \kappa$ and $\kappa_0' = \kappa$. Let G be any graph on at most κ_0 vertices such that every subgraph of G on at most κ^+ vertices has colouring number at most κ. Then* $\mathrm{Col}(G) \leqslant \kappa^+$.

Many unsolved problems remain. The simplest is the following: is it true that $\mathrm{Col}(G) \leqslant \aleph_0$ for every graph G on $\aleph_2$ vertices such that $\mathrm{Col}(H) \leqslant \aleph_0$ for every subgraph H of G on at most $\aleph_1$ vertices?

CHAPTER 6

DECOMPOSITION AND INTERSECTION PROPERTIES OF FAMILIES OF SETS

§1. Decomposition properties

The problems to be considered in this section are of the following kind. Let $\mathscr{A} = \{A_\nu; \kappa < \lambda\}$ be a family of subsets of a set S where S has power κ, such that the sets in $\mathscr{A}$ cover S closely, in the sense that for some cardinal η every η-size subset of S meets almost all the sets in $\mathscr{A}$. Can S, except for a subset of size less than κ, be written as the union of a small number of sets B_μ each with the property that for any pair x, y in B_μ there is some A in $\mathscr{A}$ with x, y in A? With the following notation the problem can be expressed more precisely.

Definition 6.1.1. The family $\mathscr{A} = \{A_\nu; \nu < \lambda\}$ is said to have *cover number* η if for every B in $[\bigcup \mathscr{A}]^\eta$,

$$|\{\nu < \lambda; B \cap A_\nu = \emptyset\}| < \lambda .$$

Definition 6.1.2. Given the family $\mathscr{A}$, a set B is said to be *connected in* $\mathscr{A}$ if for all x, y in B there is A in $\mathscr{A}$ such that $x, y \in A$.

Definition 6.1.3. The symbol $(\kappa, \lambda, \eta) \Rightarrow \theta$ means the following. Let S be any set with $|S| = \kappa$ and let $\mathscr{A} = \{A_\nu; \nu < \lambda\}$ be any decomposition of S which has cover number η. Then there is a decomposition $S = S^* \cup \bigcup\{B_\mu; \mu < \theta\}$ where $|S^*| < \kappa$ and each B_μ is connected in $\mathscr{A}$.

Results on the property $(\kappa, \lambda, \eta) \Rightarrow \theta$ (in different notation) appear in Erdös, Hajnal and Milner [35, §§11, 13, 14].

When considering the symbol $(\kappa, \lambda, \eta) \Rightarrow \theta$, to avoid trivial cases we shall always suppose that $0 < \eta \leqslant \kappa$ and that $\theta < \lambda, \kappa$ (the latter since each member of $\mathscr{A}$ is connected and also each $\{x\}$ for x in S). Simple considerations show that if for particular values of κ, λ, η and θ the relation $(\kappa, \lambda, \eta) \Rightarrow \theta$ holds, then also $(\kappa, \lambda, \eta_1) \Rightarrow \theta_1$ whenever $\eta_1 \leqslant \eta$ and $\theta_1 \geqslant \theta$.

We shall start by investigating countable decompositions $\mathscr{A}$ of a countable set S. The results are simply stated: if $\mathscr{A}$ has finite cover number then S itself is almost connected, if $\mathscr{A}$ has infinite cover number then S is almost the union of two connected sets.

Theorem 6.1.4. *Let n be finite, $n > 0$. Then $(\aleph_0, \aleph_0, n) \Rightarrow 1$.*

Proof. Let S be a set with $|S| = \aleph_0$; take any family $\mathscr{A} = \{A_\nu; \nu < \omega\}$ with $\bigcup \mathscr{A} = S$ which has cover number n. Suppose $S - S^*$ is not connected for every S^* in $[S]^{<\aleph_0}$, and seek a contradiction. Note for any subset X of S if there is M in $[\omega]^{\aleph_0}$ with $|\{\nu \in M; x \notin A_\nu\}| < \aleph_0$ for every x in X then X is connected, since given x, y in X, each miss only finitely many A_ν with ν in M and so both are in infinitely many A_ν.

Define by induction elements x_k of S and infinite subsets M_{k+1} of ω such that

$$x_j \notin A_\nu \text{ if } j \leqslant k \text{ and } \nu \in M_{k+1}\,, \tag{1}$$

as follows. Put $M_0 = \omega$. Suppose for some k that x_l and M_{l+1} have already been suitably defined whenever $l < k$. If $\{\nu \in M_k; x \notin A_\nu\}$ is finite for every x in $S - \{x_l; l < k\}$ then by the remark above, $S - \{x_l; l < k\}$ is connected, contrary to assumption. So we can choose x_k from $S - \{x_l; l < k\}$ and M_{k+1} from $[M_k]^{\aleph_0}$ such that $x_k \notin A_\nu$ for every ν in M_{k+1}. Then (1) holds.

Put $B = \{x_0, ..., x_{n-1}\}$, so $|B| = n$. Then $B \cap A_\nu = \emptyset$ whenever $\nu \in M_n$, and $|M_n| \nless \aleph_0$. This contradicts that $\mathscr{A}$ has cover number n, and so the theorem is proved.

In proving the result when the cover number is infinite, we shall make use of the following lemma.

Lemma 6.1.5. *Let the family $\mathscr{A} = \{A_\nu; \nu < \omega\}$ give a countable decomposition of a set S. If for no finite subset S^* of S is $S - S^*$ the union of two connected sets, then $\mathscr{A}$ does not have cover number $\aleph_0$.*

Proof. Let $\mathscr{A}$ be a family satisfying the conditions of the lemma. We must show that there is an infinite subset T of S such that $|\{\nu < \omega; A_\nu \cap T = \emptyset\}| = \aleph_0$.

For any subset X of S, let $G(X)$ be the graph with vertex set $S - X$ and two vertices x, y joined by an edge just when $\{x, y\} \nsubseteq A$ for every A in $\mathscr{A}$. Thus the independent subsets of $G(X)$ are exactly the subsets of $S - X$ connected in $\mathscr{A}$. This means that the assumptions on $\mathscr{A}$ ensure that for any finite subset S^* of S, the graph $G(S^*)$ has chromatic number more than 2. Hence by the theorem of de Bruijn and Erdös (Theorem 5.3.1), there is a finite subgraph of $G(S^*)$ with chromatic number more than 2.

Inductively choose finite subsets S_k of S (where $k < \omega$) such that the subgraph G_k of $G(\bigcup\{S_j; j < k\})$ spanned by S_k satisfies $\mathrm{Chr}(G_k) > 2$. For each k, then S_k has only finitely many subsets and so we can choose inductively sets M_k from $[\omega]^{\aleph_0}$ such that $A_\nu \cap S_k$ is constant, say $A_\nu \cap S_k = S_k^*$, for each ν in M_k, and $M_k \subseteq \bigcap\{M_j; j < k\}$. Since M_k is non-empty, certainly each S_k^* is independent for the graph G_k. Since $\mathrm{Chr}(G_k) > 2$ it follows that $S_k - S_k^*$ is not independent. Thus we can choose x_k, y_k in $S_k - S_k^*$ which are adjacent in G_k, and so $\{x_k, y_k\} \not\subseteq A_\nu$ for all ν.

For all k, choose $\nu(k)$ from $M_k - \{\nu(j); j < k\}$. Form the partition $[\omega]^2 = \Delta_0 \cup \Delta_1$ where for j, k in ω with $j < k$,

$$\{j, k\} \in \Delta_0 \Leftrightarrow x_k \notin A_{\nu(j)},$$

$$\{j, k\} \in \Delta_1 \Leftrightarrow y_k \notin A_{\nu(j)}.$$

By Ramsey's theorem (Theorem 2.2.9) there is H in $[\omega]^{\aleph_0}$ such that either $[H]^2 \subseteq \Delta_0$ or else $[H]^2 \subseteq \Delta_1$. Suppose $[H]^2 \subseteq \Delta_0$; the other case is similar. For j, k in H, if $j < k$ then $x_k \notin A_{\nu(j)}$ by definition of Δ_0. If $j \geqslant k$ then $\nu(j) \in M_j \subseteq M_k$ so $A_{\nu(j)} \cap S_k = S_k^*$; since $x_k \in S_k - S_k^*$ again $x_k \notin A_{\nu(j)}$. Thus if $T = \{x_k; k \in H\}$ then T is infinite and $A_\nu \cap T = \emptyset$ whenever $\nu \in \{\nu(j); j \in H\}$, and $|\{\nu(j); j \in H\}| = \aleph_0$. This suffices to prove the lemma.

Corollary 6.1.6. *For all infinite* κ, *the relation* $(\kappa, \aleph_0, \aleph_0) \Rightarrow 2$ *holds.*

The result in Corollary 6.1.6 cannot be improved to $(\aleph_0, \aleph_0, \aleph_0) \Rightarrow 1$, as the following simple example shows.

Theorem 6.1.7. *For all infinite* κ, $(\kappa, \kappa, \kappa) \not\Rightarrow 1$.

Proof. Put $S = \{0,1\} \times \kappa$ and consider the family $\mathscr{A} = \{A_\nu; \nu < \kappa\}$ where for each ν,

$$A_\nu = \{\langle 0, \mu\rangle; \mu < \nu\} \cup \{\langle 1, \mu\rangle; \nu \leqslant \mu < \kappa\},$$

so $|S| = \kappa$ and $\bigcup\mathscr{A} = S$. The family $\mathscr{A}$ has cover number κ. For take M from $[\kappa]^\kappa$ and put $A(M) = \bigcup\{A_\nu; \nu \in M\}$. For any μ with $\mu < \kappa$, there is ν in M such that $\mu < \nu$ and $\langle 0, \mu\rangle \in A_\nu$; hence $\{0\} \times \kappa \subseteq A(M)$. Take any B in $[S]^\kappa$, and put $M = \{\nu < \kappa; B \cap A_\nu = \emptyset\}$, so $B \cap A(M) = \emptyset$. If $|M| = \kappa$, we would then have to have $B \subseteq \{1\} \times \kappa$. But then for any ν from M there would be μ with $\mu \geqslant \nu$ such that $\langle 1, \mu \rangle \in B$, and thus $\langle 1, \mu \rangle \in B \cap A_\nu$ contradicting $\nu \in M$. Thus for any B in $[S]^\kappa$ it follows that $|\{\nu < \kappa; B \cap A_\nu = \emptyset\}| < \kappa$, and so $\mathscr{A}$ has cover number κ.

For any S^* in $[S]^{<\kappa}$ there are μ, ν with $\nu < \mu$ and $\langle 0, \mu \rangle, \langle 1, \nu \rangle \in S - S^*$. Since then $\{\langle 0, \mu \rangle, \langle 1, \nu \rangle\} \not\subseteq A$ for every A in $\mathscr{A}$, it follows that $S - S^*$ is not connected. Thus the family $\mathscr{A}$ gives an example that shows $(\kappa, \kappa, \kappa) \not\Rightarrow 1$.

The relation $(\kappa, \lambda, \eta) \Rightarrow$ is equivalent to a special case of the polarized partition symbol from Chapter 4, as is shown in the next theorem.

Theorem 6.1.8. *Let κ, λ and η be infinite cardinals. The relation $(\kappa, \lambda, \eta) \Rightarrow 1$ holds if and only if*

$$\binom{\kappa}{\lambda} \to \binom{\eta \quad \eta}{\lambda \quad \lambda}^{1,1} . \tag{1}$$

Proof. Suppose first that $(\kappa, \lambda, \eta) \to 1$. Take any disjoint partition $\kappa \times \lambda = \Delta_0 \cup \Delta_1$; we seek H_0 in $[\kappa]^\eta$ and H_1 in $[\lambda]^\lambda$ such that $H_0 \times H_1 \subseteq \Delta_0$ or $H_0 \times H_1 \subseteq \Delta_1$. As in the last proof, put $S = \{0, 1\} \times \kappa$ and consider the family $\mathscr{A} = \{A_\nu; \nu < \lambda\}$ where this time, for each ν,

$$A_\nu = \{\langle i, \alpha \rangle; \langle \alpha, \nu \rangle \notin \Delta_i\} .$$

Take S^* from $[S]^{<\kappa}$. There is α such that $\langle 0, \alpha \rangle, \langle 1, \alpha \rangle \in S - S^*$ and so $S - S^*$ is not connected in $\mathscr{A}$. Thus $S = \bigcup \mathscr{A}$ is not almost connected, and so by the relation $(\kappa, \lambda, \eta) \Rightarrow 1$, the family $\mathscr{A}$ can not have cover number η. Thus there is B in $[S]^\eta$ with $|\{\nu < \lambda; B \cap A_\nu = \emptyset\}| = \lambda$. There are H_0 in $[\kappa]^\eta$ and i from $\{0, 1\}$ such that $\langle i, \alpha \rangle \in B$ for all α in H_0. Put $H_1 = \{\nu < \lambda; B \cap A_\nu = \emptyset\}$, so $H_1 \in [\lambda]^\lambda$. Then if $\alpha \in H_0$ and $\nu \in H_1$, it follows that $\langle i, \alpha \rangle \notin A_\nu$ so $\langle \alpha, \nu \rangle \in \Delta_i$. Hence $H_0 \times H_1 \subseteq \Delta_i$, and the relation (1) holds.

Now suppose that (1) holds. Take a family $\mathscr{A} = \{A_\nu; \nu < \lambda\}$ where $|\bigcup \mathscr{A}| = \kappa$, which has cover number η. Suppose that $\bigcup \mathscr{A} - S^*$ is not connected for any S^* in $[\bigcup \mathscr{A}]^{<\kappa}$. Then we can define by induction sets B_α and elements x_α, y_α where $\alpha < \kappa$ as follows: B_α is a maximal connected subset of $\bigcup \mathscr{A} - (\{x_\beta; \beta < \alpha\} \cup \{y_\beta; \beta < \alpha\})$; we choose x_α from B_α and y_α from $\bigcup \mathscr{A} - (B_\alpha \cup \{y_\beta; \beta < \alpha\})$. Thus for no A in $\mathscr{A}$ is it true that $x_\alpha, y_\alpha \in A$. Form the disjoint partition $\kappa \times \lambda = \Delta_0 \cup \Delta_1$ where

$$\langle \alpha, \nu \rangle \in \Delta_0 \Leftrightarrow x_\alpha \in A_\nu .$$

By the relation (1), there are H_0 in $[\kappa]^\eta$ and H_1 in $[\lambda]^\lambda$ such that $H_0 \times H_1 \subseteq \Delta_0$

or $H_0 \times H_1 \subseteq \Delta_1$. Both $\{x_\alpha; \alpha \in H_0\}$ and $\{y_\alpha; \alpha \in H_0\}$ have power η, so since $\mathscr{A}$ has cover number η it follows that $|\{\nu < \lambda; \forall \alpha \in H_0(x_\alpha \notin A_\nu)\}| < \lambda$ and $|\{\nu < \lambda; \forall \alpha \in H_0(y_\alpha \notin A_\nu)\}| < \lambda$. Since $|H_1| = \lambda$, there must be α_1, α_2 in H_0 and ν_1, ν_2 in H_1 such that $x_{\alpha_1} \in A_{\nu_1}$ and $y_{\alpha_2} \in A_{\nu_2}$ (so certainly $x_{\alpha_2} \notin A_{\nu_2}$). Thus $\langle \alpha_1, \nu_1 \rangle \in \Delta_0$ and $\langle \alpha_2, \nu_2 \rangle \in \Delta_1$ contrary to the property of H_0, H_1. This shows that there must be S^* in $[\bigcup \mathscr{A}]^{<\kappa}$ such that $\bigcup \mathscr{A} - S^*$ is connected, and so $(\kappa, \lambda, \eta) \Rightarrow 1$.

Corollary 6.1.9 (GCH). *For infinite cardinals κ, λ the relation $(\kappa, \lambda, \kappa) \Rightarrow 1$ holds if and only if $\{\kappa, \kappa', \kappa^+, (\kappa')^+\} \cap \{\lambda, \lambda', \lambda^+, (\lambda')^+\} = \emptyset$.*

Proof. By Theorem 6.1.8 and Theorem 4.2.8.

We shall now seek to extend the earlier results on countable families to higher cardinalities. We consider first the case when the family has power equal to a successor cardinal. In Theorem 6.1.10 we shall show that $(\kappa, \lambda^+, \lambda^+) \Rightarrow \lambda$ (for arbitrary κ), a result not as strong as the relation $(\kappa, \aleph_0, \aleph_0) \Rightarrow 2$ established in Corollary 6.1.6. However, at least when $\kappa = \lambda^+$ this result is the strongest possible, for in Theorem 6.1.12 we prove $(\lambda^+, \lambda^+, \lambda^+) \not\Rightarrow \theta$ for any θ with $\theta < \lambda$.

Theorem 6.1.10. *Let λ be infinite. Then $(\kappa, \lambda^+, \lambda^+) \Rightarrow \lambda$.*

Proof. We may suppose $\kappa \geqslant \lambda^+$. Let S be a set with $|S| = \kappa$ and take any family $\mathscr{A} = \{A_\nu; \nu < \lambda^+\}$ with $\bigcup \mathscr{A} = S$ which has cover number λ^+. We shall show that $S = \bigcup\{B_\mu; \mu < \lambda\}$ where each B_μ is connected in $\mathscr{A}$; then the relation $(\kappa, \lambda^+, \lambda^+) \Rightarrow \lambda$ follows.

For x in S, put $A(x) = \{\nu < \lambda^+; x \in A_\nu\}$. Then S is the disjoint union $S = S_0 \cup S_1$ where $S_0 = \{x \in S; |A(x)| \leqslant \lambda\}$ and $S_1 = \{x \in S; |A(x)| = \lambda^+\}$.

For each x in S_1, put $D(x) = \{y \in S_1; \nexists A \in \mathscr{A}\ (x, y \in A)\}$. Suppose $|D(x)| = \lambda^+$. Then $|\{\nu < \lambda^+; D(x) \cap A_\nu = \emptyset\}| \leqslant \lambda$, because $\mathscr{A}$ has cover number λ^+. Since $|A(x)| = \lambda^+$ there must be ν in $A(x)$ with $D(x) \cap A_\nu \neq \emptyset$; thus there is y with $x, y \in A_\nu$ yet $y \in D(x)$, which is a contradiction. Hence $|D(x)| \leqslant \lambda$ for all x in S_1. For x in S_1 inductively define $D^n(x)$ for n with $n \leqslant \omega$ by

$$D^0(x) = \{x\}; D^{n+1}(x) = \bigcup\{D(y); y \in D^n(x)\}; D^\omega(x) = \bigcup\{D^n(x); n < \omega\}.$$

Then $|D^\omega(x)| \leqslant \lambda$, and $D(z) \subseteq D^\omega(x)$ whenever $z \in D^\omega(x)$.

Inductively define subsets T_α of S_1 as follows: if $S_1 \neq \bigcup\{T_\beta; \beta < \alpha\}$ choose x from $S_1 - \bigcup\{T_\beta; \beta < \alpha\}$ and put $T_\alpha = D^\omega(x) - \bigcup\{T_\beta; \beta < \alpha\}$. If $y \in T_\alpha$ and

$z \in T_\beta$ where $\beta < \alpha$ then $y \notin T_\beta$ so $y \notin D(z)$; thus there is A in $\mathcal{A}$ with $y, z \in A$. And always $|T_\alpha| \leqslant \lambda$ so we can write $T_\alpha = \{x_{\alpha\mu}; \mu < \lambda\}$. For μ with $\mu < \lambda$ let C_μ be the set of all the $x_{\alpha\mu}$. Then $S_1 = \bigcup\{C_\mu; \mu < \lambda\}$ and each C_μ is connected. So to prove the claim above, it is sufficient to show that S_0 is the union of at most λ connected sets.

By way of contraduction, suppose that this is false. Then for α with $\alpha < \lambda^+$, define inductively elements x_α of S_0 and ordinals $\nu(\alpha)$ with $\nu(\alpha) < \lambda^+$ as follows. Suppose x_β and $\nu(\beta)$ already defined whenever $\beta < \alpha$. Put

$$U_\alpha = S_0 - (\{x_\beta; \beta < \alpha\} \cup \bigcup\{A_{\nu(\beta)}; \beta < \alpha\}) .$$

Now $|\alpha| \leqslant \lambda$ and the sets $\{x_\beta\}$ and $A_{\nu(\beta)}$ are all connected, so since S_0 is not the union of λ connected subsets, certainly $U_\alpha \neq \emptyset$. Choose x_α from U_α. Since always $x_\beta \in S_0$ we know $|A(x_\beta)| \leqslant \lambda$. Hence we can choose $\nu(\alpha)$ from

$$\lambda^+ - (\{\nu(\beta); \beta < \alpha\} \cup \bigcup\{A(x_\beta); \beta \leqslant \alpha\}) .$$

This completes the definition of x_α and $\nu(\alpha)$. It follows that

$$\alpha, \beta < \lambda^+ \Rightarrow x_\alpha \notin A_{\nu(\beta)} ,$$

for if $\beta < \alpha$ this is ensured by the choice of x_α and if $\beta \geqslant \alpha$ by the choice of $\nu(\beta)$. Put $B = \{x_\alpha; \alpha < \lambda^+\}$ and $I = \{\nu(\alpha); \alpha < \lambda^+\}$. Then $|B| = |I| = \lambda^+$ and $B \cap A_\nu = \emptyset$ whenever $\nu \in I$. This contradicts that $\mathcal{A}$ has cover number λ^+. Hence S_0, and so also S, can be written as a union of λ connected sets. This proves the theorem.

Before showing that this last result is the best possible in the case $\kappa = \lambda^+$, we state a lemma. This lemma provides a strengthening of Lemma 2.7.3.

Lemma 6.1.11 (GCH). *Let S be any set with $|S| = \kappa^+$. Then there is a disjoint partition $[S]^2 = \bigcup\{\Delta_k; k < \kappa^+\}$ with the property that whenever $A \in [S]^\kappa$ and $B \in [S]^{\kappa^+}$ then there is a in A such that for every k with $k < \kappa^+$ there is b in B with $\{a, b\} \in \Delta_k$.*

Proof. Modify the proof of Lemma 2.7.3. Again identify S with κ^+. Put

$$\mathcal{A} = \{\langle A, g\rangle; A \in [\kappa^+]^\kappa \text{ and } g : A \to \kappa^+\} .$$

By GCH, $|{}^A(\kappa^+)| = \kappa^+$ for each A in $[\kappa^+]^\kappa$ and so $|\mathcal{A}| = \kappa^+$. Take a well ordering $\langle\langle A_\alpha, g_\alpha\rangle; \alpha < \kappa^+\rangle$ of $\mathcal{A}$. For each τ with $\tau < \kappa^+$, put

$$\mathcal{A}_\tau = \{\langle A_\alpha, g_\alpha\rangle; \alpha < \tau \text{ and } A_\alpha \subseteq \tau \text{ and } g[A_\alpha] \subseteq \tau\} ,$$

so $|\mathcal{A}_\tau| \leqslant \kappa$. Suppose now that for each τ with $\tau < \kappa^+$ we could find a function $f_\tau : \tau \to \tau$ such that

$$\text{if } \langle A, g\rangle \in \mathcal{A}_\tau \text{ then } f_\tau(\sigma) = g(\sigma) \text{ for some } \sigma \text{ in } A\,. \tag{1}$$

Consider the partition $\{\Delta_k; k < \kappa^+\}$ of $[\kappa^+]^2$ where for σ, τ in κ^+ with $\sigma < \tau$,

$$\{\sigma, \tau\} \in \Delta_k \Leftrightarrow f_\tau(\sigma) = k\,.$$

This partition has the new property, for take A from $[\kappa^+]^\kappa$ and B from $[\kappa^+]^{\kappa^+}$. Suppose on the contrary that for all α in A there is k such that $\{\alpha, \beta\} \in \Delta_k$ for no β in B, so there is a function $g : A \to \kappa^+$ such that always $\{\alpha, \beta\} \notin \Delta_{g(\alpha)}$ for all β in B. Let τ_0 be the least ordinal such that $\langle A, g\rangle \in \mathcal{A}_{\tau_0}$. Since $|B| = \kappa^+$ there is τ in B with $\tau \geq \tau_0$. Then also $\langle A, g\rangle \in \mathcal{A}_\tau$ so by (1) there is σ in A for which $f_\tau(\sigma) = g(\sigma)$, that is, $\{\sigma, \tau\} \in \Delta_{g(\sigma)}$. But since $\sigma \in A$ and $\tau \in B$, by choice of g we know $\{\sigma, \tau\} \notin \Delta_{g(\sigma)}$. This contradiction proves the result.

In order to prove the lemma then, it suffices to find functions $f_\tau : \tau \to \tau$ with the property (1). We may suppose $\mathcal{A}_\tau \neq \emptyset$, for otherwise any function f_τ satisfies (1). Thus $\tau \geq \kappa$, so $|\tau| = \kappa$, and we may write $\mathcal{A}_\tau = \{\langle A_{\tau\alpha}, g_{\tau\alpha}\rangle; \alpha < \kappa\}$. Since always $|A_{\tau\alpha}| = \kappa$ we can choose inductively $\sigma_{\tau\alpha}$ for α with $\alpha < \kappa$ such that $\sigma_{\tau\alpha} \in A_{\tau\alpha} - \{\sigma_{\tau\beta}; \beta < \alpha\}$. Define $f_\tau : \tau \to \tau$ as follows: if $\sigma = \sigma_{\tau\alpha}$ then $f_\tau(\sigma) = g_{\tau\alpha}(\sigma)$; otherwise $f_\tau(\sigma)$ is arbitrary. Then clearly (1) holds, since if $\langle A, g\rangle \in \mathcal{A}_\tau$ then $\langle A, g\rangle = \langle A_{\tau\alpha}, g_{\tau\alpha}\rangle$ for some α, and then $\sigma_{\tau\alpha}$ is an element of A on which both f_τ and g agree, as required by (1). This completes the proof of the lemma.

Theorem 6.1.12 (GCH). *Let λ be an infinite cardinal with $\theta < \lambda$. Then* $(\lambda^+, \lambda^+, \lambda^+) \not\to \theta$.

Proof. Using Lemma 6.1.11, take a disjoint partition $[\lambda^+]^2 = \bigcup\{\Delta_k; k < \lambda^+\}$ such that

$$A \in [\lambda^+]^\lambda \text{ and } B \in [\lambda^+]^{\lambda^+} \Rightarrow \exists\, \alpha \in A\ \forall k < \lambda^+\ \exists\, \beta \in B(\langle\alpha,\beta\rangle \in \Delta_k)\,.$$

Put $S = \lambda^+ \times \lambda$ so $|S| = \lambda^+$, and for β with $\beta < \lambda^+$ put

$$A_\beta = \{\langle\alpha, k\rangle \in S;\ \{\alpha, \beta\} \in \Delta_k\}\,.$$

We consider the decomposition of S given by the family $\mathcal{A} = \{A_\beta; \beta < \lambda^+\}$.

First observe that $\mathcal{A}$ has cover number λ^+. For take any X in $[S]^{\lambda^+}$, and for a contradiction suppose that if

$$I = \{\beta < \lambda^+; X \cap A_\beta = \emptyset\}$$

then $|I| = \lambda^+$. Put $X' = \{\alpha < \lambda^+; \exists\, k < \lambda\,(\langle\alpha, k\rangle \in X)\}$; since $|X| = \lambda^+$ also

$|X'| = \lambda^+$. By the choice of the partition, there is α in X' such that for all k with $k < \lambda^+$ there is β in I with $\{\alpha, \beta\} \in \Delta_k$. Since $\alpha \in X'$ there is $k_0 < \lambda$ with $\langle \alpha, k_0 \rangle \in X$ and then there is β_0 in I with $\{\alpha, \beta_0\} \in \Delta_{k_0}$, that is, $\langle \alpha, k_0 \rangle \in A_{\beta_0}$. Hence $\langle \alpha, k_0 \rangle \in X \cap A_{\beta_0}$, contradicting that $\beta_0 \in I$. Thus indeed, whenever $X \in [S]^{\lambda^+}$ then $|\{\beta < \lambda^+; X \cap A_\beta = \emptyset\}| < \lambda^+$, so $\mathcal{A}$ has cover number λ^+ as claimed.

However, S is not almost the union of θ connected sets. Note that since the partition is disjoint, whenever $\{\alpha, \beta\} \in \Delta_k$ then it follows that $\{\alpha, \beta\} \notin \Delta_l$ if $l \neq k$. Thus if $\langle \alpha, k \rangle \in A_\beta$ then $\langle \alpha, l \rangle \notin A_\beta$ whenever $l \neq k$, so no connected sets includes pairs $\langle \alpha, k \rangle$, $\langle \alpha, l \rangle$ with $l \neq k$. Now take any S^* from $[S]^{<\lambda^+}$. There must be α such that $\langle \alpha, k \rangle \in S - S^*$ for all k with $k < \lambda$. Suppose $S - S^* = \bigcup\{B_\mu; \mu < \theta\}$. There must be some μ such that for distinct k, l both $\langle \alpha, k \rangle \in B_\mu$ and $\langle \alpha, l \rangle \in B_\mu$, so this B_μ is not connected. Hence S is not almost the union of θ connected sets. Thus the family $\mathcal{A}$ provides an example to show that $(\lambda^+, \lambda^+, \lambda^+) \not\Rightarrow \theta$.

In almost the same way, we can prove the following result where the covering number is reduced to λ.

Theorem 6.1.13 (GCH). *Let λ be an infinite cardinal with $\theta^+ < \lambda$. Then $(\lambda^+, \lambda^+, \lambda) \not\Rightarrow \theta$.*

Proof. Almost as for Theorem 6.1.12, only using the set $S = \lambda^+ \times \theta^+$.

If the cardinals λ and η in the relation $(\kappa, \lambda, \eta) \Rightarrow \theta$ are far apart, then positive results can be proved. If $\lambda' > \eta^+$ then by Theorem 4.2.8

$$\begin{pmatrix} \eta \\ \lambda \end{pmatrix} \to \begin{pmatrix} \eta & \eta \\ \lambda & \lambda \end{pmatrix}^{1,1},$$

and hence whenever $\kappa \geqslant \eta$,

$$\begin{pmatrix} \kappa \\ \lambda \end{pmatrix} \to \begin{pmatrix} \eta & \eta \\ \lambda & \lambda \end{pmatrix}^{1,1}.$$

If we now apply Theorem 6.1.8, the following result is immediate.

Theorem 6.1.14 (GCH). *Let κ, λ, η be cardinals such that $\kappa \geqslant \eta$ and $\lambda' > \eta^+$. Then $(\kappa, \lambda, \eta) \Rightarrow 1$.*

If η is replaced by η^+ in the above theorem, then GCH is not needed in the proof. In fact we have the following.

Theorem 6.1.15. *Let κ, λ, η be cardinals such that $\kappa > \eta$ and $\lambda' > \eta^+$. Then* $(\kappa, \lambda, \eta^+) \Rightarrow 1$.

Proof. Suppose $\lambda' > \eta^+$, let S be a set with $|S| = \kappa$ and take a family $\mathscr{A} = \{A_\nu; \nu < \lambda\}$ with $\bigcup \mathscr{A} = S$ which has cover number η^+. For x in S, put $A(x) = \{\nu < \lambda; x \in A_\nu\}$. Then S is the disjoint union, $S = S_0 \cup S_1$ where $S_0 = \{x \in S; |A(x)| < \lambda\}$ and $S_1 = \{x \in S; |A(x)| = \lambda\}$. If $|S_0| = \kappa$, we could take S' from $[S_0]^{\eta^+}$, and noting that then $|\bigcup\{A(x); x \in S'\}| < \lambda$, we would have $|\{\nu < \lambda; S' \cap A_\nu = \emptyset\}| = \lambda$, contradicting that $\mathscr{A}$ has cover number η^+. Hence $|S_0| < \kappa$. Thus to prove the theorem, it suffices to show that S_1, but for fewer than κ elements, is connected. Let B be a maximal connected subset of S_1. Then I claim that $|S_1 - B| \leqslant \eta$. For if not, take B' from $[S_1 - B]^{\eta^+}$. Since $\mathscr{A}$ has cover number η^+, then $|\{\nu < \lambda; B' \cap A_\nu = \emptyset\}| < \lambda$. Take any x in B. By the maximality of B, for any ν if $x \in A_\nu$ then $A_\nu \subseteq B$, so if $x \in A_\nu$ then $B' \cap A_\nu = \emptyset$. Thus $|\{\nu < \lambda; x \in A_\nu\}| < \lambda$, that is, $|A(x)| < \lambda$, contrary to x being in S_1. Thus indeed $|S_1 - B| \leqslant \eta$. Since $\eta < \kappa$, this shows that S_1 is almost connected, as required.

In the case that λ is a singular cardinal with $\lambda' = \aleph_0$, there are techniques special to cofinality $\aleph_0$ which enable us to obtain results which correspond to unsolved problems for singular cardinals of higher cofinality. The first of these that we shall establish is that $(\kappa, \lambda, \lambda) \Rightarrow \lambda$, for arbitrary κ. This should be compared with Corollary 6.1.6 and Theorem 6.1.10.

Theorem 6.1.16. *Let λ be a singular cardinal with $\lambda' = \aleph_0$. Then $(\kappa, \lambda, \lambda) \Rightarrow \lambda$, for all κ.*

Proof. We may suppose $\kappa > \lambda$. Let $\langle \lambda_n, n < \omega\rangle$ be an increasing sequence of regular cardinals with always $\aleph_0 < \lambda_n < \lambda$ and $\lambda = \Sigma(\lambda_n, n < \omega)$. Let S be any set with $|S| = \kappa$, and take a family $\mathscr{A} = \{A_\nu; \nu < \lambda\}$ where $\bigcup \mathscr{A} = S$ and $\mathscr{A}$ has cover number λ.

Take any set S' from $[S]^{\geqslant\lambda}$. We start by observing that if there is I in $[\lambda]^\lambda$ for which

$$B \in [S']^{\lambda_n} \Rightarrow |\{\nu \in I; A_\nu \cap B = \emptyset\}| < \lambda\,, \tag{1}$$

then S' can be written as the union of λ_n connected sets. For suppose (1) is true. For x in S', put $A(x) = \{\nu \in I; x \in A_\nu\}$ and write S' as the disjoint union $S' = S'_0 \cup S'_1$ where $S'_0 = \{x \in S'; |A(x)| < \lambda\}$ and $S'_1 = \{x \in S'; |A(x)| = \lambda\}$. For x in S'_1, put $D(x) = \{y \in S'_1; \nexists\, \nu \in I(x, y \in A_\nu)\}$. It follows that if $x \in A_\nu$ then $A_\nu \cap D(x) = \emptyset$, so $A(x) \subseteq \{\nu \in I; A_\nu \cap D(x) = \emptyset\}$. Hence $|D(x)| < \lambda_n$, for if $|D(x)| \geqslant \lambda_n$ by (1) then $|\{\nu \in I; A_\nu \cap D(x) = \emptyset\}| < \lambda$ contradicting that $|A(x)| = \lambda$. For x in S_1', inductively define $D^n(x)$ for n with $n \leqslant \omega$ by

$$D^0(x) = \{x\}; D^{n+1}(x) = \bigcup\{D(y); y \in D^n(x)\};$$

$$D^\omega(x) = \bigcup\{D_n(x); n < \omega\}.$$

Then $|D^\omega(x)| < \lambda_n$. Inductively define subsets T_α of S_1' as follows: if $S_1 \neq \bigcup\{T_\beta; \beta < \alpha\}$, choose x in $S_1' - \bigcup\{T_\beta; \beta < \alpha\}$ and put $T_\alpha = D^\omega(x) - \bigcup\{T_\beta; \beta < \alpha\}$. Then always $|T_\alpha| < \lambda_n$. It now follows, just as in the proof of Theorem 6.1.10, that S_1' can be written as a union of λ_n connected sets. To complete the proof of the observation above, we must show that the same is true of S_0'. If $|S_0'| < \lambda_n$ this is trivial, so suppose $|S_0'| \geqslant \lambda_n$. For any x in S_0' there is an integer $m(x)$ such that $|A(x)| < \lambda_{m(x)}$. Since $\lambda_n' = \lambda_n > \aleph_0$, there is a subset S_0'' of S_0' with $|S_0''| = \lambda_n$ such that $m(x)$ is constant for x in S_0'', say $m(x) = m$. Thus $|\bigcup\{A(x); x \in S_0''\}| \leqslant \lambda_n \cdot \lambda_m < \lambda$, so $|\{\nu \in I; S_0'' \cap A_\nu \neq \emptyset\}| < \lambda$. Since $|S_0''| = \lambda_n$ this contradicts (1), and so it is impossible that $|S_0'| \geqslant \lambda_n$. Thus the observation is established.

We shall show that S can be decomposed into λ connected sets; from this the theorem follows. For a contradiction, suppose this to be false. Use induction on the integer n to define sets I_n in $[\lambda]^\lambda$, S_n^* in $[S]^{\lambda_n}$ and I_n^* in $[\lambda]^{\lambda_n}$ as follows: Put $I_0 = \lambda$ and choose I_0^* from $[I_0]^{\lambda_0}$. Write

$$S_n = S - \bigcup\{A_\nu; \nu \in I_m^* \text{ for some } m \text{ with } m \leqslant n\}.$$

Now $|\bigcup\{I_m^*; m < n\}| \leqslant \Sigma(\lambda_m; m < n) < \lambda$, so since S is not the union of λ connected sets, neither is S_n. In particular $|S_n| > \lambda$ so (1), applied in particular to S_n and I_n, must be false. Thus there are S_n^* in $[S_n]^{\lambda_n}$ and I_{n+1} in $[I_n]^\lambda$ such that $A_\nu \cap S_n^* = \emptyset$ whenever $\nu \in I_{n+1}$. Choose I_{n+1}^* from $[I_{n+1}]^{\lambda_{n+1}}$. This defines S_n^*, I_{n+1} and I_{n+1}^*, and completes the definition.

Put $S^* = \bigcup\{S_n^*; n < \omega\}$ and $I^* = \bigcup\{I_n^*; n < \omega\}$ so $|S^*| = \lambda$ and $|I^*| = \lambda$. However,

$$S^* \cap A_\nu = \emptyset \text{ if } \nu \in I^*, \tag{2}$$

for suppose $\nu \in I_m^*$. If $m \leqslant n$ by definition of S_n then $S_n^* \cap A_\nu = \emptyset$. If $m > n$, then since $I_m^* \subseteq I_m \subseteq I_{n+1}$, by choice of S_n^* and I_{n+1} we know $A_\nu \cap S_n^* = \emptyset$. In both cases then, $S_n^* \cap A_\nu = \emptyset$ and so indeed $S^* \cap A_\nu = \emptyset$. Since $|S^*| = \lambda$, (2) contradicts that $\mathscr{A}$ has cover number λ. This contradiction shows that S can be decomposed into λ connected sets, as required.

For small values of κ, the result in Theorem 6.1.16 can be improved substantially. From Theorem 4.2.5, when $\lambda' = \aleph_0$ then

$$\binom{\lambda^+}{\lambda} \to \binom{\lambda \quad \lambda}{\lambda \quad \lambda}^{1,1},$$

so by Theorem 6.1.8, the relation $(\lambda^+, \lambda, \lambda) \Rightarrow 1$ holds. When $\kappa = \lambda$, we shall show that $(\kappa, \lambda, \lambda) \Rightarrow 2$. In view of Theorem 6.1.7, this is the strongest positive result possible.

Theorem 6.1.17 (GCH). *If* $\lambda' = \aleph_0$ *then* $(\lambda^+, \lambda, \lambda) \Rightarrow 1$ *and* $(\lambda, \lambda, \lambda) \Rightarrow 2$.

Proof. Only the second relation remains to be proved, and in view of Corollary 6.1.6 we need only consider the case where $\lambda > \aleph_0$. Let $\langle \lambda_n; n < \omega \rangle$ be an increasing sequence of regular cardinals with always $\aleph_1 < \lambda_n < \lambda$ and $\lambda = \Sigma(\lambda_n; n < \omega)$. Let S be any set with $|S| = \lambda$, and consider any decomposition $\mathscr{A} = \{A_\nu; \nu < \lambda\}$ of S which has cover number λ. For a contradiction, suppose that whenever $S^* \in [S]^{<\lambda}$ then $S - S^*$ is not the union of two connected sets. The argument that follows is similar to that in the proof of Lemma 6.1.5.

For a subset X of S, let $G(X)$ be the graph on $S - X$ with vertices x, y joined by an edge in $G(X)$ just when $\{x, y\} \not\subseteq A$ for every A in $\mathscr{A}$. Then when $S^* \in [S]^{<\lambda}$ the graph $G(S^*)$ has chromatic number more than 2, so by Theorem 5.3.1 some finite subgraph of $G(S^*)$ has chromatic number more than 2. Thus we can choose inductively finite subsets S_α of S for α with $\alpha < \lambda$ so that the subgraph G_α of $G(\bigcup\{S_\beta; \beta < \alpha\})$ spanned by S_α has chromatic number more than 2.

For each integer n, we can choose I_n' from $[\lambda]^{\lambda_n}$ such that $|S_\alpha|$ is constant for α in I_n', say $|S_\alpha| = k(n)$. For such α, let $\mathfrak{P} S_\alpha = \{X_{\alpha i}; i < 2^{k(n)}\}$. Consider the partition $I_n' \times \lambda = \bigcup\{\Delta_i; i < 2^{k(n)}\}$ where

$$\langle \alpha, \nu \rangle \in \Delta_i \Leftrightarrow S_\alpha \cap A_\nu = X_{\alpha i}\,.$$

Since $\aleph_1 < \lambda_n = \lambda_n' < \lambda_n^+ < \lambda$, from Theorem 4.2.8 it follows that the relation

$$\binom{\lambda_n}{\lambda} \to \binom{\lambda_n}{\lambda}^{1,1}_{\gamma}$$

holds, where $\gamma = 2^{k(n)}$. Thus there are I_n'' in $[I_n']^{\lambda_n}$, M_n'' in $[\lambda]^\lambda$ and $i(n) < \gamma$ such that

$$\alpha \in I_n'' \text{ and } \nu \in M_n'' \Rightarrow S_\alpha \cap A_\nu = X_{\alpha i(n)}\,.$$

Further, we can arrange inductively that $M_n'' \subseteq \bigcap\{M_m''; m < n\}$. Choose inductively I_n from $[I_n'' - \bigcup\{I_m; m < n\}]^{\lambda_n}$ and M_n from $[M_n'' - \bigcup\{M_m; m < n\}]^{\lambda_n}$ and let I and M be the disjoint unions $I = \bigcup\{I_n; n < \omega\}$, $M = \bigcup\{M_n; n < \omega\}$.

Fix α in I. There is a unique n such that $\alpha \in I_n$. The set $X_{\alpha i(n)}$, being contained in some A_ν, is independent in the graph G_α. Since $\mathrm{Chr}(G_\alpha) > 2$ it follows that $S_\alpha - X_{\alpha i(n)}$ is not independent, so we can choose x_α, y_α from $S_\alpha - X_{\alpha i(n)}$ which are adjacent in G_α; thus $\{x_\alpha, y_\alpha\} \not\subseteq A$ for all A in $\mathcal{A}$. Take a disjoint partition $I \times M = \Delta_0 \cup \Delta_1$ such that

$$\langle \alpha, \nu \rangle \in \Delta_0 \Rightarrow x_\alpha \notin A_\nu; \langle \alpha, \nu \rangle \in \Delta_1 \Rightarrow y_\alpha \notin A_\nu .$$

From the proof of the polarized canonization lemma, Lemma 4.1.7, there are sets I_n^* in $[I]^{\lambda_n}$, M_n^* in $[M]^{\lambda_n}$ and a function $h : \omega \times \omega \to \{0,1\}$ such that if $m, n < \omega$, then

$$I_m^* \times M_n^* \subseteq \Delta_{h(m,n)} ,$$

and moreover there are increasing functions $f, g : \omega \to \omega$ with always $f(m) \leqslant g(m)$ such that $I_n^* \subseteq I_{f(n)}$ and $M_n^* \subseteq M_{g(n)}$. By Ramsey's theorem (Theorem 2.2.9) there are H in $[\omega]^\omega$ and i in $\{0,1\}$ such that whenever $m, n \in H$ with $m > n$ then $h(m, n) = i$. Suppose without loss of generality that $i = 0$. For m, n in H with $m > n$ then $I_m^* \times M_n^* \subseteq \Delta_0$; thus if $\alpha \in I_m^*$ and $\nu \in M_n^*$ then $x_\alpha \notin A_\nu$. On the other hand, if $m \leqslant n$ then $I_m^* \subseteq I_{f(m)} \subseteq I''_{f(m)}$ and $M_n^* \subseteq M_{g(n)} \subseteq M_{g(m)} \subseteq M_{f(m)}$ so for α in I_m^* and ν in M_n^* it follows that $S_\alpha \cap A_\nu = X_{\alpha i(f(m))}$; since $x_\alpha \in S_\alpha - X_{\alpha i(f(m))}$ thus $x_\alpha \notin A_\nu$. Put $T = \{x_\alpha; \alpha \in \bigcup\{I_n^*; n < \omega\}\}$ and $N = \bigcup\{M_n^*; n < \omega\}$. Then $|T| = \lambda$, $|N| = \lambda$ and $T \cap A_\nu = \emptyset$ whenever $\nu \in N$. Thus $|\{\nu < \lambda; T \cap A_\nu = \emptyset\}| = \lambda$, which contradicts that $\mathcal{A}$ has cover number λ. This contradiction proves the theorem.

For more results and unsolved problems connected with the relation $(\kappa, \lambda, \eta) \Rightarrow \theta$ the reader is referred to Erdös, Hajnal and Milner [35].

§2. Delta-systems

Let $\mathcal{A} = (A_i; i \in I)$ be an indexed family of sets. The family $\mathcal{A}$ is said to *contain* the family $\mathcal{B} = (B_j; j \in J)$ if there is a one-to-one map f from J into I such that always $B_j = A_{f(j)}$. For the purposes of this chapter, if $\mathcal{A}$ contains $\mathcal{B}$ and also $\mathcal{B}$ contains $\mathcal{A}$ then we may consider the indexed families $\mathcal{A}$ and $\mathcal{B}$ as being the same. In particular, we may always suppose that an indexed family $\mathcal{A}$ is indexed by the ordinals less than some cardinal, so $\mathcal{A} = (A_\nu; \nu < \lambda)$. A family $\mathcal{A} = (A_i; i \in I)$ is said to be a *(λ, κ)-family* if $|I| = \lambda$ and $|A_i| = \kappa$ for each i. Expressions such as $(\lambda, <\kappa)$-family have the obvious meanings. We shall be concerned with indexed families for which the intersection of any two sets from the family is always the same. An indexed family

$\mathcal{A} = (A_i; i \in I)$ is said to be a $\Delta(\lambda)$-*family* if $|I| = \lambda$ and $A_i \cap A_j = A_k \cap A_l$ whenever $\{i, j\}, \{k, l\} \in [I]^2$.

Definition 6.2.1. Let θ, κ, λ be cardinal numbers. The symbol $(\lambda, \kappa) \to \Delta(\theta)$ means the following: every (λ, κ)-family contains a $\Delta(\theta)$-family.

We shall ask for which values of θ, κ, λ the relation $(\lambda, \kappa) \to \Delta(\theta)$ is true. The first result of this kind in the literature seems to be due to Mazur [67], where the relation $(\aleph_1, \leqslant n) \to \Delta(\aleph_1)$ is proved. The general problem was first discussed by Erdös and Rado [31]. Michael [68] independently discovered some of their results. Erdös and Rado returned to the problem in [33], and there completely answered the question of finding for given κ and θ (not both finite) the least λ such that $(\lambda, <\kappa) \to \Delta(\theta)$.

In Theorem 6.2.5 below we shall solve the easier problem of deciding the truth of the relation $(\lambda, \kappa) \to \Delta(\theta)$, under the assumption of the GCH. (This relation is easily seen to be equivalent to $(\lambda, <\kappa^+) \to \Delta(\theta)$.) For a discussion not depending on the GCH, the reader is referred to Erdös and Rado [33] where the GCH is not assumed.

A couple of simple remarks are in order. Every (λ, κ)-family where $\lambda \geqslant 2$ contains a $\Delta(2)$-family, so in considering the relation $(\lambda, \kappa) \to \Delta(\theta)$ only values of θ with $3 \leqslant \theta \leqslant \lambda$ need be considered, and $(\lambda, \kappa) \to \Delta(\lambda)$ is the strongest positive result while $(\lambda, \kappa) \not\to \Delta(3)$ is the strongest negative result. Clearly if the relation $(\lambda, \kappa) \to \Delta(\theta)$ is true, then if $\lambda_1 \geqslant \lambda$ and $\theta_1 \leqslant \theta$ also the relation $(\lambda_1, \kappa) \to \Delta(\theta_1)$ holds; and moreover if $\kappa_1 \leqslant \kappa$ also the relation $(\lambda, \kappa_1) \to \Delta(\theta)$ holds. To verify the last claim, suppose given a (λ, κ_1)-family $\mathcal{A}$. Choose pairwise disjoint sets C_A, one for each A in $\mathcal{A}$, such that $|A \cup C_A| = \kappa$ and consider the (λ, κ)-family $\mathcal{C} = \{A \cup C_A; A \in \mathcal{A}\}$. Since for A, B from $\mathcal{A}$ we have $(A \cup C_A) \cap (B \cup C_B) = A \cap B$, any $\Delta(\theta)$-family contained in $\mathcal{C}$ gives a $\Delta(\theta)$-family contained in $\mathcal{A}$.

The discussion is based on the following two lemmas. The proof for the first follows Davies [8].

Lemma 6.2.2. *Suppose that λ is an infinite regular cardinal, that $\kappa \geqslant 1$ and that $\eta^\kappa < \lambda$ whenever $\eta < \lambda$. Then $(\lambda, \kappa) \to \Delta(\lambda)$.*

Proof. Take any indexed family $\mathcal{A} = (A_\nu; \nu < \lambda)$ where always $|A_\nu| = \kappa$. If λ of the A_ν are the same trivially $\mathcal{A}$ contains a $\Delta(\lambda)$-family, so we may assume this does not happen and in fact since λ is regular we may assume all the A_ν are different.

Put $S = \bigcup\{A_\nu; \nu < \lambda\}$, so $|S| \leqslant \kappa \cdot \lambda = \lambda$. In fact $|S| = \lambda$, for if $|S| = \eta < \lambda$

then S has only η^κ different subsets of power κ, and $\eta^\kappa < \lambda$. Take any well ordering $\prec$ of S with order type λ. For each ν, let $\langle x(\nu,\alpha); \alpha < \epsilon_\nu\rangle$ enumerate A_ν in increasing $\prec$-order, so $|\epsilon_\nu| = \kappa$. For α with $\alpha < \kappa^+$, put $S_\alpha = \{x(\nu,\alpha); \nu < \lambda\}$. Since $S = \bigcup\{S_\alpha; \alpha < \kappa^+\}$ and $|S| = \lambda$ with λ regular and $\kappa^+ \leqslant 2^\kappa < \lambda$, there must be some S_α with $|S_\alpha| = \lambda$. In fact, let β be the least ordinal with $|S_\beta| = \lambda$. Put $T = \bigcup\{S_\alpha; \alpha < \beta\}$ so $|T| < \lambda$, by the regularity of λ. Then also $|[T]^{\leqslant\kappa}| < \lambda$. For each x in S_β, choose one ν such that $x = x(\nu,\beta)$. Consider the sets $X_\nu = \{x(\nu,\alpha); \alpha < \beta\}$ for each of these ν. Then $|X_\nu| = |\beta| \leqslant \kappa$, so $X_\nu \in [T]^{\leqslant\kappa}$. Since $|S_\beta| = \lambda$ and λ is regular, there must be λ of the ν having the same X_ν. Thus there is L in $[\lambda]^\lambda$ such that for ν, μ in L we have $X_\nu = X_\mu = X$ say, and $x(\nu,\beta) \neq x(\mu,\beta)$.

Inductively choose ν_ξ from L, for ξ with $\xi < \lambda$, such that for all x in $\bigcup\{A_{\nu_\zeta}; \zeta < \xi\}$ we have $x \prec x(\nu_\xi,\beta)$. Since $|\bigcup\{A_{\nu_\zeta}; \zeta < \xi\}| < \lambda$ (again by the regularity) such ν_ξ can be found. Then always $A_{\nu_\zeta} \cap A_{\nu_\xi} = X$ so $(A_{\nu_\xi}; \xi < \lambda)$ is a $\Delta(\lambda)$-family contained in $\mathscr{A}$, and the lemma is proved.

Lemma 6.2.3. *Let θ_ν for ν with $\nu < \kappa$ be cardinals such that $\theta_\nu \dot{+} 1 < \theta$. Put $\lambda = \Sigma(\Pi(\theta_\nu; \nu < \alpha); \alpha < \kappa)$. Then $(\lambda, <\kappa) \not\to \Delta(\theta)$.*

Proof. For each α, let P_α be the cartesian product of the θ_ν with $\nu < \alpha$, and put $\mathscr{A} = \bigcup\{P_\alpha; \alpha < \kappa\}$. Consider the functions in $\mathscr{A}$ as sets of ordered pairs. Certainly $|\mathscr{A}| = \lambda$ and $|f| < \kappa$ for each f in $\mathscr{A}$. Consider $\mathscr{A}$ as an indexed family of distinct sets, indexed by λ; then $\mathscr{A}$ is a $(\lambda, <\kappa)$-family and we shall show that $\mathscr{A}$ contains no $\Delta(\theta)$-family.

Let $\mathscr{B} = (f_i; i \in I)$ be an indexed family contained in $\mathscr{A}$ such that the pairs of elements from $\mathscr{B}$ have constant intersection; that is, there is a subset F of $\bigcup\mathscr{A}$ such that $f \cap g = F$ for all pairs f, g from $\mathscr{B}$. Then $|F| < \kappa$, so if σ is the least ordinal not in the domain of F then $\sigma < \kappa$. Also let τ be the least ordinal such that $\beta < \tau$ for all β in dom(F), so $\tau < \kappa$.

If $\sigma < \tau$ then $\sigma \in \text{dom}(f)$ for all f from $\mathscr{B}$. The choice of σ ensures that $f(\sigma) \neq g(\sigma)$ for distinct f, g from $\mathscr{B}$, so if $\sigma < \tau$ then $|I| \leqslant \theta_\sigma < \theta$. On the other hand, suppose $\tau \leqslant \sigma$. Then dom$(F) = \tau$ so $F \in \mathscr{A}$. Moreover, for any f, g in $\mathscr{B}$ with $f, g \neq F$ we have $\tau \in \text{dom}(f) \cap \text{dom}(g)$ and $f(\tau) \neq g(\tau)$. Hence $|I| \leqslant \theta_\tau \dot{+} 1 < \theta$. So in either event $|I| < \theta$. This proves the lemma.

Lemma 6.2.4. *Suppose κ is infinite and $\lambda \leqslant \kappa^+$. Then $(\lambda, \kappa) \not\to \Delta(3)$.*

Proof. The relation $(\kappa^+, \kappa) \not\to \Delta(3)$ is an immediate corollary of Lemma 6.2.3. It can also be seen directly by considering the family $(\alpha; \kappa \leqslant \alpha < \kappa^+)$. Then certainly if $\lambda \leqslant \kappa^+$, also $(\lambda, \kappa) \not\to \Delta(3)$.

By Lemma 6.2.4, in considering the relation $(\lambda, \kappa) \to \Delta(\theta)$ we may henceforth suppose that $\kappa^+ < \lambda$. The next theorem covers all non-trivial cases in which $\lambda \dotplus \kappa \geqslant \aleph_0$. (Unsolved problems remain if both λ and κ are finite. See Erdös and Rado [31].)

Theorem 6.2.5 (GCH). *Suppose that λ is infinite and $\kappa \geqslant 1$. Then*

(i) *if λ is inaccessible and $\kappa < \lambda$, then $(\lambda, \kappa) \to \Delta(\lambda)$;*

(ii) *if λ is singular and $\kappa, \theta < \lambda$, then $(\lambda, \kappa) \to \Delta(\theta)$;*

(iii) *if λ is singular and $\kappa < \lambda$, then $(\lambda, \kappa) \not\to \Delta(\lambda)$;*

(iv) *if $\kappa < \lambda'$, then $(\lambda^+, \kappa) \to \Delta(\lambda^+)$;*

(v) *if λ is singular and $\kappa, \theta < \lambda$, then $(\lambda^+, \kappa) \to \Delta(\theta)$;*

(vi) *if λ is singular and $\lambda' \leqslant \kappa < \lambda$, then $(\lambda^+, \kappa) \not\to \Delta(\lambda)$.*

Proof. Part (i) is immediate from Lemma 6.2.2. For (ii), take κ and θ with $\kappa, \theta < \lambda$. By Lemma 6.2.2, $((\theta^\kappa)^+, \kappa) \to \Delta((\theta^\kappa)^+)$. Since (by GCH) $(\theta^\kappa)^+ < \lambda$, and $\theta < (\theta^\kappa)^+$, certainly $(\lambda, \kappa) \to \Delta(\theta)$. From (ii), (v) follows trivially. Lemma 6.2.2 implies (iv), noting that λ^+ is regular and that by GCH, $\lambda^\kappa = \lambda < \lambda^+$ when $\kappa < \lambda'$.

For (vi), take κ with $\lambda' \leqslant \kappa < \lambda$. Take a sequence $\langle \lambda_\nu; \nu < \kappa^+ \rangle$ of cardinals such that always $\lambda_\nu < \lambda$ and $\langle \lambda_\nu; \nu < \lambda' \rangle$ is a sequence cofinal in λ. For α with $\alpha < \kappa^+$, always $\Pi(\lambda_\nu; \nu < \alpha) \leqslant \lambda^{|\alpha|} \leqslant \lambda^+$ by GCH, and $\Pi(\lambda_\nu; \nu < \lambda') = \lambda^+$ by König's Theorem. Hence

$$\lambda^+ \leqslant \Sigma(\Pi(\lambda_\nu; \nu < \alpha); \alpha < \kappa^+) \leqslant \lambda^+ \cdot \kappa^+ = \lambda^+ ,$$

so by Lemma 6.2.3 it follows that $(\lambda^+, <\kappa^+) \not\to \Delta(\lambda)$; hence $(\lambda^+, \kappa) \not\to \Delta(\lambda)$.

Only (iii) remains, so suppose λ is singular. Write λ as a disjoint union, $\lambda = \bigcup\{S_\sigma; \sigma < \lambda'\}$ where always $|S_\sigma| < \lambda$. Take pairwise disjoint sets A_σ for σ with $\sigma < \lambda'$ such that always $|A_\sigma| = \kappa$. If $\sigma < \lambda'$ and $x \in S_\sigma$, put $A_{\sigma x} = A_\sigma$. Consider the indexed family $\mathcal{A} = (A_{\sigma x}; x \in S_\sigma$ and $\sigma < \lambda')$. Clearly $\mathcal{A}$ is a (λ, κ)-family which contains no $\Delta(\lambda)$-family, and so (iii) holds. This completes the proof.

Lemma 6.2.4 shows that a (κ^+, κ)-family may fail to contain any non-trivial Δ-family. However, if some restriction is placed on the (κ^+, κ)-family then positive results are possible. We shall investigate (κ^+, κ)-families $\mathcal{A}$ such that every two sets from $\mathcal{A}$ have an intersection of power less that θ, for some θ with $\theta < \kappa$. Since the sets from such an indexed family $\mathcal{A} = (A_i; i \in I)$ are pairwise distinct, we may identify $\mathcal{A}$ with the non-indexed family $\{A_i; i \in I\}$ and refer to the degree of disjunction $\delta(\mathcal{A})$ as defined in Chapter 4. The following theorem is from Erdös, Milner and Rado [39].

Theorem 6.2.6 (GCH). *Let $\mathscr{A}$ be a (κ^+, κ)-family with $\delta(\mathscr{A}) = \theta$ where $\theta < \kappa$.*
(i) *If $\theta < \kappa'$ then $\mathscr{A}$ contains a $\Delta(\kappa^+)$-family.*
(ii) *If $\kappa' \leqslant \theta < \kappa$ and $\eta < \kappa$ then $\mathscr{A}$ contains a $\Delta(\eta)$-family.*

Proof. Let the family $\mathscr{A} = \{A_\nu; \nu < \kappa^+\}$ be given, with $\delta(\mathscr{A}) = \theta$ and $|A_\nu| = \kappa$ for each ν. Use transfinite induction to choose subsets I_α of κ^+ for α with $\alpha < \kappa^{++}$ so that I_α is a subset of $\kappa^+ - \bigcup\{I_\beta; \beta < \alpha\}$ maximal with the property that $\{A_\nu - \bigcup\{A_\mu; \mu \in I_\beta$ for some β with $\beta < \alpha\}; \nu \in I_\alpha\}$ is a pairwise disjoint family. For brevity, put

$$I(\alpha) = \bigcup\{I_\beta; \beta < \alpha\}, \ A(\alpha) = \bigcup\{A_\mu; \mu \in I(\alpha)\};$$

thus I_α is a subset of $\kappa^+ - I(\alpha)$ maximal such that $\{A_\nu - A(\alpha); \nu \in I_\alpha\}$ is pairwise disjoint. Thus in particular

$$\{\mu, \nu\} \in [I_\alpha]^2 \Rightarrow A_\mu \cap A_\nu \subseteq A(\alpha), \tag{1}$$

since by choice of I_α it follows that $(A_\mu - A(\alpha)) \cap (A_\nu - A(\alpha)) = \emptyset$. Further,

$$\nu \in \kappa^+ - I(\alpha) \Rightarrow |A_\nu \cap A(\alpha)| \geqslant |\alpha|. \tag{2}$$

For take any β with $\beta < \alpha$. Since $\nu \notin I_\beta$ by the maximality of I_β there is $\mu(\beta)$ in I_β such that $(A_\nu - A(\beta)) \cap (A_{\mu(\beta)} - A(\beta)) \neq \emptyset$. Hence we can choose x_β from $(A_\nu \cap A_{\mu(\beta)}) - A(\beta)$. And if $\gamma < \beta$ then $x_\gamma \neq x_\beta$ since $x_\gamma \in A_{\mu(\gamma)} \subseteq A(\beta)$. Thus $\{x_\beta; \beta < \alpha\} \subseteq A_\alpha \cap A(\alpha)$, so indeed $|A_\alpha \cap A(\alpha)| \geqslant |\alpha|$.

Now we can prove (i). So suppose $\theta < \kappa'$, and we wish to show that $\mathscr{A}$ contains a $\Delta(\kappa^+)$-family. Assume first that $|I_\alpha| \leqslant \kappa$ for all α with $\alpha < \theta$. Then $|I(\theta)| \leqslant \theta \cdot \kappa = \kappa$, so $|A(\theta)| \leqslant \kappa \cdot \kappa = \kappa$ and $|\kappa^+ - I(\theta)| = \kappa^+$. By (2), $|A_\nu \cap A(\theta)| \geqslant \theta$ for each ν in $\kappa^+ - I(\theta)$, so we can choose B_ν from $[A_\nu \cap A(\theta)]^\theta$. However, $[A(\theta)]^\theta \leqslant \kappa^\theta = \kappa$ and so there must be distinct μ, ν in $\kappa^+ - I(\theta)$ for which $B_\mu = B_\nu$. But then $|A_\mu \cap A_\nu| \geqslant |B_\mu| = \theta$, contradicting that $\delta(\mathscr{A}) = \theta$.

Thus there must be α with $\alpha < \theta$ for which $|I_\alpha| = \kappa^+$, and in fact let α be the least such ordinal. Then $|A(\alpha)| \leqslant |\alpha| \cdot \kappa \cdot \kappa = \kappa$. Put

$$P = \{\nu \in I_\alpha; |A_\nu \cap A(\alpha)| \geqslant \theta\}; Q = \{\nu \in I_\alpha; |A_\nu \cap A(\alpha)| < \theta\}.$$

Consider what would happen if $|P| = \kappa^+$. Then for each ν in P, choose B_ν from $[A_\nu \cap A(\alpha)]^\theta$. Since $|[A(\alpha)]^\theta| \leqslant \kappa^\theta = \kappa$, there must be distinct μ, ν in P for which $B_\mu = B_\nu$. But then $|A_\mu \cap A_\nu| \geqslant |B_\mu| = \theta$, contrary to $\delta(\mathscr{A}) = \theta$. Thus in fact $|P| < \kappa^+$, and so $|Q| = \kappa^+$. Since $|[A(\alpha)]^{<\theta}| \leqslant \kappa^\theta = \kappa$ there must be A in $[A(\alpha)]^{<\theta}$ and Q' in $[Q]^{\kappa^+}$ such that $A_\nu \cap A(\alpha) = A$ whenever $\nu \in Q'$. Thus $(A_\nu \cap A(\alpha); \nu \in Q')$ is a $\Delta(\kappa^+)$-family and hence by (1), $\{A_\nu; \nu \in Q'\}$ is a $\Delta(\kappa^+)$-family, and is contained in $\mathscr{A}$. Thus (i) is proved.

Now consider (ii). Thus κ is singular, $\kappa' \leqslant \theta < \kappa$ and $\eta < \kappa$. By increasing

θ if necessary, we may assume that $\theta' \neq \kappa'$. Suppose $|I(\kappa)| \leqslant \kappa$. Then $|A(\kappa)| \leqslant \kappa \cdot \kappa = \kappa$ and $|\kappa^+ - I(\kappa)| = \kappa^+$. By (2), $|A_\nu \cap A(\kappa)| \geqslant \kappa$ whenever $\nu \in \kappa^+ - I(\kappa)$. Thus there is a decomposition $\mathcal{B} = \{A_\nu \cap A(\kappa); \nu \in \kappa^+ - I(\kappa)\}$ of a set of power κ into κ^+ sets each of power at least κ, with $\delta(\mathcal{B}) \leqslant \theta$. By Theorem 1.1.6 this is impossible. Hence $|I(\kappa)| = \kappa^+$. Since $I(\kappa) = \bigcup\{I_\alpha; \alpha < \kappa\}$, there must be α with $\alpha < \kappa$ such that $|I_\alpha| = \kappa^+$, and we may suppose that α is the least such ordinal. Then $|A(\alpha)| \leqslant \kappa \cdot \kappa \cdot \kappa = \kappa$. Again by Theorem 1.1.6 there cannot be I in $[I_\alpha]^{\kappa^+}$ such that $|A_\nu \cap A(\alpha)| = \kappa$ for each ν in I. Thus there must be I in $[I_\alpha]^{\kappa^+}$ such that $|A_\nu \cap A(\alpha)| < \kappa$ for each ν in I. There is a cardinal κ_1 with $\kappa_1 < \kappa$ and a set I' in $[I]^{\kappa^+}$ such that $|A_\nu \cap A(\alpha)| = \kappa_1$ for all ν in I'. By Theorem 6.2.5(v), $(\kappa^+, \kappa_1) \to \Delta(\eta)$. Thus there is I'' in $[I']^\eta$ such that $(A_\nu \cap A(\alpha); \nu \in I'')$ is a $\Delta(\eta)$-family, and so by (1) also $\{A_\nu; \nu \in I''\}$ is a $\Delta(\eta)$-family. This proves (ii).

The result in (ii) in the theorem above is the best possible in the sense that if κ is singular then there is a (κ^+, κ)-family $\mathcal{A}$ with $\delta(\mathcal{A}) = \kappa'$ which contains no $\Delta(\kappa)$-family. For take a sequence $\langle \kappa_\sigma; \sigma < \kappa' \rangle$ of cardinals below κ with $\kappa = \Sigma(\kappa_\sigma; \sigma < \kappa')$, and let P be the cartesian product of the sets κ_σ. For each f in P, put $B_f = \{f \restriction \sigma; \sigma < \kappa'\}$. Take pairwise disjoint sets C_f with $|C_f| = \kappa$, disjoint from all the B_f, and put $A_f = B_f \cup C_f$. Then $\mathcal{A}. = \{A_f; f \in P\}$ is a (κ^+, κ)-family with $\delta(\mathcal{A}) = \kappa'$. Suppose $\{A_f; f \in Q\}$ where $Q \subseteq P$ is a $\Delta(|Q|)$-family. For distinct f, g from Q then $A_f \cap A_g = B_f \cap B_g$, and there is σ with $\sigma < \kappa'$ such that $f \restriction \sigma \notin B_f \cap B_g$ (for otherwise $B_f = B_g$ and so $f = g$). Clearly the least such σ is a successor ordinal, $\sigma = \tau + 1$, and $f(\tau) \neq g(\tau)$. Further, σ is the same for all pairs f, g from Q. Hence $|Q| \leqslant \kappa_\tau < \kappa$.

The restriction that $\theta < \kappa$ in Theorem 6.2.6 is justified by the following example of a $(2^\kappa, \kappa)$-family $\mathcal{A}$ with $\delta(\mathcal{A}) = \kappa$ which contains no $\Delta(3)$-family. For each f in ${}^\kappa 2$ put $A_f = \{f \restriction \alpha; \alpha < \kappa\}$. Then the family $\mathcal{A} = \{A_f; f \in {}^\kappa 2\}$ is a $(2^\kappa, \kappa)$-family with $\delta(\mathcal{A}) = \kappa$. Suppose $\{A_f; f \in I\}$ where $I \subseteq {}^\kappa 2$ is a $\Delta(|I|)$-family. For distinct f, g from I there is α with $\alpha < \kappa$ such that $f \restriction \alpha \notin A_f \cap A_g$. The least such α is a successor ordinal, $\alpha = \beta + 1$, and moreover is constant for all pairs f, g from I. Thus $f(\beta) \neq g(\beta)$ for distinct f, g from I, so that $|I| < 3$. Hence $\mathcal{A}$ contains no $\Delta(3)$-family.

Lemma 6.2.4 gives also the result $(\lambda, \kappa) \not\to \Delta(3)$ if $\lambda \leqslant \kappa$, and one may ask if a restriction to the degree of disjunction of the (λ, κ)-family will lead to a positive relation, as happened in Theorem 6.2.6 when $\lambda = \kappa^+$. However, no such positive relation is forthcoming; the following gives an example of a (κ, κ)-family $\mathcal{A}$ with $\delta(\mathcal{A}) = 2$ which contains no non-trivial Δ-family. For α with $\alpha < \kappa$, put $A_\alpha = \{\langle \alpha, \beta \rangle; \beta < \kappa\} \cup \{\langle \mu, \alpha \rangle; \mu < \alpha\}$. Then if $\gamma < \alpha < \kappa$, clearly $A_\gamma \cap A_\alpha = \{\langle \gamma, \alpha \rangle\}$. Hence $(A_\alpha; \alpha < \kappa)$ has the required properties.

§3. Weak delta-systems

An indexed family $\mathscr{A} = (A_i; i \in I)$ is said to be a weak $\Delta(\lambda)$-family if $|I| = \lambda$ and $|A_i \cap A_j| = |A_k \cap A_l|$ whenever $\{i, j\}, \{k, l\} \in [I]^2$. Clearly every $\Delta(\lambda)$-family is also a weak $\Delta(\lambda)$-family, but not conversely. We shall consider those cases when the relation $(\lambda, \kappa) \to \Delta(\theta)$ from the last section is false, and ask about the truth of the weaker requirement that every (λ, κ)-family contain only a weak $\Delta(\theta)$-family.

Definition 6.3.1. Let θ, κ, λ be cardinal numbers. The symbol $(\lambda, \kappa) \to \mathrm{wk}\Delta(\theta)$ means the following: every (λ, κ)-family contains a weak $\Delta(\theta)$-family.

The relation $(\lambda, \kappa) \to \mathrm{wk}\ \Delta(\theta)$ was introduced by Erdös, Milner and Rado [39], and for infinite λ and κ they gave a complete discussion (assuming GCH).

The same simple remarks that were made in the last section shortly after the definition of the relation $(\kappa, \lambda) \to \Delta(\theta)$ apply also to the relation $(\kappa, \lambda) \to \mathrm{wk}\ \Delta(\theta)$.

Lemma 6.2.4 shows that $(\lambda, \kappa) \not\to \Delta(3)$ whenever $\lambda \leqslant \kappa^+$, so a discussion of the relation $(\lambda, \kappa) \to \mathrm{wk}\ \Delta(\theta)$ is pertinent. Many special cases need to be considered, and there is no simple result analogous to $(\lambda, \kappa) \not\to \Delta(3)$. We first establish several lemmas.

For the rest of this section, write $\psi(\kappa)$ for the number of cardinals less than or equal to κ, so

$$\psi(\kappa) = |\{\eta; \eta \text{ is a cardinal and } 0 \leqslant \eta \leqslant \kappa\}| .$$

Thus if $\kappa = \aleph_\alpha$ then $\psi(\kappa) = |\alpha| \dotplus \aleph_0$.

Lemma 6.3.2. *Suppose* $\lambda \to (\theta)^2_{\psi(\kappa)}$. *Then* $(\lambda, \kappa) \to \mathrm{wk}\ \Delta(\theta)$.

Proof. Let $\mathscr{A} = (A_\nu; \nu < \lambda)$ be any (λ, κ)-family. For each cardinal η with $\eta \leqslant \kappa$, put

$$\Gamma_\eta = \{\{\mu, \nu\} \in [\lambda]^2; |A_\mu \cap A_\nu| = \eta\} ,$$

so $[\lambda]^2$ is the union of the Γ_η. By the partition relation $\lambda \to (\theta)^2_{\psi(\kappa)}$, there is H in $[\lambda]^\theta$ such that $[H]^2 \subseteq \Gamma_\eta$ for some η. Then $(A_\nu; \nu \in H)$ is a weak $\Delta(\theta)$-family contained in $\mathscr{A}$.

Lemma 6.3.3. *For all infinite cardinals* κ, *the relation* $(2^{\psi(\kappa)}, \kappa) \not\to \mathrm{wk}\ \Delta(3)$ *holds.*

Proof. Let $\langle \lambda_\beta; \beta < \psi(\kappa)\rangle$ enumerate in order all cardinals (finite and infinite) strictly less than κ. Start with a pairwise disjoint family $\{X(g); g \in {}^\beta 2$ for some β with $\beta < \psi(\kappa)\}$ where if $g \in {}^\beta 2$ then $|X(g)| = \lambda_\beta^+$. For f in ${}^{\psi(\kappa)}2$ put

$$A_f = \bigcup\{X(f \upharpoonright \beta); \beta < \psi(\kappa)\} ,$$

so always $|A_f| = \Sigma(\lambda_\beta^+; \beta < \psi(\kappa)) = \kappa$. Consider the $(2^{\psi(\kappa)}, \kappa)$-family $\mathcal{A} = (A_f; f \in {}^{\psi(\kappa)}2)$. For distinct f, g from ${}^{\psi(\kappa)}2$, as before write $\delta(f, g)$ for the least ordinal α such that $f(\alpha) \neq g(\alpha)$, and note that $A_f \cap A_g = \bigcup\{X(f \upharpoonright \beta); \beta < \delta(f, g)\}$ so

$$|A_f \cap A_g| = \Sigma(\lambda_\beta^+; \beta < \delta(f, g)) .$$

Thus $|A_f \cap A_g| = \lambda_{\delta(f,g)}$ if $\delta(f, g)$ is infinite, and $|A_f \cap A_g| = \frac{1}{2}\delta(f, g)(\delta(f,g)+1)$ if $\delta(f, g)$ is finite. Take distinct functions f, g, h from ${}^{\psi(\kappa)}2$. We may suppose $\delta(f, g) \neq \delta(g, h)$, and hence $|A_f \cap A_g| \neq |A_g \cap A_h|$. Thus the family $\mathcal{A}$ contains no weak $\Delta(3)$-family, and the lemma is proved.

Lemma 6.3.4. *Let* $\lambda' = \kappa$; *then* $(\lambda^+, \aleph_\kappa) \nrightarrow$ wk $\Delta(\lambda)$. *If* $\lambda' = \aleph_0$ *then* $(\lambda^+, \aleph_0) \nrightarrow$ wk $\Delta(\lambda)$.

Proof. Take a sequence $\langle \lambda_\sigma; \sigma < \kappa\rangle$ of cardinals such that always $1 < \lambda_\sigma < \lambda$ and $\lambda = \Sigma(\lambda_\sigma; \sigma < \kappa)$. Choose a pairwise disjoint family $\{X(g); g \in \mathsf{X}(\lambda_\sigma; \sigma < \beta)$ for some β with $\beta < \kappa\}$, where if $g \in \mathsf{X}(\lambda_\sigma; \sigma < \beta)$ then $|X(g)| = \aleph_{\beta+1}$. Write $P = \mathsf{X}(\lambda_\sigma; \sigma < \kappa)$ and for f in P put

$$A_f = \bigcup\{X(f \upharpoonright \beta); \beta < \kappa\} ,$$

so always $|A_f| = \Sigma(\aleph_{\beta+1}; \beta < \kappa) = \aleph_\kappa$. And from König's Lemma, $|P| > \lambda$, so the family $\mathcal{A} = (A_f; f \in P)$ is a $(\geqslant \lambda^+, \aleph_\kappa)$-family. For distinct f, g from P we have $A_f \cap A_g = \bigcup\{X(f \upharpoonright \beta); \beta < \delta(f, g)\}$ so that

$$|A_f \cap A_g| = \Sigma(\aleph_{\beta+1}; \beta < \delta(f, g)) = \aleph_{\delta(f,g)} .$$

Let $\mathcal{B} = (A_f; f \in I)$ be a weak Δ-family contained in $\mathcal{A}$, where we may suppose $I \subseteq P$. Then $\delta(f, g)$ must be constant for all pairs $\{f, g\}$ from $[I]^2$, say with value σ (so $\sigma < \kappa$). Then $|I| \leqslant \lambda_\sigma < \lambda$. Hence $\mathcal{A}$ contains no weak $\Delta(\lambda)$-family.

If $\lambda' = \aleph_0$, modify the above construction by starting with $|X(g)| = \beta + 1$ when $g \in \mathsf{X}(\lambda_\sigma; \sigma < \beta)$, and noting that β is finite.

Lemma 6.3.5. *Let κ be regular, and let f be a function, $f: [\kappa^+]^2 \to \{0, 1\}$. There is a (κ^+, κ)-family $(A_\nu; \nu < \kappa^+)$ such that whenever $\mu < \nu < \kappa^+$ then*

$$|A_\mu \cap A_\nu| < \kappa \textit{ if } f(\{\mu, \nu\}) = 0; \ |A_\mu \cap A_\nu| = \kappa \textit{ if } f(\{\mu, \nu\}) = 1 .$$

Proof. Start with a pairwise disjoint family $\{S_{\alpha\beta}; \alpha < \kappa^+ \text{ and } \beta < \kappa\}$ where each $S_{\alpha\beta}$ has power κ. Since κ is regular, by Theorem 1.3.3 any family $\mathcal{B}$ of κ pairwise almost disjoint sets each of power κ has a κ-transversal, that is, there is a set T with $1 \leqslant |B \cap T| < \kappa$ for each B in $\mathcal{B}$. Thus we can inductively choose sets B_μ of power κ for μ with $\mu < \kappa^+$ such that B_μ is a κ-transversal of

$$\{S_{\alpha\beta}; \alpha < \mu \text{ and } \beta < \kappa\} \cup \{B_\sigma; \sigma < \mu\} .$$

Put $S_\alpha = \bigcup\{S_{\alpha\beta}; \beta < \kappa\}$, for α with $\alpha < \kappa^+$. Then the S_α are pairwise disjoint. An easy induction shows that always $B_\mu \subseteq \bigcup\{S_\alpha; \alpha < \mu\}$. Put $A_{\alpha\mu} = S_\alpha \cap B_\mu$, so $B_\mu = \bigcup\{A_{\alpha\mu}; \alpha < \mu\}$. Also $|A_{\alpha\mu}| = \kappa$ whenever $\alpha < \mu$, for then $|S_{\alpha\beta} \cap B_\mu| \geqslant 1$ for each β (with $\beta < \kappa$) and $S_{\alpha\beta} \cap B_\mu \subseteq S_\alpha \cap B_\mu = A_{\alpha\mu}$.

For ν with $\nu < \kappa^+$, put

$$A_\nu = S_\nu \cup \bigcup\{A_{\mu\nu}; \mu < \nu \text{ and } f(\{\mu, \nu\}) = 1\} .$$

Then $S_\nu \subseteq A_\nu \subseteq \bigcup\{S_\alpha; \alpha \leqslant \nu\}$ so $|A_\nu| = \kappa$. Thus if $\mathcal{A} = (A_\nu; \nu < \kappa^+)$ then $\mathcal{A}$ is a (κ^+, κ)-family. Further, $\mathcal{A}$ has the required property. For take μ, ν with $\mu < \nu < \kappa^+$. If $f(\{\mu, \nu\}) = 1$ then $A_{\mu\nu} \subseteq S_\mu \cap A_\nu \subseteq A_\mu \cap A_\nu$ so $|A_\mu \cap A_\nu| = \kappa$. If $f(\{\mu, \nu\}) = 0$ then

$$A_\mu \cap A_\nu \subseteq \bigcup\{A_{\sigma\mu}; \sigma < \mu\} \cap \bigcup\{A_{\sigma\nu}; \sigma < \nu\} = B_\mu \cap B_\nu$$

so $|A_\mu \cap A_\nu| < \kappa$, since $|B_\mu \cap B_\nu| < \kappa$ by choice of B_ν. This proves the lemma.

Theorem 6.3.6 (GCH). *Suppose κ is an infinite successor cardinal. Then $(\kappa^+, \kappa) \nrightarrow \text{wk } \Delta(\kappa^+)$ and $(\kappa^+, \kappa) \to \text{wk } \Delta(\kappa)$.*

Proof. To see that $(\kappa^+, \kappa) \to \text{wk } \Delta(\kappa)$, note that if $\kappa = \eta^+$ then the number γ of cardinals $\leqslant \kappa$ is the same as the number of cardinals $\leqslant \eta$, so $\gamma \leqslant \eta < \kappa$. By Theorem 2.2.4 the partition relation $\kappa^+ \to (\kappa)^2_\gamma$ holds, so $(\kappa^+, \kappa) \to \text{wk } \Delta(\kappa)$ follows from Lemma 6.3.2.

To establish that $(\kappa^+, \kappa) \nrightarrow \text{wk } \Delta(\kappa^+)$, note that $\kappa^+ \nrightarrow (\kappa^+)^2_2$ by Theorem 2.5.7. Thus there is a function $f: [\kappa^+]^2 \to \{0, 1\}$ such that if $H \subseteq \kappa^+$ and f is constant on $[H]^2$, then $|H| < \kappa^+$. Take the (κ^+, κ)-family $\mathcal{A}$ constructed in Lemma 6.3.5 for this particular function f. Then $\mathcal{A}$ contains no wk $\Delta(\kappa^+)$-family.

Theorem 6.3.7 (GCH). *Suppose κ is not a successor cardinal. Then*
(i) $(\kappa^+, \kappa) \nrightarrow \text{wk } \Delta(\kappa)$,
(ii) *if* $\aleph_0 < \kappa < \aleph_\kappa$ *and* $\theta < \kappa$ *then* $(\kappa^+, \kappa) \to \text{wk } \Delta(\theta)$,
(iii) *if* $\kappa = \aleph_0$ *or if* $\kappa = \aleph_\kappa$ *then* $(\kappa^+, \kappa) \nrightarrow \text{wk } \Delta(3)$.

Proof. For (i), given any function f in ${}^\kappa 2$, put $A_f = \{f \upharpoonright \alpha; \alpha < \kappa\}$ and consider the family $\mathcal{A} = (A_f; f \in {}^\kappa 2)$. Then $\mathcal{A}$ is a (κ^+, κ)-family of pairwise different sets. For distinct f, g in ${}^\kappa 2$ we have $A_f \cap A_g = \{f \upharpoonright \alpha; \alpha < \delta(f, g)\}$ so $|A_f \cap A_g| = |\delta(f, g)|$. Let $(A_f; f \in I)$ where $I \subseteq {}^\kappa 2$ be a weak Δ-family contained in $\mathcal{A}$, say $|A_f \cap A_g| = \eta$ for distinct f, g in I. Then $\eta = |\delta(f, g)| < \kappa$. Thus for all $\{f, g\}$ from $[I]^2$ we have $|\delta(f, g)| = \eta$ so $\delta(f, g) < \eta^+$. Thus

$$|I| \leqslant |\{f \upharpoonright \alpha; f \in {}^\kappa 2 \text{ and } \alpha < \eta^+\}| = 2^{\eta^+} = \eta^{++} < \kappa\ .$$

Hence $\mathcal{A}$ contains no weak $\Delta(\kappa)$-family.

To show (ii), suppose $\aleph_0 < \kappa < \aleph_\kappa$ and take θ with $\theta < \kappa$. The restriction on κ ensures that $\psi(\kappa) < \kappa$. Because κ is a limit cardinal with $\psi(\kappa), \theta < \kappa$ then $\kappa \to (\theta)^2_{\psi(\kappa)}$ (from Theorem 2.5.10). Hence $(\kappa, \kappa) \to \text{wk } \Delta(\theta)$ by Lemma 6.3.2, so surely $(\kappa^+, \kappa) \to \text{wk } \Delta(\theta)$.

Finally for (iii), suppose $\kappa = \aleph_0$ or $\kappa = \aleph_\kappa$. Thus $\kappa = \psi(\kappa)$, so Lemma 6.3.3 ensures that $(\kappa^+, \kappa) \nrightarrow \text{wk } \Delta(3)$. This completes the proof.

Together, Theorems 6.3.6 and 6.3.7 provide a complete discussion of the relation $(\kappa^+, \kappa) \to \text{wk } \Delta(\theta)$. We shall now consider the relation $(\lambda, \kappa) \to \text{wk } \Delta(\theta)$ in the case $\aleph_0 \leqslant \lambda \leqslant \kappa$. The discussion breaks into several cases. (The GCH is assumed throughout.)

If $\lambda \leqslant \psi(\kappa)^+$, it follows from Lemma 6.3.3 that $(\lambda, \kappa) \nrightarrow \text{wk } \Delta(3)$, so we may suppose that $\psi(\kappa)^+ < \lambda \leqslant \kappa$. Suppose first that λ is not a successor cardinal. By Theorem 6.3.7(i), $(\lambda^+, \lambda) \nrightarrow \text{wk } \Delta(\lambda)$, and hence $(\lambda, \kappa) \nrightarrow \text{wk } \Delta(\lambda)$. If $\theta < \lambda$ then $\lambda \to (\theta)^2_{\psi(\kappa)}$ (from Theorem 2.5.10) and hence $(\lambda, \kappa) \to \text{wk } \Delta(\theta)$ by Lemma 6.3.2. Now take the case that λ is a successor cardinal, say $\lambda = \eta^+$. If η is not a successor cardinal, by Theorem 6.3.7(i), $(\eta^+, \eta) \nrightarrow \text{wk } \Delta(\eta)$ and hence $(\lambda, \kappa) \nrightarrow \text{wk } \Delta(\eta)$. However, if $\theta < \eta$ then $\eta^+ \to (\theta)^2_{\psi(\kappa)}$ since $\psi(\kappa) < \eta$ and so $(\lambda, \kappa) \to \text{wk } \Delta(\theta)$ by Lemma 6.3.2. The final case to consider is that $\lambda = \eta^+$ where η is a successor cardinal. Here $(\eta^+, \eta) \nrightarrow \text{wk } \Delta(\eta^+)$ from Theorem 6.3.6, so $(\lambda, \kappa) \nrightarrow \text{wk } \Delta(\lambda)$. And if $\theta < \lambda$, since $\psi(\kappa) < \eta = \eta'$ it follows from Theorem 2.2.4 that $\eta^+ \to (\theta)^2_{\psi(\kappa)}$ so $(\lambda, \kappa) \to \text{wk } \Delta(\theta)$ by Lemma 6.3.2. All cases have now been covered.

The remaining situations where the strong relation $(\lambda, \kappa) \to \Delta(\theta)$ fails are given by Theorem 6.2.5(iii), (vi). Theorem 6.2.5(iii) states that $(\lambda, \kappa) \nrightarrow \Delta(\lambda)$ if λ is singular (and $\kappa < \lambda$). In fact the example given to establish this result is an example of a (λ, κ)-family which contains no weak $\Delta(\lambda)$-family, so the stronger negative relation $(\lambda, \kappa) \nrightarrow \text{wk } \Delta(\lambda)$ holds. There remains the result in Theorem 6.2.5(vi): if λ is singular with $\lambda' \leqslant \kappa < \lambda$ then $(\lambda^+, \kappa) \nrightarrow \Delta(\lambda)$. In this situation one can ask if either of the relations $(\lambda^+, \kappa) \to \text{wk } \Delta(\lambda^+)$ or $(\lambda^+, \kappa) \to \text{wk } \Delta(\lambda)$ hold. The outcome is summarized in the following theorem.

Theorem 6.3.8 (GCH). *Let λ be singular with $\lambda' \leqslant \kappa < \lambda$. If λ' is countable or $\aleph_{\lambda'} \leqslant \kappa$ then $(\lambda^+, \kappa) \nrightarrow \text{wk } \Delta(\lambda)$; otherwise $(\lambda^+, \kappa) \to \text{wk } \Delta(\lambda^+)$.*

Proof. By Lemma 6.3.4, $(\lambda^+, \aleph_{\lambda'}) \nrightarrow \text{wk } \Delta(\lambda)$ so if $\aleph_{\lambda'} \leqslant \kappa$ then certainly $(\lambda^+, \kappa) \nrightarrow \text{wk } \Delta(\lambda)$. If $\lambda' = \aleph_0$ then even $(\lambda^+, \aleph_0) \nrightarrow \text{wk } \Delta(\lambda)$ so $(\lambda^+, \kappa) \nrightarrow \text{wk } \Delta(\lambda)$ for all κ. So suppose $\lambda' \leqslant \kappa < \aleph_{\lambda'}$ with $\lambda' \neq \aleph_0$. We shall show that in this situation, $(\lambda^+, \kappa) \to \text{wk } \Delta(\lambda^+)$.

Take a (λ^+, κ)-family $\mathcal{A} = (A_\nu; \nu < \lambda^+)$ and for a contradiction we shall suppose that $\mathcal{A}$ contains no weak $\Delta(\lambda^+)$-family. There is a least cardinal η such that $\mathcal{A}$ contains a (λ^+, κ)-family $\mathcal{B}$ with $\delta(\mathcal{B}) = \eta$ (so $0 < \eta \leqslant \kappa^+$), and we may in fact suppose that $\mathcal{A} = \mathcal{B}$.

Suppose for the moment that η is a successor cardinal, say $\eta = \theta^+$. Take any A in $\mathcal{A}$, and put

$$J(A) = \{\nu < \lambda^+; |A_\nu \cap A| = \theta\}.$$

Since $|[A]^\theta| = \kappa^\theta \leqslant \kappa^\kappa < \lambda$, if $|J(A)| = \lambda^+$ then there would be B in $[A]^\theta$ and J in $[J(A)]^{\lambda^+}$ such that $A_\nu \cap A = B$ for all ν in J. Then for distinct μ, ν from J since $B \subseteq A_\mu \cap A_\nu$ it follows that $\theta = |B| \leqslant |A_\mu \cap A_\nu| < \delta(\mathcal{A}) = \theta^+$; hence $|A_\mu \cap A_\nu| = \theta$. Thus $(A_\nu; \nu \in J)$ would be a weak $\Delta(\lambda^+)$-family contained in $\mathcal{A}$, contrary to the choice of $\mathcal{A}$. Hence $|J(A)| \leqslant \lambda$ for each A from $\mathcal{A}$. This means that we can choose inductively ordinals $\nu(\alpha)$ for α with $\alpha < \lambda^+$ so that

$$\nu(\alpha) \in \lambda^+ - \bigcup\{J(A_{\nu(\beta)}); \beta < \alpha\}.$$

Then $|A_{\nu(\alpha)} \cap A_{\nu(\beta)}| < \theta$ whenever $\beta < \alpha$, so that $\mathcal{B} = (A_{\nu(\alpha)}; \alpha < \lambda^+)$ is a (λ^+, κ)-family contained in $\mathcal{A}$ with $\delta(\mathcal{B}) \leqslant \theta$. This contradicts the minimum property of η. Consequently η cannot be a successor cardinal. So further $0 < \eta \leqslant \kappa$.

It follows that $\eta' \neq \lambda'$. If $\eta = \aleph_0$ this is trivial since $\lambda' \neq \aleph_0$. Otherwise since $\eta \leqslant \kappa < \aleph_{\lambda'}$, then $\eta = \aleph_\alpha$ for some α where $0 < \alpha < \lambda'$. Since η is not a successor, $\eta = \Sigma(\aleph_\beta; \beta < \alpha)$ and hence $\eta' \leqslant |\alpha| \leqslant \alpha < \lambda'$. Thus indeed $\eta' \neq \lambda'$.

Use transfinite induction to choose non-empty subsets I_α of λ^+ with always $|I_\alpha| \leqslant \lambda$, for α with $\alpha \leqslant \kappa^+$, as follows. Suppose for some α that I_β has already been defined for all β with $\beta < \alpha$. Put

$$I(\alpha) = \bigcup\{I_\beta; \beta < \alpha\}; A(\alpha) = \bigcup\{A_\mu; \mu \in I(\alpha)\}\,.$$

Then $|I(\alpha)| \leqslant \lambda \cdot \kappa^+ = \lambda$ and $|A(\alpha)| \leqslant \kappa \cdot \lambda = \lambda$. Define $I^*(\alpha)$ by

$$I^*(\alpha) = \{\nu < \lambda^+; |A_\nu \cap A(\alpha)| \geqslant \eta\}\,.$$

Put $\mathscr{A}^*(\alpha) = \{A_\nu \cap A(\alpha); \nu \in I^*(\alpha)\}$; then $\mathscr{A}^*(\alpha)$ is a family of the at least η size subsets of $A(\alpha)$ with $\delta(A^*(\alpha)) \leqslant \eta$. Since $|A(\alpha)| \leqslant \lambda$ and $\lambda' \neq \eta'$ it follows from Theorem 1.1.6 that $|I^*(\alpha)| \leqslant \lambda$. Thus $\lambda^+ - I^*(\alpha) \neq \emptyset$. Choose for I_α a subset of $\lambda^+ - I^*(\alpha)$ maximal with the property that $\{A_\nu - A(\alpha); \nu \in I_\alpha\}$ is pairwise disjoint. We must verify that $|I_\alpha| \leqslant \lambda$. Suppose, on the contrary, that $|I_\alpha| = \lambda^+$. For each ν in I_α since $\nu \notin I^*(\alpha)$ it follows that $|A_\nu \cap A(\alpha)| < \eta$. Since the number γ of cardinals less than η satisfies $\gamma \leqslant \eta \leqslant \kappa < \lambda$, there must be a cardinal θ with $\theta < \eta$ and a set I in $[I_\alpha]^{\lambda^+}$ such that always $|A_\nu \cap A(\alpha)| = \theta$ for ν in I. Now $\{A_\nu - A(\alpha); \nu \in I_\alpha\}$ is pairwise disjoint so $A_\mu \cap A_\nu \subseteq A(\alpha)$ for distinct μ, ν from I_α and hence $|A_\mu \cap A_\nu| \leqslant \theta$. Thus $\mathscr{B}^* = (A_\nu; \nu \in I_\alpha)$ gives a (λ^+, κ)-family contained in $\mathscr{A}$ with $\delta(\mathscr{B}^*) \leqslant \theta^+$. Since η is a limit cardinal, $\theta^+ < \eta$, and this contradicts the minimum property of η. Thus in fact $|I_\alpha| \leqslant \lambda$.

In particular, $I_{\kappa^+} \neq \emptyset$. Choose μ from I_{κ^+}. Then $\mu \notin I_\alpha$ whenever $\alpha < \kappa^+$. By the maximal property of I_α there is $\mu(\alpha)$ in I_α such that $(A_\mu \cap A_{\mu(\alpha)}) - A(\alpha) \neq \emptyset$. Choose x_α from $(A_\mu \cap A_{\mu(\alpha)}) - A(\alpha)$. If $\alpha < \beta < \kappa^+$ then $x_\alpha \in A_{\mu(\alpha)} \subseteq A(\beta)$ while $x_\beta \notin A(\beta)$ and hence $x_\alpha \neq x_\beta$. Thus $|A_\mu| \geqslant |\{x_\alpha; \alpha < \kappa^+\}| = \kappa^+$; yet $|A_\mu| = \kappa$. This is the required contradiction. Thus $\mathscr{A}$ must contain a weak $\Delta(\lambda^+)$-family, and the relation $(\lambda^+, \kappa) \to \text{wk}\ \Delta(\lambda^+)$ is established.

CHAPTER 7

ORDINARY PARTITION RELATIONS FOR ORDINAL NUMBERS

§1. Introductory remarks

As noted in Chapter 2, the definition of the ordinary partition relation for cardinal numbers can be extended to order types. In this chapter we shall consider the partition symbol for well ordered types, that is, for ordinal numbers. The definition is as follows.

Definition 7.1.1. Let α, γ, α_k (where $k < \gamma$) be ordinal numbers and let n be a positive integer. The *ordinary partition symbol* $\alpha \to (\alpha_k; k < \gamma)^n$ means the following. Let S be a set ordered with order type α. For all partitions $\mathbf{\Delta} = \{\Delta_k; k < \gamma\}$ of $[S]^n$ into γ parts, there exist k with $k < \gamma$ and a subset H of S having order type α_k such that $[H]^n \subseteq \Delta_k$.

The various conventions concerning the use of the partition symbol adopted for cardinal numbers in Chapter 2 will be followed without further comment for ordinal numbers.

The problems concerning the symbol for ordinal numbers are considerably more ramified than those for cardinal numbers. We shall confine most of the discussion to the case $n = 2$, and frequently $\gamma = 2$ as well. Even so there are many unsolved problems, and we shall not attempt to cover even all the cases where progress has been made. We shall limit our treatment to a few special cases where a reasonably complete discussion is possible. Thus §2 is devoted to partitions of $[\omega^\alpha]^2$ into two classes, mainly for finite α. In §3, we prove Chang's theorem for ω^ω. And in §4 we shall consider relations of the form $\omega_1 \to (\alpha_1, ..., \alpha_k)^2$, and prove that $\omega_1 \to (\alpha)_k^2$ for countable α and finite k.

In this chapter we shall need to distinguish clearly between the order type of a well ordered set and its cardinality. The problem is particularly acute with the initial ordinals. When order type is to be emphasized, we shall write

$\omega, \omega_1, \omega_2, ..., \omega_\alpha, ...$ for the sequence of infinite initial ordinals, although the sequence $\aleph_0, \aleph_1, \aleph_2, ..., \aleph_\alpha, ...$ of infinite cardinals denotes the same sequence (as a sequence of sets). It will be left to the context to distinguish between the symbols α^β for ordinal exponentiation and κ^λ for cardinal exponentiation. In particular, symbols such as $\omega_\alpha^2, \omega_\alpha^\omega, ...$ stand for the ordinal operation, whereas $\aleph_\alpha^2, \aleph_\alpha^{\aleph_0}, ...$ indicate cardinal exponentiation. The symbols Σ_0, Π_0 are used to indicate the ordinal sum and the ordinal product of a well ordered sequence of ordinals.

It is easy to see that a partition relation between cardinal numbers is equivalent to the same relation between the corresponding initial ordinals.

Unlike the situation with cardinal numbers, for ordinal numbers results with $n = 1$ are not trivial. In [73], Milner and Rado consider this situation. They give an algorithm to determine in finitely many steps for any sequence $\langle \alpha_k; k < \gamma \rangle$ of ordinals the least α such that $\alpha \to (\alpha_k; k < \gamma)^1$. We mention a couple of results from [73], but otherwise refer the reader to the original. There is the following lemma.

Lemma 7.1.2. (i) *Suppose* $\alpha \to (\alpha_k; k < \gamma)^1$ *and* $\beta \to (\beta_k; k < \gamma)^1$. *Then* $\alpha\beta \to (\alpha_k\, \beta_k; k < \gamma)^1$.

(ii) *Let* γ *be finite. Suppose* $\alpha_\mu \to (\alpha_{\mu k}; k < \gamma)^1$ *for* μ *with* $\mu < \rho$. *Put* $\alpha(\rho) = \Pi_0(\alpha_\mu; \mu < \rho)$ *and* $\alpha_k(\rho) = \Pi_0(\alpha_{\mu k}; \mu < \rho)$. *Then* $\alpha(\rho) \to (\alpha_k(\rho); k < \gamma)^1$.

Proof. A special case of (i) was stated as Lemma 5.1.6, and the proof of the general case hardly differs from the proof of Lemma 5.1.6. Let S be a set well ordered with order type $\alpha\beta$, so we may suppose $S = \beta \times \alpha$ under the lexicographic ordering. Take any partition $S = \bigcup\{\Delta_k; k < \gamma\}$. For x with $x < \beta$ put $\Delta_k(x) = \{y < \alpha; \langle x, y \rangle \in \Delta_k\}$. For each x, we have $\alpha = \bigcup\{\Delta_k(x); k < \gamma\}$, and so there is $k(x)$ such that $\mathrm{tp}(\Delta_{k(x)}(x)) \geqslant \alpha_{k(x)}$. Put $\Gamma_k = \{x < \beta; k(x) = k\}$, so $\beta = \bigcup\{\Gamma_k; k < \gamma\}$. Then there is k_0 such that $\mathrm{tp}(\Gamma_{k_0}) = \beta_{k_0}$. Put $R = \{\langle x, y \rangle \in S; x \in \Gamma_{k_0} \text{ and } y \in \Delta_{k_0}(x)\}$. Then $R \subseteq \Delta_{k_0}$ and $\mathrm{tp}(R) \geqslant \alpha_{k_0}\beta_{k_0}$, so $\mathrm{tp}(\Delta_{k_0}) \geqslant \alpha_{k_0}\beta_{k_0}$. This proves (i).

To prove (ii), for each ν with $\nu \leqslant \rho$, put $\alpha(\nu) = \Pi_0(\alpha_\mu; \mu < \nu)$ and $\alpha_k(\nu) = \Pi_0(\alpha_{\mu k}; \mu < \nu)$. Use transfinite induction on ρ. The case $\rho = 1$ is trivial. Suppose ρ is a successor ordinal, say $\rho = \sigma + 1$. Then $\alpha(\sigma) \to (\alpha_k(\sigma); k < \gamma)^1$ by the inductive hypothesis, and $\alpha_\sigma \to (\alpha_{\sigma k}; k < \gamma)^1$. So by (i), $\alpha(\rho) \to (\alpha_k(\rho); k < \gamma)^1$. Now suppose that ρ is a limit ordinal. Let S be a set well ordered by a relation $<$ with order type $\alpha(\rho)$, and suppose S is partitioned, $S = \bigcup\{\Delta_k; k < \gamma\}$. For each μ with $\mu < \rho$ there is a subset T_μ of S with $\mathrm{tp}(T_\mu) = \alpha(\mu)$. By the inductive hypothesis there are always $k(\mu)$ and a subset H_μ of T_μ with $\mathrm{tp}(H_\mu) \geqslant \alpha_{k(\mu)}(\mu)$ and $H_\mu \subseteq \Delta_{k(\mu)}$. Put $\Gamma_k = \{\mu < \rho; k(\mu) = k\}$, so $\rho = \bigcup\{\Gamma_k; k < \gamma\}$. Since γ is finite and ρ is a limit ordinal there is some k_0 such that Γ_{k_0} is cofinal in ρ. Then for each μ in Γ_{k_0} we have

$$\mathrm{tp}(\Delta_{k_0}) = \mathrm{tp}(\Delta_{k(\mu)}) \geqslant \mathrm{tp}(H_\mu) \geqslant \alpha_{k(\mu)}(\mu) = \alpha_{k_0}(\mu)\,.$$

Hence $\mathrm{tp}(\Delta_{k_0}) \geqslant \sup(\alpha_{k_0}(\mu); \mu \in \Gamma_{k_0}) = \alpha_{k_0}(\rho)$. This completes the proof.

Theorem 7.1.3. *If m is finite then $\omega^\alpha \to (\omega^\alpha)^1_m$.*

Proof. From Lemma 7.1.2(ii), noting that $\omega \to (\omega)^1_m$.

The second, somewhat surprising, result from [73] concerns partitions into infinitely many classes. It has been referred to as the Milner-Rado paradox.

Theorem 7.1.4. *For all β, if $\alpha < \omega_{\beta+1}$ then $\alpha \nrightarrow (\omega_\beta^k; k < \omega)^1$.*

Proof. (Note that trivially $\omega_{\beta+1} \to (\omega_{\beta+1})^1_\omega$, so certainly $\omega_{\beta+1} \to (\omega_\beta^k; k < \omega)^1$.) It suffices to show that if $\alpha < \omega_{\beta+1}$ then $\omega_\beta^\alpha \nrightarrow (\omega_\beta^k; k < \omega)^1$. This we prove by induction on α. The case $\alpha = 1$ is trivial. Suppose α is a successor ordinal, $\alpha = \gamma + 1$, and $\omega_\beta^\gamma \nrightarrow (\omega_\beta^k; k < \omega)^1$. Take any set S ordered by a relation $<$ with order type ω_β^α, so $S = \bigcup\{S_\mu; \mu < \omega_\beta\}$ where $\mathrm{tp}(S_\mu) = \omega_\beta^\gamma$ and $S_\mu < S_\nu$ whenever $\mu < \nu < \omega_\beta$. By the inductive hypothesis, for each μ there is a decomposition $S_\mu = \bigcup\{\Delta_{\mu k}; k < \omega\}$ where $\mathrm{tp}(\Delta_{\mu k}) < \omega_\beta^k$. Put $\Delta_0 = \Delta_1 = \emptyset$ and for k with $k \geqslant 0$,

$$\Delta_{k+2} = \bigcup\{\Delta_{\mu k}; \mu < \omega_\beta\}\,.$$

Then $S = \bigcup\{\Delta_k; k < \omega\}$ and $\mathrm{tp}(\Delta_0) = \mathrm{tp}(\Delta_1) = 0$,

$$\mathrm{tp}(\Delta_{k+2}) \leqslant \Sigma_0(\omega_\beta^k; \mu < \omega_\beta) = \omega_\beta^{k+1} < \omega_\beta^{k+2}\,.$$

This gives a partition of S which demonstrates $\omega_\beta^\alpha \nrightarrow (\omega_\beta^k; k < \omega)^1$. Now suppose α is a limit ordinal and that $\omega_\beta^\gamma \nrightarrow (\omega_\beta^k; k < \omega)^1$ whenever $\gamma < \alpha$. There is a sequence $\langle\alpha(\mu); \mu < \omega_\beta\rangle$ of ordinals below α with $\alpha(0) \leqslant \alpha(1) \leqslant \alpha(2)$ ... such that $\alpha = \sup\{\alpha(\mu); \mu < \omega_\beta\}$. Let S be a set ordered with order type $\Sigma_0(\omega_\beta^{\alpha(\mu)}; \mu < \omega_\beta)$, so $\mathrm{tp}(S) \geqslant \omega_\beta^\alpha$, and write $S = \bigcup\{S_\mu; \mu < \omega_\beta\}$ where $\mathrm{tp}(S_\mu) = \omega_\beta^{\alpha(\mu)}$ and $S_\mu < S_\nu$ whenever $\mu < \nu < \omega_\beta$. By the inductive hypothesis, for each μ there is a decomposition $S_\mu = \bigcup\{\Delta_{\mu k}; k < \omega\}$ where $\mathrm{tp}(\Delta_{\mu k}) < \omega_\beta^k$. As before, put $\Delta_0 = \Delta_1 = \emptyset$ and for k with $k \geqslant 0$,

$$\Delta_{k+2} = \bigcup\{\Delta_{\mu k}; \mu < \omega_\beta\}\,;$$

then this decomposition of S shows $\omega_\beta^\alpha \nrightarrow (\omega_\beta^k; k < \omega)^1$. This completes the induction, and proves Theorem 7.1.4.

We shall conclude this section by noting a couple of negative relations.

Theorem 7.1.5. *For all β, if $\alpha < \omega_{\beta+1}$ then $\alpha \nrightarrow (\omega_\beta + 1, \omega)^2$.*

Proof. If $|\alpha| < \aleph_\beta$ this is clear, so suppose $|\alpha| = \aleph_\beta$. Let $\lessdot$ be a well ordering of α of order type ω_β. Define a partition $[\alpha]^2 = \Delta_0 \cup \Delta_1$ by: if $\sigma, \tau < \alpha$ with $\sigma < \tau$,

$$\{\sigma, \tau\} \in \Delta_0 \Leftrightarrow \sigma \lessdot \tau; \ \{\sigma, \tau\} \in \Delta_1 \Leftrightarrow \sigma \gtrdot \tau .$$

Take a subset H of α. If $[H]^2 \subseteq \Delta_0$ then both $<$ and $\lessdot$ agree on H, so $\mathrm{tp}(H, <) = \mathrm{tp}(H, \lessdot)$; hence $\mathrm{tp}(H, <) \leqslant \omega_\beta$. If $[H]^2 \subseteq \Delta_1$ then H enumerated in increasing $\lessdot$-order gives a descending $<$-chain of ordinals; hence H is finite. Thus this partition of $[\alpha]^2$ suffices to prove the theorem.

Results of Kruse [62] extend Theorem 7.1.5 as follows for values of n with $n \geqslant 3$.

Theorem 7.1.6. *Suppose $n \geqslant 3$. For all β, if $\alpha < \omega_{\beta+1}$ then $\alpha \nrightarrow (\omega_\beta + 1, n + 1)^n$.*

Proof. By Theorem 7.1.5, $\alpha \nrightarrow (\omega_\beta + 1, \omega)^{n-1}$. Thus there is a disjoint partition $\mathbf{\Delta} = \{\Delta_0, \Delta_1\}$ of $[\alpha]^{n-1}$ such that there is no subset A of α of order type $\omega_\beta + 1$ with $[A]^{n-1} \subseteq \Delta_0$, nor an infinite subset B with $[B]^{n-1} \subseteq \Delta_1$. Define a disjoint partition $\mathbf{\Gamma} = \{\Gamma_0, \Gamma_1\}$ of $[\alpha]^n$ as follows: if $\sigma_1 < \sigma_2 < ... < \sigma_n < \alpha$ then

$$\{\sigma_1, ..., \sigma_n\} \in \Gamma_0 \Leftrightarrow \{\sigma_1, ..., \sigma_{n-1}\} \in \Delta_1 \text{ or } \{\sigma_2, ..., \sigma_n\} \in \Delta_0 ,$$

$$\{\sigma_1, ..., \sigma_n\} \in \Gamma_1 \Leftrightarrow \{\sigma_1, ..., \sigma_{n-1}\} \in \Delta_0 \text{ and } \{\sigma_2, ..., \sigma_n\} \in \Delta_1 .$$

Take H from $[\alpha]^{n+1}$, say $H = \{\sigma_1, ..., \sigma_{n+1}\}$, listed in increasing order. If both $\{\sigma_1, ..., \sigma_n\} \in \Gamma_1$ and $\{\sigma_2, ..., \sigma_{n+1}\} \in \Gamma_1$ we would have the contradiction $\{\sigma_2, ..., \sigma_n\} \in \Delta_0 \cap \Delta_1$, so $[H]^2 \not\subseteq \Gamma_1$.

Suppose there is a subset H of α with $\mathrm{tp}(H) = \omega_\beta + 1$ such that $[H]^2 \subseteq \Gamma_0$. Put

$$B = \{\sigma \in H; \text{ for all } \{\sigma_1, ..., \sigma_{n-1}\} \text{ from } [H]^{n-1}, \text{ if}$$
$$\sigma_1, ..., \sigma_{n-1} \leqslant \sigma \text{ then } \{\sigma_1, ..., \sigma_{n-1}\} \in \Delta_1\} ,$$

so $[B]^{n-1} \subseteq \Delta_1$. Put $A = H - B$; then $B < A$. And in fact $[A]^{n-1} \subseteq \Delta_0$. For take $\tau_1, ..., \tau_{n-1}$ from A where $\tau_1 < \tau_2 < ... < \tau_{n-1}$. There are $\sigma_1, ..., \sigma_{n-1}$ in B with $\sigma_1 < ... < \sigma_{n-1} \leqslant \tau_1$ and $\{\sigma_1, ..., \sigma_{n-1}\} \notin \Delta_1$. Let $\sigma_1, ..., \sigma_m$ list $\{\sigma_1, ..., \sigma_{n-1}\} \cup \{\tau_1, ..., \tau_{n-1}\}$ in order (so $m = 2n - 2$ or $2n - 3$ depending on whether $\sigma_{n-1} < \tau_1$ or $\sigma_{n-1} = \tau_1$). Always $\{\sigma_{i+1}, ..., \sigma_{i+n}\} \in \Gamma_0$, and $\{\sigma_1, ..., \sigma_{n-1}\} \notin \Delta_1$. Hence an easy induction on i shows that always $\{\sigma_{i+2}, ..., \sigma_{i+n}\} \in \Delta_0$. In particular $\{\tau_1, ..., \tau_{n-1}\} \in \Delta_0$, so indeed $[A]^{n-1} \subseteq \Delta_0$ as claimed. Now $\mathrm{tp}(B) + \mathrm{tp}(A) = \mathrm{tp}(H) = \omega_\beta + 1$, so either $\mathrm{tp}(B) \geqslant \omega$ or $\mathrm{tp}(A) \geqslant \omega_\beta + 1$. This contradicts the choice of the partition $\mathbf{\Delta}$. Hence if $H \subseteq \alpha$ with $\mathrm{tp}(H) = \omega_\beta + 1$ then $[H]^2 \not\subseteq \Gamma_0$. Thus the partition $\mathbf{\Gamma}$ of $[\alpha]^n$ suffices to prove the theorem.

§2. Countable ordinals

In this section we shall discuss partition relations of the form $\alpha \to (\alpha_0, \alpha_1)^n$ where α is a denumerable ordinal. From Ramsey's theorem, $\alpha \to (\omega, \omega)^n$. By Theorems 7.1.5 and 7.1.6 (with $\beta = 0$), $\alpha \not\to (\omega + 1, \omega)^2$ and $\alpha \not\to (\omega + 1, n+1)^n$ if $n \geqslant 3$. Hence the only relations of interest are those of the form $\alpha \to (\alpha_0, m)^2$ where m is finite.

Some of the first such relations to be established are in Erdös and Rado [29], where it is shown that $\omega m \to (\omega + l, m)^2$ and $\omega m \not\to (\omega + 1, m + 1)^2$. Moreover, for each k and m it is shown that there is a least integer $l_0(m, k)$ such that $\omega l_0(m, k) \to (\omega k, m)^2$, and that $\alpha \not\to (\omega k, m)^2$ if $\alpha < \omega l_0(m, k)$. In [32], this result is generalized to arbitrary β by showing that there is a least integer $l_\beta(m, k)$ such that $\omega_\beta l_\beta(m, k) \to (\omega_\beta k, m)^2$, and that $\alpha \not\to (\omega_\beta k, m)^2$ if $\alpha < \omega_\beta l_\beta(m, k)$. Erdös and Rado conjecture in [32] that in fact $l_\beta(m, k) = l_0(m, k)$ for all β. This was later proved correct by Baumgartner [3]. The reader is referred to the papers mentioned for the proofs of these results.

We shall concentrate on partition relations for ω^α, the ordinal powers of ω, with countable α. The first results will be concerned with ω^n for finite n.

For each positive integer n, put

$$W(n) = \{\langle a_0, ..., a_{n-1}\rangle \in {}^n\omega; a_0 < a_1 < ... < a_{n-1}\},$$

and let $\prec$ be the lexicographic ordering of $W(n)$, so

$$\langle a_0, ..., a_{n-1}\rangle \prec \langle b_0, ..., b_{n-1}\rangle \Leftrightarrow a_0 < b_0$$
$$\text{or } (a_0 = b_0 \text{ and } a_1 < b_1) \text{ or } ... .$$

Under this ordering, $W(n)$ has order type ω^n. For a sequence $\boldsymbol{a}$ from $W(n)$, the i-th component of $\boldsymbol{a}$ will be denoted either by a_i or $\boldsymbol{a}(i)$, and similarly with $\boldsymbol{b}, \boldsymbol{c}, ...$.

Definition 7.2.1. A subset S of $W(n)$ is said to be *free in the i-th component* if for all $\langle a_0, ..., a_{n-1}\rangle$ in S there is an infinite subset A of ω such that for all a in A there are $a'_{i+1}, ..., a'_{n-1}$ in ω such that $\langle a_0, ..., a_{i-1}, a, a'_{i+1}, ..., a'_{n-1}\rangle \in S$.

Lemma 7.2.2. *Let S be a subset of $W(n)$ and take m with $m \leqslant n$. Then* $\mathrm{tp}(S) \geqslant \omega^m$ *if and only if there is a subset T of S which is free in m different components.*

Proof. Suppose there is T with $T \subseteq S$ such that T is free in m components. We shall show $\mathrm{tp}(T) \geqslant \omega^m$ (and so also $\mathrm{tp}(S) \geqslant \omega^m$) by induction on m. This is trivial for $m = 1$. So take T, free in $m + 1$ components. Let i be least such that T is free in the i-th component, and let A be the infinite set of i-th components given by the definition. For each $\boldsymbol{a}$ in $W(i + 1)$ which is an initial segment of a sequence in T, the set

$$T(\boldsymbol{a}) = \{\boldsymbol{b} \in T; \boldsymbol{b} \restriction (i + 1) = \boldsymbol{a}\}$$

is free in m components and so by inductive assumption $\mathrm{tp}(T(\boldsymbol{a})) \geqslant \omega^m$. Since $\mathrm{tp}(T) \geqslant \Sigma_0(\mathrm{tp}(T(\boldsymbol{a})); \boldsymbol{a}(i) \in A)$ thus $\mathrm{tp}(T) \geqslant \omega^m\omega = \omega^{m+1}$.

Now suppose $\mathrm{tp}(S) \geqslant \omega^m$, and use induction to show that there is a subset of S free in m components. We may suppose $\mathrm{tp}(S) = \omega^m$. If $m = 1$, let i be maximal such that there is some sequence $\boldsymbol{a}$ of length i for which the set

$$T(\boldsymbol{a}) = \{\boldsymbol{b} \in S; \boldsymbol{b} \restriction i = \boldsymbol{a}\}$$

is infinite (where $i = 0$ and $\boldsymbol{a} = \emptyset$ is allowed). For such an $\boldsymbol{a}$, put $A = \{b_i; \boldsymbol{b} \in T(\boldsymbol{a})\}$; then the choice of i ensures that A is infinite. Thus $T(\boldsymbol{a})$ is a subset of S free in the i-th component.

Now suppose the statement is true for m, and take a subset S of $W(n)$ with $\mathrm{tp}(S) = \omega^{m+1}$. Let i be the largest integer such that there is a sequence $\boldsymbol{a}$ in $W(i)$ for which the set

$$T(\boldsymbol{a}) = \{\boldsymbol{b} \in S; \boldsymbol{b} \restriction i = \boldsymbol{a}\}$$

has order type ω^{m+1}. (Again $i = 0$ is allowed). So by the choice of i, for such an $\boldsymbol{a}$, if

$$T_x(\boldsymbol{a}) = \{\boldsymbol{b} \in T(\boldsymbol{a}); b_i = x\}$$

then $\mathrm{tp}(T_x(\boldsymbol{a})) \leqslant \omega^m$ for each x in ω. Put

$$X = \{x \in \omega; \mathrm{tp}(T_x(\boldsymbol{a})) = \omega^m\},$$

then X is infinite (since $\omega^{m+1} = \mathrm{tp}(T(\boldsymbol{a})) = \Sigma_0(\mathrm{tp}(T_x(\boldsymbol{a})); x \in \omega))$. By the inductive hypothesis, each $T_x(\boldsymbol{a})$ for x in X is free in m components. Hence there is an infinite subset A of X such that all the $T_x(\boldsymbol{a})$ for x in A are free in the same m components. But then if $T = \bigcup\{T_x(a); x \in A\}$, it follows that T is free in the i-th component and m later components; thus T is a subset of S free in $m + 1$ components. This completes the proof.

The first paper in which partition relations for ω^n are proved is Specker [93], where it is shown that $\omega^2 \to (\omega^2, m)^2$ for every finite m, and $\omega^n \nrightarrow (\omega^n, 3)^2$ if $n \geqslant 3$. The following simple proof of the first of these relations is from Haddad and Sabbagh [50].

Theorem 7.2.3. *For all finite m, the relation $\omega^2 \to (\omega^2, m)^2$ holds.*

Proof. Since $\mathrm{tp}(W(2)) = \omega^2$, it suffices to consider partitions of $[W(2)]^2$. So let any partition $[W(2)]^2 = \Delta_0 \cup \Delta_1$ be given. Define a partition of $[\omega]^4$ into 16 classes,

$$[\omega]^4 = \bigcup\{\Delta(i_0, ..., i_3); i_0, ..., i_3 = 0, 1\}\ ,$$

by: for a, b, c, d in ω with $a < b < c < d$,

$$\begin{aligned}\{a, b, c, d\} \in \Delta(i_0, ..., i_3) \Leftrightarrow\ & \{\langle a, b\rangle, \langle c, d\rangle\} \in \Delta_{i_0} \text{ and} \\ & \{\langle a, c\rangle, \langle b, d\rangle\} \in \Delta_{i_1} \text{ and} \\ & \{\langle a, d\rangle, \langle b, c\rangle\} \in \Delta_{i_2} \text{ and} \\ & \{\langle a, b\rangle, \langle a, c\rangle\} \in \Delta_{i_3}\ .\end{aligned}$$

By Ramsey's theorem there is H in $[\omega]^{\aleph_0}$ such that H is homogeneous for this partition of $[\omega]^4$, say $[H]^4 \subseteq \Delta(i_0, ..., i_3)$. Let $\langle h_k; k < \omega\rangle$ enumerate H in increasing order. If $i_j = 1$ for any j, the following gives a subset I of $W(2)$ of power m with $[I]^2 \subseteq \Delta_1$:

$$\{\langle h_{2k}, h_{2k+1}\rangle; k < m\} \text{ if } i_0 = 1\ ,$$

$$\{\langle h_k, h_{m+k}\rangle; k < m\} \text{ if } i_1 = 1\ ,$$

$$\{\langle h_k, h_{2m-k}\rangle; k < m\} \text{ if } i_2 = 1\ ,$$

$$\{\langle h_0, h_{1+k}\rangle; k < m\} \text{ if } i_3 = 1\ .$$

On the other hand, if $i_j = 0$ for all j, write H as a disjoint union, $H = \bigcup\{H_k; k < \omega\}$ where each H_k is infinite, and put

$$I = \{\langle h, h'\rangle; h \in H_0 \text{ and } h' \in H_{h+1} \text{ and } h < h'\}\ .$$

Then $I \subseteq W(2)$, $\mathrm{tp}(I) = \omega^2$ and $[I]^2 \subseteq \Delta_0$. Thus the theorem is proved.

Lemma 7.2.4. *The relation $\omega^3 \to (\omega^3, 3)^2$ is false.*

Proof. Consider the following disjoint partition $[W(3)]^3 = \Gamma_0 \cup \Gamma_1$ where

$$\{\boldsymbol{a}, \boldsymbol{b}\} \in \Gamma_1 \Leftrightarrow a_0 < a_1 < b_0 < a_2 < b_1 < b_2 \,.$$

Suppose $[\{\boldsymbol{a}, \boldsymbol{b}, \boldsymbol{c}\}]^2 \subseteq \Gamma_1$, where $\boldsymbol{a} \prec \boldsymbol{b} \prec \boldsymbol{c}$. Since $\{\boldsymbol{a}, \boldsymbol{b}\} \in \Gamma_1$ certainly $a_2 < b_1$, from $\{\boldsymbol{b}, \boldsymbol{c}\} \in \Gamma_1$ follows $b_1 < c_0$ and so $a_2 < c_0$; yet $\{\boldsymbol{a}, \boldsymbol{c}\} \in \Gamma_1$ requires $c_0 < a_2$, so this is impossible. On the other hand, take any subset H of $W(3)$ with $\mathrm{tp}(H) = \omega^3$. By Lemma 7.2.2, H has a subset free in all components. Thus for any $\boldsymbol{a}$ in H, there are $\boldsymbol{b}, \boldsymbol{c}, \boldsymbol{d}$ in H such that $b_0 > a_1$ and $c_0 = a_0$, $c_1 = a_1$, $c_2 > b_0$ and $d_0 = b_0$, $d_1 > c_2$. Then $\langle \boldsymbol{c}, \boldsymbol{d} \rangle \in [H]^2 \cap \Gamma_1$ so $[H]^2 \not\subseteq \Gamma_0$. Thus there is no H of order type ω^3 with $[H]^2 \subseteq \Gamma_0$. Thus this partition provides an example which proves the lemma.

Theorem 7.2.5. *If* $3 \leqslant m < \omega$ *then* $\omega^m \nrightarrow (\omega^m, 3)^2$.

Proof. The case $m = 3$ is Lemma 7.2.4. For m with $m > 3$, it is enough to prove the following:
(1) there is a one-to-one map $f: \omega^m \to \omega^3$ such that if X is a subset of ω^m with $\mathrm{tp}(X) = \omega^m$ then $\mathrm{tp}(f[X]) = \omega^3$.
(In this situation, one says that f *pins* ω^m to ω^3.) For suppose (1) established. Take a partition $\boldsymbol{\Delta} = \{\Delta_0, \Delta_1\}$ of $[\omega^3]^2$ such that there is no H in $[\omega^3]^3$ with $[H]^2 \subseteq \Delta_1$, nor H with $\mathrm{tp}(H) = \omega^3$ and $[H]^2 \subseteq \Delta_0$. Define a partition $\boldsymbol{\Delta}^* = \{\Delta_0^*, \Delta_1^*\}$ of ω^m by, for x, y in ω^m,

$$\{x, y\} \in \Delta_i^* \Leftrightarrow \{f(x), f(y)\} \in \Delta_i \,.$$

Then H in $[\omega^m]^3$ with $[H]^2 \subseteq \Delta_1^*$ would give $f[H]$ in $[\omega^3]^3$ with $[f[H]]^2 \subseteq \Delta_1$, whereas if $H \subseteq \omega^m$ and $\mathrm{tp}(H) = \omega^m$ with $[H]^2 \subseteq \Delta_0^*$ then $f[H] \subseteq \omega^3$, $\mathrm{tp}(f[H]) = \omega^3$ and $[f[H]]^2 \subseteq \Delta_0$. Thus $\boldsymbol{\Delta}^*$ shows $\omega^m \nrightarrow (\omega^m, 3)^2$.

We prove (1) by observing that for any integers k, l:
(2) if ω^k can be pinned to ω^l, then ω^{k+1} can be pinned to ω^{l+1}.
Since any one-to-one map $f: \omega^{m-2} \to \omega$ pins ω^{m-2} to ω, two applications of (2) establish (1). To prove (2), suppose indeed that ω^k can be pinned to ω^l. Write ω^{k+1} as a disjoint union, $\omega^{k+1} = \bigcup\{S_i; i < \omega\}$ where $\mathrm{tp}(S_i) = \omega^k$ and $S_i < S_j$ whenever $i < j$, and similarly $\omega^{l+1} = \bigcup\{T_i; i < \omega\}$ where $\mathrm{tp}(T_i) = \omega^l$ and $T_i < T_j$ if $i < j$. Then for each i there is $f_i : S_i \to T_i$ which pins S_i to T_i, that is, for each subset X of S_i with $\mathrm{tp}(X) = \omega^k$ we have $\mathrm{tp}(f_i[X]) = \omega^l$. Put $f = \bigcup\{f_i; i < \omega\}$; then $f: \omega^{k+1} \to \omega^{l+1}$ and f pins ω^{k+1} to ω^{l+1}. For take a subset X of ω^{k+1} with $\mathrm{tp}(X) = \omega^{k+1}$. Then $\{i < \omega; \mathrm{tp}(S_i \cap X) = \omega^k\}$ is infinite, and since $f[S_i \cap X] = f_i[S_i \cap X] = T_i \cap f[X]$ in fact $\{i < \omega; \mathrm{tp}(T_i \cap f[X]) = \omega^l\}$ is infinite. Hence $\mathrm{tp}(f[X]) = \omega^{l+1}$, as required. This completes the proof.

Galvin and Larson [46] investigate just which countable ordinals can be pinned to ω^3 (the use of "pin" in this context is due to them), and as a consequence they show: if α is a countable ordinal and $\alpha \to (\alpha, 3)^2$ then $\alpha = 0, 1, \omega^2$ or $\alpha = \omega^{\omega^\beta}$ for some β.

It was noted by Erdös and Hajnal [24, see Problem 6] that for each n and k ($k \geqslant 3$) there is a least integer $f(k, n)$ such that $\omega^n \not\to (\omega^k, f(k, n))^2$. (So from Lemma 7.2.4, $f(3, 3) = 3$.) The exact value of $f(k, n)$ is not known in general. However, Nosal [75] has shown $f(3, n) = 2^{n-2} + 1$. This result appears as Theorem 7.2.9 below. The proof depends on the existence of cartain canonical partitions of $W(n)$.

Definition 7.2.6. Two pairs $\{\boldsymbol{a}, \boldsymbol{b}\}$, $\{\boldsymbol{a}', \boldsymbol{b}'\}$ from $[W(n)]^2$ are said to be *similar* if for all i, j with $i, j < n$,

$$a_i < b_j \Leftrightarrow a_i' < b_j' \text{ and } a_i > b_j \Leftrightarrow a_i' > b_j' .$$

A partition $\boldsymbol{\Delta}$ of $[W(n)]^2$ is said to be *canonical* if for all $\{\boldsymbol{a}, \boldsymbol{b}\}$, $\{\boldsymbol{a}', \boldsymbol{b}'\}$ from $[W(n)]^2$ whenever $\{\boldsymbol{a}, \boldsymbol{b}\}$ is similar to $\{\boldsymbol{a}', \boldsymbol{b}'\}$ then $\{\boldsymbol{a}, \boldsymbol{b}\} \equiv \{\boldsymbol{a}', \boldsymbol{b}'\} \pmod{\boldsymbol{\Delta}}$.

The existence of canonical partitions was first proved by Hajnal, and independently later by Galvin.

Theorem 7.2.7. *Given any disjoint partition* $\boldsymbol{\Delta}$ *of* $[W(n)]^2$ *into finitely many classes there is an infinite subset* H *of* ω *such that the induced partition on* $[H^n \cap W(n)]^2$ *is canonical.*

Proof. Let $\mathcal{F}$ be the set of all strictly increasing functions $f: n \to 2n$. For f in $\mathcal{F}$ and x from $[\omega]^{2n}$, say $x = \{x_0, x_1, ..., x_{2n-1}\}_<$ put $\boldsymbol{x}_f = \langle x_{f(0)}, ..., x_{f(n-1)}\rangle$. Suppose $\boldsymbol{\Delta} = \{\Delta_0, ..., \Delta_{m-1}\}$. For each x in $[\omega]^{2n}$ define a function $F_x : [\mathcal{F}]^2 \to m$ by

$$F_x(\{f, g\}) = i \Leftrightarrow \{\boldsymbol{x}_f, \boldsymbol{x}_g\} \in \Delta_i .$$

Since there are only finitely many functions mapping from $[\mathcal{F}]^2$ to m, by Ramsey's theorem there is an infinite subset H of ω such that all the sets x from $[H]^{2n}$ have the same F_x, say $F_x = F$. Take any similar pairs $\{\boldsymbol{a}, \boldsymbol{b}\}$, $\{\boldsymbol{a}', \boldsymbol{b}'\}$ from $[H^n \cap W(n)]^2$. There are x, y in $[H]^{2n}$ and f, g in $\mathcal{F}$ such that $\boldsymbol{a} = \boldsymbol{x}_f$, $\boldsymbol{b} = \boldsymbol{x}_g$, $\boldsymbol{a}' = \boldsymbol{y}_f$ and $\boldsymbol{b}' = \boldsymbol{y}_g$. Since $F_x = F_y$ thus $\{\boldsymbol{a}, \boldsymbol{b}\} = \{\boldsymbol{x}_f, \boldsymbol{x}_g\} \in \Delta_{F(\{f,g\})}$ and $\{\boldsymbol{a}', \boldsymbol{b}'\} = \{\boldsymbol{y}_f, \boldsymbol{y}_g\} \in \Delta_{F(\{f,g\})}$, so $\{\boldsymbol{a}, \boldsymbol{b}\} \equiv \{\boldsymbol{a}', \boldsymbol{b}'\} \pmod{\boldsymbol{\Delta}}$ as required.

Let n be fixed, with $n \geqslant 3$. We define the following subsets of $W(n)$: for i,

j, k with $0 \leqslant i < j < k < n$, put

$$E^n_{ijk} = \{\{a, b\} \in W(n); a_0 = b_0 \text{ and } \dots \text{ and } a_{i-1} = b_{i-1} \text{ and}$$
$$a_i < \dots < a_{k-1} < b_i < \dots < b_{j-1} < a_k < \dots < a_{n-1} < b_j < \dots < b_{n-1}\}.$$

Consider the disjoint partition $\mathbf{\Gamma}^n = \{\Gamma^n_0, \Gamma^n_1\}$ of $W(n)$, where

$$\Gamma^n_1 = \bigcup\{E^n_{ijk}; 0 \leqslant i < j < k < n\}.$$

This partition has several useful properties, which we collect in the following lemma.

Lemma 7.2.8. *For this partition* $\mathbf{\Gamma}^n$ *of* $W(n)$,

(i) *there is no subset* H *of* $W(n)$ *of order type* ω^3 *such that* $[H]^2 \subseteq \Gamma^n_0$,

(ii) *there is no subset* H *of* $W(n)$ *with* $|H| = 2^{n-2} + 1$ *such that* $[H]^2 \subseteq \Gamma^n_1$,

(iii) *there is a subset* H *of* $W(n)$ *with* $|H| = 2^{n-2}$ *such that* $[H]^2 \subseteq \Gamma^n_1$.

Proof. For (i), take any subset H of $W(n)$ with $\text{tp}(H) = \omega^3$. By Lemma 7.2.2 we may suppose that H is free in three components, say i, j, k where $0 \leqslant i < j < k$. Thus given $\boldsymbol{a}$ in H, there is $\boldsymbol{b}$ in H such that $\boldsymbol{b} \upharpoonright i = \boldsymbol{a} \upharpoonright i$ and $\boldsymbol{b}(i) > \boldsymbol{a}(k-1)$. There is $\boldsymbol{c}$ in H such that $\boldsymbol{c} \upharpoonright k = \boldsymbol{a} \upharpoonright k$ and $\boldsymbol{c}(k) > \boldsymbol{b}(j-1)$. There is $\boldsymbol{d}$ in H such that $\boldsymbol{d} \upharpoonright j = \boldsymbol{b} \upharpoonright j$ and $\boldsymbol{d}(j) > \boldsymbol{c}(n-1)$. Then $\{\boldsymbol{c}, \boldsymbol{d}\} \in E^n_{ijk}$, and so $[H]^2 \cap \Gamma^n_1 \neq \emptyset$. This proves (i).

To prove (ii), use induction on n. When $n = 3$, the partition $\mathbf{\Gamma}^3$ is the Specker partition used to prove Lemma 7.2.4, so (ii) is true when $n = 3$. Suppose (ii) is true for a particular value of n. Take a set A from $[W(n+1)]^{q+1}$ such that $[A]^2 \subseteq \Gamma^{n+1}_1$; we wish to show that $q < 2^{n-1}$. Let $A = \{\boldsymbol{a}_0, \dots, \boldsymbol{a}_q\}$ listed in lexicographic order. Note that for $\boldsymbol{a}, \boldsymbol{b}$ with $\boldsymbol{a} \prec \boldsymbol{b}$ and $\{\boldsymbol{a}, \boldsymbol{b}\} \in \Gamma^{n+1}_1$, always $\boldsymbol{a}(i) \leqslant \boldsymbol{b}(i)$. Further the definition of any E^{n+1}_{ijk} ensures that $\boldsymbol{b}(1) \neq \boldsymbol{a}(n)$, for if $j = 1$ then $\boldsymbol{a}(n) < \boldsymbol{b}(j) = \boldsymbol{b}(1)$ and if $j > 1$ then $\boldsymbol{b}(1) \leqslant \boldsymbol{b}(j-1) < \boldsymbol{a}(k) \leqslant \boldsymbol{a}(n)$. Hence there is l, with $l \leqslant q$, such that

$$\boldsymbol{a}_0(1) \leqslant \boldsymbol{a}_1(1) \leqslant \dots \leqslant \boldsymbol{a}_l(1) < \boldsymbol{a}_0(n) < \boldsymbol{a}_{l+1}(1) \leqslant \dots \leqslant \boldsymbol{a}_q(1). \tag{1}$$

We show first that $l < 2^{n-2}$. Observe for $\boldsymbol{a}, \boldsymbol{b}$ with $\boldsymbol{a} \prec \boldsymbol{b}$ and $\{\boldsymbol{a}, \boldsymbol{b}\} \in \Gamma^{n+1}_1$ that if $\boldsymbol{a}(0) \leqslant \boldsymbol{b}(0) < \boldsymbol{b}(1) < \boldsymbol{a}(n)$ then also $\{\boldsymbol{a}, \boldsymbol{b}'\} \in \Gamma^{n+1}_1$ where $\boldsymbol{b}' = \langle \boldsymbol{a}(0), \boldsymbol{b}(1), \boldsymbol{b}(2), \dots, \boldsymbol{b}(n) \rangle$. For if $\boldsymbol{a}(0) = \boldsymbol{b}(0)$ this is trivial, and if $\boldsymbol{a}(0) < \boldsymbol{b}(0)$ then $\{\boldsymbol{a}, \boldsymbol{b}\} \in E^{n+1}_{0jk}$ for some j, k, where $j > 1$. But then $\{\boldsymbol{a}, \boldsymbol{b}'\} \in E^{n+1}_{1jk}$, so indeed $\{\boldsymbol{a}, \boldsymbol{b}'\} \in \Gamma^{n+1}_1$. Thus if

$$A' = \{\langle \boldsymbol{a}_0(0), \boldsymbol{a}_m(1), \dots, \boldsymbol{a}_m(n) \rangle; m \leqslant l\}$$

then $[A']^2 \subseteq \Gamma^{n+1}_1$. Consequently, if

$$A'' = \{\langle \boldsymbol{a}_m(1), \dots, \boldsymbol{a}_m(n) \rangle; m \leqslant l\}$$

then $[A'']^2 \subseteq \Gamma_1^n$. So by the inductive hypothesis, $l + 1 \leqslant 2^{n-2}$.

Return now to (1). Thus if $l = q$, nothing more is necessary. So suppose $l < q$. Observe first that

$$a_{l+1}(0) = a_{l+2}(0) = \ldots = a_q(0) . \tag{2}$$

For suppose (2) is false. Then there is m with $l + 1 < m \leqslant q$ such that $a_{l+1}(0) < a_m(0)$, and so $\{a_{l+1}, a_m\} \in E_{0jk}^{n+1}$ for some j, k. Since then $a_{l+1}(1) \leqslant a_{l+1}(k-1) < a_m(0)$, in fact $a_{l+1}(1) < a_m(0)$ and so $a_0(n) < a_{l+1}(1) < a_m(0)$. However, $a_0(n) < a_m(0)$ is impossible with $\{a_0, a_m\} \in \Gamma_1^{n+1}$. Hence (2) must hold. Thus if

$$A^* = \{\langle a_m(1), \ldots, a_m(n)\rangle; l + 1 \leqslant m \leqslant q\} ,$$

as above $[A^*]^2 \subseteq \Gamma_1^n$. So again by the inductive hypothesis $|A^*| \leqslant 2^{n-2}$, that is, $q - l \leqslant 2^{n-2}$. Since $l < 2^{n-2}$, thus $q < 2^{n-1}$. This completes the inductive step, and proves (ii).

Finally, we establish (iii). Again use induction; the case $n = 3$ is trivial. So suppose for some n that (iii) holds. The partition of $\{a \in W(n + 1); a(0) = 0\}$ induced by $\mathbf{\Gamma}^{n+1}$ is isomorphic to the partition $\mathbf{\Gamma}^n$ of $W(n)$, so by hypothesis there is a subset H of $W(n + 1)$ with $|H| = 2^{n-2}$ and $a(0) = 0$ for all a in H such that $[H]^2 \subseteq \Gamma_1^{n+1}$. Let $H = \{a_0, \ldots, a_p\}$, listed in lexicographic order (so $p = 2^{n-2} - 1$). Then

$$0 = a_0(0) = \ldots = a_p(0) < a_0(1) \leqslant a_1(1) \leqslant \ldots \leqslant a_p(1) < a_0(n)$$
$$< a_1(n) < \ldots < a_p(n) .$$

Define b_l and c_l in $W(n + 1)$ for l with $0 \leqslant l \leqslant p$ by

$$b_l(m) = \begin{cases} a_l(m) \text{ if } a_l(m) < a_0(n) \\ a_l(m) + 1 \text{ if } a_l(m) \geqslant a_0(n) , \end{cases}$$

$$c_l(m) = \begin{cases} a_0(n) \text{ if } m = 0 \\ a_p(n) + a_l(m) + 1 \text{ if } m > 0 . \end{cases}$$

Put $B = \{b_l; 0 \leqslant l \leqslant p\}$ and $C = \{c_l; 0 \leqslant l \leqslant p\}$. It is easy to see that $\{a_l, a_m\}$ is similar to both $\{b_l, b_m\}$ and $\{c_l, c_m\}$. The definition of $\mathbf{\Gamma}^{n+1}$ ensures that $\mathbf{\Gamma}^{n+1}$ is a canonical partition of $[W(n + 1)]^2$, so since $[H]^2 \subseteq \Gamma_1^{n+1}$ also $[B]^2 \subseteq \Gamma_1^{n+1}$ and $[C]^2 \subseteq \Gamma_1^{n+1}$. Moreover, for any l, m with l, $m \leqslant p$ there is k such that

$$b_l(0) < \ldots < b_l(k - 1) < a_0(n) = c_m(0) < b_l(k) < \ldots < b_l(n)$$
$$\leqslant a_p(n) + 1 < c_m(1) < \ldots < c_m(n) , \tag{1}$$

and hence $\{b_l, c_m\} \in E^{n+1}_{01k}$, so $\{b_l, c_m\} \in \Gamma^{n+1}_1$. Hence $[B \cup C]^2 \subseteq \Gamma^{n+1}_1$, and since $|B \cup C| = 2^{n-2} + 2^{n-2} = 2^{n-1}$, the induction step is complete. This proves (iii), and completes the proof of the lemma.

Theorem 7.2.9. *If* $n \leqslant 3$ *then* $\omega^n \nrightarrow (\omega^3, 2^{n-2}+1)^2$ *and* $\omega^n \rightarrow (\omega^3, 2^{n-2})^2$.

Proof. From (i) and (ii) of Lemma 7.2.8, the negative relation is established by the partition $\mathbf{\Gamma}^n$ of $W(n)$.

For the positive relation, take any disjoint partition $\mathbf{\Delta} = \{\Delta_0, \Delta_1\}$ of $[W(n)]^2$ such that there is no subset H of $W(n)$ of order type ω^3 with $[H]^2 \subseteq \Delta_0$ and no 2^{n-2}-element subset H of $W(n)$ with $[H]^2 \subseteq \Delta_1$, and seek a contradiction. By Theorem 7.2.7 there is an infinite subset of ω on which the induced partition is canonical, and we may suppose that $\mathbf{\Delta}$ itself is canonical. For i, j, k with $0 \leqslant i < j < k < n$ and integers a_0, a_1, a_2 put

$$f_{ijk}(a_0, a_1, a_2) = \langle 0, 1, \ldots, i-1, i+a_0, \ldots, j-1+a_0, j+a_1, \ldots,$$
$$k-1+a_1, k+a_2, \ldots, n-1+a_2\rangle,$$

so $f_{ijk}(a_0, a_1, a_2) \prec f_{ijk}(b_0, b_1, b_2) \Leftrightarrow \langle a_0, a_1, a_2\rangle \prec \langle b_0, b_1, b_2\rangle$. Define a disjoint partition $\mathbf{\Delta}^* = \{\Delta^*_0, \Delta^*_1\}$ of $[W(3)]^2$ by

$$\langle a_1, a_2, a_3\rangle \in \Delta^*_0 \Leftrightarrow f_{ijk}(a_1, a_2, a_3) \in \Delta_0\ .$$

Clearly $\mathbf{\Delta}^*$ is canonical since $\mathbf{\Delta}$ is. A subset of $W(3)$ homogeneous for $\mathbf{\Delta}^*$ gives a subset of $W(n)$ homogeneous for $\mathbf{\Delta}$ in the obvious way. Hence $\mathbf{\Delta}^*$ is a canonical partition of $[W(3)]^2$ having no H of order type ω^3 with $[H]^2 \subseteq \Delta^*_0$ and no H of size 2^{n-2} with $[H]^2 \subseteq \Delta^*_1$. A check of the possible canonical partitions of $[W(3)]^2$ (see Milner [72]) shows that apart from the Specker partition $\mathbf{\Gamma}^3$ all have either H of order type ω^3 with $[H]^2$ in class 0 or else H of arbitrary large finite size with $[H]^2$ in class 1. Hence $\mathbf{\Delta}^*$ must be the Specker partition $\mathbf{\Gamma}^3$. Thus if $\langle a_0, a_1, a_2\rangle \prec \langle b_0, b_1, b_2\rangle$ then

$$\{f_{ijk}(a_0, a_1, a_2), f_{ijk}(b_0, b_1, b_2)\} \in \Delta_1 \Leftrightarrow a_0 < a_1 < b_0 < a_2 < b_1 < b_2\ .$$

An easy check shows $\{f_{ijk}(0, 1, n+1), f_{ijk}(n, 2n, 2n+1)\} \in E^n_{ijk}$, and hence $E^n_{ijk} \cap \Delta_1 \neq \emptyset$. Since $\mathbf{\Delta}$ is canonical, it follows that $E^n_{ijk} \subseteq \Delta_1$. Since i, j, k were arbitrary, thus $\Gamma^n_1 \subseteq \Delta_1$. Hence by Lemma 7.2.8(iii) there is a 2^{n-2}-element subset H of $W(n)$ with $[H]^2 \subseteq \Delta_1$, contrary to the choice of $\mathbf{\Delta}$. This is the required contradiction, and the proof is complete.

In particular, if we put $n = 4$ in Theorem 7.2.9, then we find $\omega^4 \rightarrow (\omega^3, 4)^2$ and $\omega^4 \nrightarrow (\omega^3, 5)^2$. These particular results were also obtained by Galvin, by

Hajnal (see [24]) and by Haddad and Sabbagh [51]. Earlier results for ω^4 were obtained by Milner [71], where it was shown that $\omega^4 \to (\omega^3, 3)^2$, $\omega^4 \not\to (\omega^3 + 1, 3)^2$ and $\alpha \not\to (\omega^3, 3)^2$ if $\alpha < \omega^4$.

We conclude this section by proving the following theorem of Erdös and Milner [26]: $\omega^{1+\mu m} \to (\omega^{1+\mu}, 2^m)^2$ where $m < \omega$ and $\mu < \omega_1$. This result dates back to 1959; a proof also occurs in Milner [69]. The theorem does not give best possible results – for example with $\mu = 2$, it gives $\omega^{2m+1} \to (\omega^3, 2^m)^2$ whereas by Theorem 7.2.9 in fact $\omega^{m+2} \to (\omega^3, 2^m)^2$ – but it seems to be the best general result of this type known.

Theorem 7.2.10. *Let α, β be countable, let k be finite. If $\omega^\alpha \to (\omega^{1+\beta}, k)^2$ then $\omega^{\alpha+\beta} \to (\omega^{1+\beta}, 2k)^2$.*

Corollary 7.2.11. *If $m < \omega$ and $\mu < \omega_1$ then $\omega^{1+\mu m} \to (\omega^{1+\mu}, 2^m)^2$.*

Proof. By induction on m using Theorem 7.2.10, noting that trivially $\omega^{1+\mu} \to (\omega^{1+\mu}, 2)^2$.

Proof of Theorem 7.2.10. Suppose $\omega^\alpha \to (\omega^{1+\beta}, k)^2$. Take any set S ordered by a relation $<$ with order type $\omega^{\alpha+\beta}$, and let a partition $[S]^2 = \Delta_0 \cup \Delta_1$ be given. Assume there is no H in $[S]^{\omega^{1+\beta}}$ with $[H]^2 \subseteq \Delta_0$ and no H in $[S]^{2k}$ with $[H]^2 \subseteq \Delta_1$, and seek a contradiction.

For any x in S, put

$$\Delta_i(x) = \{y \in S;\ \{x, y\} \in \Delta_i\}\,.$$

We shall use the following observation. Suppose given a family $\{A_\nu; \nu < \delta\}$ where each A_ν is a subset of S of order type ω^α. For x in S, put

$$M(x) = \{\nu < \delta;\ \mathrm{tp}(\Delta_0(x) \cap A_\nu) = \omega^\alpha\}\,.$$

Then

$$A \in [S]^{\omega^\alpha} \Rightarrow \mathrm{tp}\,\{x \in A;\ \mathrm{tp}(M(x)) = \delta\} = \omega^\alpha\,. \tag{1}$$

Consider first the case when δ is an ordinal power of ω, say $\delta = \omega^\gamma$. Suppose (1) is false for a particular set A from $[S]^{\omega^\alpha}$. Put $A' = \{x \in A;\ \mathrm{tp}(M(x)) < \omega^\gamma\}$, so $\mathrm{tp}(A') = \omega^\alpha$. By the relation $\omega^\alpha \to (\omega^{1+\beta}, k)^2$ and the choice of the partition, there is H in $[A']^k$ such that $[H]^2 \subseteq \Delta_1$. Since $\omega^\gamma \to (\omega^\gamma)^1_k$ (by Theorem 7.1.3) and $\mathrm{tp}(M(x)) < \omega^\gamma$ if $i \in H$, there is ν with $\nu < \omega^\gamma$ such that $\nu \notin \bigcup\{M(x); x \in H\}$. Thus for all x in H, $\mathrm{tp}(\Delta_0(x) \cap A_\nu) < \omega^\alpha$. Hence $\mathrm{tp}(A_\nu \cap \bigcap\{\Delta_1(x); x \in H\}) = \omega^\alpha$. Again by the relation $\omega^\alpha \to (\omega^{1+\beta}, k)^2$ there is I in $[\bigcap\{\Delta_1(x); x \in H\} - H]^k$ such that $[I]^2 \subseteq \Delta_1$. Then $|H \cup I| = 2k$

and $[H \cup I]^2 \subseteq \Delta_1$, contrary to the choice of the partition. Thus (1) holds when $\delta = \omega^\gamma$. For arbitrary δ, write δ as a finite sum of powers of ω and successively use the above result finitely often to prove (1).

From (1) we shall prove the following. Suppose given a family $\{A_\nu; \nu < \omega^\beta\}$ where each A_ν is a subset of S of order type ω^α and moreover $A_\mu < A_\nu$ whenever $\mu < \nu < \omega^\beta$. Suppose also finitely many values $\nu_0, ..., \nu_n$ less than ω^β are given. Then

for any A in $[S]^{\omega^\alpha}$ there is x in A and an order preserving function $g : \omega^\beta \to \omega^\beta$ such that $g(\nu_0) = \nu_0, ..., g(\nu_n) = \nu_n$ and $\text{tp}(\Delta_0(x) \cap A_{g(\nu)} = \omega^\alpha$ whenever $\nu < \omega^\beta$. (2)

For we may suppose $\nu_0 < \nu_1 < ... < \nu_n$. Let A in $[S]^{\omega^\alpha}$ be given. Define subsets $B_0, ..., B_{n+1}$ of S as follows: $B_0 = A$,

$$B_{m+1} = \{x \in B_m; \text{tp}(\Delta_0(x) \cap A_{\nu_m}) = \omega^\alpha\}, (m = 0, 1, ..., n) .$$

Successive applications of (1) with $\delta = 1$ show that always $\text{tp}(B_{m+1}) = \omega^\alpha$. In particular, if $x \in B_{n+1}$ then $\text{tp}(\Delta_0(x) \cap A_{\nu_m}) = \omega^\alpha$ for each m. Define subsets $C_0, ..., C_{n+1}$ of S as follows. Put

$$M_0(x) = \{\nu; \nu < \nu_0 \text{ and } \text{tp}(\Delta_0(x) \cap A_\nu) = \omega^\alpha\} ,$$

$$C_0 = \{x \in B_{n+1}; \text{tp}(M_0(x)) = \nu_0\} ,$$

and for m with $m = 0, 1, ..., n$ put

$$M_{m+1}(x) = \{\nu; \nu_m < \nu < \nu_{m+1} \text{ and } \text{tp}(\Delta_0(x) \cap A_\nu) = \omega^\alpha\} ,$$

$$C_{m+1} = \{x \in C_m; \nu_m + 1 + \text{tp}(M_{m+1}(x)) = \nu_{m+1}\} ,$$

(with the convention $\nu_{n+1} = \omega^\beta$). Successive applications of (1) show that always $\text{tp}(C_m) = \omega^\alpha$. In particular, $\text{tp}(C_{n+1}) = \omega^\alpha$ so $C_{n+1} \neq \emptyset$. Choose x from C_{n+1}. Then $\text{tp}(M_0(x)) = \nu_0$ and $\nu_m + 1 + \text{tp}(M_{m+1}(x)) = \nu_{m+1}$ for each m. Hence if

$$M(x) = \{\nu < \omega^\beta; \text{tp}(\Delta_0(x) \cap A_\nu) = \omega^\alpha\} ,$$

then $M(x)$ is the ordered union

$$M(x) = M_0(x) \cup \{\nu_0\} \cup M_1(x) \cup \{\nu_1\} \cup ... \cup \{\nu_n\} \cup M_{n+1}(x) ,$$

and so $\text{tp}(M(x)) = \omega^\beta$. Let $g : \omega^\beta \to M(x)$ be the enumeration of $M(x)$ in order. Then $g(\nu_0) = \nu_0, ..., g(\nu_n) = \nu_n$, and (2) is proved.

Since $|\omega^\beta| = \aleph_0$, there is a sequence $\langle \gamma_n; n < \omega \rangle$ of ordinals below ω^β in which every ν with $\nu < \omega^\beta$ is repeated infinitely often. We shall define by induction on the integer n elements x_n of S and subsets $A(n, \nu)$ of S for ν with

$\nu < \omega^\beta$. Write

$$S = \bigcup\{A(0,\nu); \nu < \omega^\beta\}$$

where always $\text{tp}(A(0,\nu)) = \omega^\alpha$ and $A(0,\mu) < A(0,\nu)$ whenever $\mu < \nu < \omega^\beta$. Suppose for some integer n that we have already defined the x_m whenever $m < n$ and the sets $A(n,\nu)$ such that always $\text{tp}(A(n,\nu)) = \omega^\alpha$ and $A(n,\mu) < A(n,\nu)$ if $\mu < \nu < \omega^\beta$. Put

$$A^*(n,\nu) = \{x \in A(n,\nu); x > x_m \text{ for all } m \text{ with } m < n\},$$

then $\text{tp}(A^*(n,\nu)) = \omega^\alpha$. Use (2) on the family $\{A(n,\nu); \nu < \omega^\beta\}$ with $\nu_0 = \gamma_0, \ldots, \nu_n = \gamma_n$ and $A = A^*(n,\gamma_n)$ to find x_n in $A^*(n,\gamma_n)$ and an order preserving map $g_n : \omega^\beta \to \omega^\beta$ with $g_n(\gamma_m) = \gamma_m$ if $m \leqslant n$ such that $\text{tp}(\Delta_0(x_n) \cap A(n, g_n(\nu))) = \omega^\alpha$ whenever $\nu < \omega^\beta$. Put $A(n+1,\nu) = \Delta_0(x_n) \cap A(n, g_n(\nu))$. Then $\text{tp}(A(n+1,\nu)) = \omega^\alpha$ and since g_n is order preserving also $A(n+1,\mu) < A(n+1,\nu)$ whenever $\mu < \nu < \omega^\beta$. This completes the construction.

Put $H = \{x_n; n < \omega\}$. The construction ensures that if $n > m$ then $A(n,\nu) \subseteq \Delta_0(x_m)$ for all ν; in particular since $x_n \in A(n,\gamma_n)$ thus $x_n \in \Delta_0(x_m)$. Hence $[H]^2 \subseteq \Delta_0$. We shall show that $\text{tp}(H) = \omega^{1+\beta}$, contrary to the choice of the original partition. This will provide the contradiction required to prove the theorem. Take integers m, n; say $m < n$. Since

$$A(n,\nu) \subseteq A(n-1, g_{n-1}(\nu)) \subseteq A(n-2, g_{n-2}g_{n-1}(\nu)) \subseteq \ldots$$
$$\subseteq A(m, g_m \ldots g_{n-2}g_{n-1}(\nu)),$$

in particular

$$A(n,\gamma_n) \subseteq A(m, g_m \ldots g_{n-2}g_{n-1}(\gamma_n)), \tag{3}$$

$$A(n,\gamma_m) \subseteq A(m, g_m \ldots g_{n-2}g_{n-1}(\gamma_m)). \tag{4}$$

Further

$$g_m \ldots g_{n-2}g_{n-1}(\gamma_m) = \gamma_m.$$

Since each g_i is order preserving, it follows that

$$\gamma_m < \gamma_n \Leftrightarrow \gamma_m < g_m \ldots g_{n-2}g_{n-1}(\gamma_n).$$

Hence from (3),

$$\gamma_m < \gamma_n \Rightarrow A(m,\gamma_m) < A(n,\gamma_n) \Rightarrow x_m < x_n. \tag{5}$$

And from (4), $A(n,\gamma_m) \subseteq A(m,\gamma_m)$, so

$$m < n \text{ and } \gamma_n = \gamma_m \Rightarrow A(n,\gamma_n) \subseteq A(m,\gamma_m) \Rightarrow x_n \in A(m,\gamma_m).$$

Now given m, the set $\{n < \omega; m < n \text{ and } \gamma_n = \gamma_m\}$ is infinite. Further the definition of x_n ensures that if $p < n$ and $\gamma_n = \gamma_p$ then $x_n > x_p$. Thus $\text{tp}(H \cap A(m, \gamma_m)) = \omega$. Since $\{\gamma_m; m < \omega\} = \omega^\beta$, it now follows from (5) that $\text{tp}(H) = \omega\omega^\beta = \omega^{1+\beta}$, as claimed. This completes the proof.

§3. Chang's Theorem for ω^ω

One of the long-standing problems in the partition calculus for countable ordinals has been to decide whether or not the relation $\omega^\omega \to (\omega^\omega, 3)^2$ holds (see Specker [93]). A positive answer was finally given by Chang [7] in a lengthy tour de force. His result was subsequently extended by Milner who showed that $\omega^\omega \to (\omega^\omega, m)^2$ for all finite m. In [65], Larson gave a new short proof of this. We shall follow Larson's method.

We shall make use of the following lemma, a consequence of Ramsey's theorem.

Lemma 7.3.1. *Let $S \in [\omega]^{\aleph_0}$. Suppose for each a from $[\omega]^{<\aleph_0}$ there are given an integer $m(a)$ and a partition $\Delta(a) = \{\Delta_0(a), \Delta_1(a)\}$ of $[S]^{m(a)}$. Then there is an infinite subset H of S such that for each a from $[H]^{<\aleph_0}$, the set $\{h \in H; h > \max(a)\}$ is homogeneous for $\Delta(a)$.*

Proof. Inductively define sets H_k from $[S]^{\aleph_0}$ and elements h_k of H_k, for finite k, as follows. Put $H_0 = S$, and let h_0 be the least element of H_0. If H_k and $h_0, ..., h_k$ have already been defined, list all subsets a of $\{h_0, ..., h_k\}$ with $\max(a) = h_k$, say $a_0, a_1, ..., a_l$. Put $H_{k0} = H_k$ and use Ramsey's theorem to choose successively infinite subsets H_{ki+1} from H_{ki} so that H_{ki+1} is homogeneous for $\Delta(a_i)$. Put $H_{k+1} = \{h \in H_{kl+1}; h > h_k\}$; then H_{k+1} is homogeneous for each $\Delta(a_i)$. Let h_{k+1} be the least element of H_{k+1}.

Put $H = \{h_k; k < \omega\}$. Take a from $[H]^{<\aleph_0}$, and suppose $\max(a) = h_k$. Then $\{h \in H; h > h_k\}$ is a subset of H_{k+1}, and hence is homogeneous for $\Delta(a)$, as required.

Theorem 7.3.2. *Let m be finite. Then $\omega^\omega \to (\omega^\omega, m)^2$.*

As in the previous section, for positive integer n put

$$W(n) = \{\langle a_0, ..., a_{n-1}\rangle \in {}^n\omega; a_0 < a_1 < ... < a_{n-1}\},$$

and order $W(n)$ lexicographically. Put $W = \bigcup\{W(n); 0 < n < \omega\}$ and order the elements of W first by length of sequence, and then lexicographically in each $W(n)$. Under this ordering, W has order type ω^ω. So to prove Theorem 7.3.2, it suffices to take any partition $[W]^2 = \Delta_0 \cup \Delta_1$ and show that, given m, either there is a subset H of W with $\text{tp}(H) = \omega^\omega$ such that $[H]^2 \subseteq \Delta_0$ or else there is a subset H of W with $|H| = m$ such that $[H]^2 \subseteq \Delta_1$.

Let m be fixed. Take any partition $\boldsymbol{\Delta} = \{\Delta_0, \Delta_1\}$ of $[W]^2$ for which there is no set H in $[W]^m$ with $[H]^2 \subseteq \Delta_1$. We must find H in $[W]^{\omega^\omega}$ with

$[H]^2 \subseteq \Delta_0$. By Corollary 7.2.11, certainly the relation $\omega^{mn} \to (\omega^n, 2^m)^2$ holds. By the relation applied to the partition of $W(mn)$ induced by $\boldsymbol{\Delta}$, there must be a subset $W'(n)$ of $W(mn)$ with $\text{tp}(W'(n)) = \omega^n$ such that $[W'(n)]^2 \subseteq \Delta_0$. Put $W' = \bigcup\{W'(n); 0 \leqslant n < \omega\}$, so $\text{tp}(W') = \omega^\omega$ and for $\{\boldsymbol{a}, \boldsymbol{b}\}$ from $[W']^2$ with $\text{ln}(\boldsymbol{a}) = \text{ln}(\boldsymbol{b})$ we know $\{\boldsymbol{a}, \boldsymbol{b}\} \in \Delta_0$. It will suffice to find a subset H of W' with $\text{tp}(H) = \omega^\omega$ such that for $\{\boldsymbol{a}, \boldsymbol{b}\}$ from $[H]^2$ with $\text{ln}(\boldsymbol{a}) \neq \text{ln}(\boldsymbol{b})$ still $\{\boldsymbol{a}, \boldsymbol{b}\} \in \Delta_0$. By redefining the partition $\boldsymbol{\Delta}$, we may suppose that $W' = W$.

For the rest of this section we shall use the convention that if $\boldsymbol{a}$ is a finite increasing sequence of integers then a is the set of entries of $\boldsymbol{a}$ and conversely, if a is a finite set of integers then $\boldsymbol{a}$ is the sequence of the elements of a arranged in increasing order. The concatenation of sequences $\boldsymbol{a}$ and $\boldsymbol{b}$ is $\boldsymbol{a} {}^\frown \boldsymbol{b}$.

Definition 7.3.3. For $\boldsymbol{a}, \boldsymbol{b}$ from W with $\text{ln}(\boldsymbol{a}) < \text{ln}(\boldsymbol{b})$ and for integer k, say $\{\boldsymbol{a}, \boldsymbol{b}\}$ has *form* $2k$ [or *form* $2k+1$] if there are non-empty finite sets of integers $a_0, a_1, \ldots, a_k$ and $b_0, b_1, \ldots, b_{k-1}$ [b_k] such that

(i) $a_0 < b_0 < a_1 < b_1 < \ldots < a_{k-1} < b_{k-1} < a_k\, [< b_k]$;

(ii) $\boldsymbol{a} = \boldsymbol{a}_0 {}^\frown \boldsymbol{a}_1 {}^\frown \boldsymbol{a}_1 \ldots {}^\frown \boldsymbol{a}_k$;

$\boldsymbol{b} = \boldsymbol{b}_0 {}^\frown \boldsymbol{b}_1 {}^\frown \ldots {}^\frown \boldsymbol{b}_{k-1}$ for form $2k$; $\boldsymbol{b} = \boldsymbol{b}_0 {}^\frown \boldsymbol{b}_1 {}^\frown \ldots {}^\frown \boldsymbol{b}_k$ for form $2k+1$;

(iii) if $\boldsymbol{c} = \langle |a_0|, |a_0 \cup a_1|, \ldots, |a_0 \cup \ldots \cup a_k| \rangle$ and

$\boldsymbol{d} = \langle |b_0|, |b_0 \cup b_1|, \ldots, |b_0 \cup \ldots \cup b_{k-1} [\cup b_k]| \rangle$ then

$c < a_0 < d < b_0$.

The pair $\{\boldsymbol{c}, \boldsymbol{d}\}$ will be called the *shape* of $\{\boldsymbol{a}, \boldsymbol{b}\}$.

We shall need several lemmas.

Lemma 7.3.4. *Let $S \in [\omega]^{\aleph_0}$. There is H in $[S]^{\aleph_0}$, $H = \{h_1, h_2, h_3, \ldots\}$ listed in increasing order, such that for all positive integers l there is $i(l) = 0, 1$ so that if $\{\boldsymbol{a}, \boldsymbol{b}\}$ is any pair of sequences of form l and shape $\{\boldsymbol{c}, \boldsymbol{d}\}$ with $a, b, c, d \subseteq \{h \in H; h > h_l\}$ then $\{\boldsymbol{a}, \boldsymbol{b}\} \in \Delta_{i(l)}$.*

Proof. For each possible shape $\{\boldsymbol{c}, \boldsymbol{d}\}$ with $c, d \subseteq S$ and each a_0 in $[S]^{<\aleph_0}$ with $c < a_0 < d$, choose a partition $\boldsymbol{\Delta}(c, a_0, d)$ of suitable size subsets of S into two classes so that: whenever $x = a_0 \cup b_0 \cup a_1 \cup b_1 \cup \ldots$ where if $\boldsymbol{a} = \boldsymbol{a}_0 \hat{} \boldsymbol{a}_1 \hat{} \ldots, \boldsymbol{b} = \boldsymbol{b}_0 \hat{} \boldsymbol{b}_1 \hat{} \ldots$ then $\{\boldsymbol{a}, \boldsymbol{b}\}$ has shape $\{\boldsymbol{c}, \boldsymbol{d}\}$ and $\{\boldsymbol{a}, \boldsymbol{b}\} \in \Delta_i$, then $x \in \Delta_i(c, a_0, d)$.

By Lemma 7.3.1. there is H_0 in $[S]^{\aleph_0}$ such that for c, a_0, d from H_0 there is $i(c, a_0, d)$ so that if $a \cup b \subseteq \{h \in H_0; h > \max(d)\}$ and $\{\boldsymbol{a}_0 \hat{} \boldsymbol{a}, \boldsymbol{b}\}$ has shape $\{\boldsymbol{c}, \boldsymbol{d}\}$ then

$$\{\boldsymbol{a}_0 \hat{} \boldsymbol{a}, \boldsymbol{b}\} \in \Delta_{i(c, a_0, d)} .$$

For each finite subset c of H_0, choose two partitions $\boldsymbol{\Delta}(c)$, $\boldsymbol{\Delta}^*(c)$ into two classes of the $\boldsymbol{c}(0) + |c|$, $\boldsymbol{c}(0) + |c| - 1$ size subsets of H_0 so that:

$$\text{if } x = a_0 \cup d \text{ where } |a_0| = \boldsymbol{c}(0),\ |d| = |c|,\ c < a_0 < d \text{ and}$$
$$i(c, a_0, d) = i, \text{ then } x \in \Delta_i(c) ;$$
$$\text{if } x = a_0 \cup d \text{ where } |a_0| = \boldsymbol{c}(0),\ |d| = |c| - 1,\ c < a_0 < d \text{ and}$$
$$i(c, a_0, d) = i, \text{ then } x \in \Delta_i^*(c) .$$

Apply Lemma 7.3.1 twice to find H_1 in $[H_0]^{\aleph_0}$ such that for c from $[H_1]^{<\aleph_0}$ there are $i(c)$, $i^*(c)$ such that for appropriate sized subsets a_0, d of $\{h \in H_1; h_1 > \max(c)\}$,

$$\text{if } |d| = |c| \text{ then } i(c, a_0, d) = i(c) ;$$
$$\text{if } |d| = |c| - 1 \text{ then } i(c, a_0, d) = i^*(c) .$$

Finally, define partitions $\boldsymbol{\Delta}(2k)$, $\boldsymbol{\Delta}(2k+1)$ of $[H_1]^{k+1}$ by

$$c \in \Delta_i(2k) \Leftrightarrow i^*(c) = i ; c \in \Delta_i(2k+1) \Leftrightarrow i(c) = i .$$

Use Ramsey's theorem repeatedly to find infinite subsets H_{l+1} of H_l minus its least element such that H_{l+1} is homogeneous for $\boldsymbol{\Delta}(l)$ (where $l \geqslant 1$). Let h_l be the least element of H_{l+1}, and put $H = \{h_l; 0 < l < \omega\}$. Then for all positive l there is $i(l)$ such that if $c \in [\{h \in H; h > h_l\}]^{k+1}$ then $i^*(c) = i(l)$ or $i(c) = i(l)$, depending on whether l equals $2k$ or $2k+1$. It follows that H has the property required. For take a pair $\{\boldsymbol{a}, \boldsymbol{b}\}$ of form l and shape $\{\boldsymbol{c}, \boldsymbol{d}\}$ where $a, b, c, d \subseteq \{h \in H; h > h_l\}$. Since $H \subseteq H_1 \subseteq H_0$, then $\{\boldsymbol{a}, \boldsymbol{b}\} \in \Delta_{i(c, a_0, d)}$ (where $\boldsymbol{a}_0$ is the sequence of the first $\boldsymbol{c}(0)$ elements of $\boldsymbol{a}$), and

$$\begin{aligned} i(c, a_0, d) &= i^*(c) \text{ or } i(c) \text{ (depending on whether } l \text{ is even or odd)} \\ &= i(l) . \end{aligned}$$

This completes the proof.

Lemma 7.3.5. *Given any infinite subset H of ω, for all positive integers l, m there is M in $[W]^m$ such that whenever $\boldsymbol{a}, \boldsymbol{b} \in M$ then $a, b \subseteq H$, $\{\boldsymbol{a}, \boldsymbol{b}\}$ has form l and if $\{\boldsymbol{c}, \boldsymbol{d}\}$ is the shape of $\{\boldsymbol{a}, \boldsymbol{b}\}$ then $c, d \subseteq H$.*

Proof. Suppose $l = 2k$ or $l = 2k + 1$. Choose increasing finite sequences $\boldsymbol{c}_i, \boldsymbol{a}_{ij}$ from H for i, j with $1 \leqslant i \leqslant m$, $0 \leqslant j \leqslant k + 1$ in the order that follows, so that always the last element of one sequence is less than the least element of the next:

$$\boldsymbol{c}_1, \boldsymbol{a}_{10}, \boldsymbol{c}_2, \boldsymbol{a}_{20}, \ldots, \boldsymbol{c}_m, \boldsymbol{a}_{m0}, \boldsymbol{a}_{11}, \boldsymbol{a}_{21}, \ldots, \boldsymbol{a}_{m1},$$

$$\boldsymbol{a}_{12}, \boldsymbol{a}_{22}, \ldots, \boldsymbol{a}_{m2}, \ldots, \boldsymbol{a}_{1k}, \ldots, \boldsymbol{a}_{mk}, \boldsymbol{a}_{mk+1}, \ldots, \boldsymbol{a}_{2k+1}, \boldsymbol{a}_{1k+1},$$

such that $\ln(\boldsymbol{c}_i) = k + 1$, $\ln(\boldsymbol{a}_{i0}) = \boldsymbol{c}_i(0)$ and $\ln(\boldsymbol{a}_{ij}) = \boldsymbol{c}_i(j) - \boldsymbol{c}_i(j-1)$ (for i, j with $1 \leqslant i \leqslant m$, $1 \leqslant j \leqslant k$), $\ln(\boldsymbol{a}_{ik+1}) = c_i(k) - c_i(k-1)$.

If $l = 2k + 1$, put $\boldsymbol{a}_i = \boldsymbol{a}_{i0} {}^\frown \boldsymbol{a}_{i1} {}^\frown \ldots {}^\frown \boldsymbol{a}_{ik}$ whereas if $l = 2k$, put $\boldsymbol{a}_i = \boldsymbol{a}_{i0} {}^\frown \boldsymbol{a}_{i1} {}^\frown \ldots {}^\frown \boldsymbol{a}_{ik-1} {}^\frown \boldsymbol{a}_{ik+1}$ and put

$$M = \{\boldsymbol{a}_i;\ 1 \leqslant i \leqslant m\}.$$

If $l = 2k + 1$ then $\{\boldsymbol{a}_i, \boldsymbol{a}_j\}$ has form l and shape $\{\boldsymbol{c}_i, \boldsymbol{c}_j\}$, whereas if $l = 2k$ then $\{\boldsymbol{a}_i, \boldsymbol{a}_j\}$ has form l and shape $\{\boldsymbol{c}_i, \boldsymbol{d}_j\}$ where $\boldsymbol{d}_j = \langle c_j(0), \ldots, c_j(k-2), c_j(k+1)\rangle$. Hence M has the property required in the lemma.

Lemma 7.3.6. *Given any infinite subset H of ω, say $H = \{h_1, h_2, h_3, \ldots\}$ listed in increasing order, there is a subset X of W with $\mathrm{tp}(X) = \omega^\omega$ such that whenever $\boldsymbol{a}, \boldsymbol{b} \in X$ with $\ln(\boldsymbol{a}) < \ln(\boldsymbol{b})$ then $a, b \subseteq H$ and $\{\boldsymbol{a}, \boldsymbol{b}\}$ has form l for some positive integer l; moreover if $\{\boldsymbol{c}, \boldsymbol{d}\}$ is the shape of $\{\boldsymbol{a}, \boldsymbol{b}\}$ then $h_l < \boldsymbol{c}(0)$ and $c, d \subseteq H$.*

Proof. Choose increasing finite sequences $\boldsymbol{c}_i, \boldsymbol{a}_i, \boldsymbol{a}(i, j, k)$ from H for i, j, k with $1 \leqslant i < \omega$ and $1 \leqslant j \leqslant i \leqslant k < \omega$ as follows. Suppose for given k that $\boldsymbol{c}_i, \boldsymbol{a}_i, \boldsymbol{a}(i, j, l)$ have already been chosen whenever $1 \leqslant j \leqslant i \leqslant l < k$. We choose $\boldsymbol{c}_k, \boldsymbol{a}_k, \boldsymbol{a}(i, j, k)$ where $1 \leqslant j \leqslant i \leqslant k$ in the order that follows so that always the last element of one sequence is less than the least element of the next. Ensure also that the least element of $\boldsymbol{c}_k$ is greater that h_{2k+1} and greater than every element in the sequences already defined. The order is:

$$\boldsymbol{c}_k, \boldsymbol{a}_k, \boldsymbol{a}(1, 1, k), \boldsymbol{a}(2, 1, k), \boldsymbol{a}(2, 2, k), \boldsymbol{a}(3, 1, k), \ldots, \boldsymbol{a}(k, k, k)$$

(that is, the $\boldsymbol{a}(i, j, k)$ are chosen according to the lexicographic ordering of $\langle i, j\rangle$, where $1 \leqslant j \leqslant i \leqslant k$). The sequences have lengths as follows: $\ln(\boldsymbol{c}_k) = k + 1$, $\ln(\boldsymbol{a}_k) = \boldsymbol{c}_k(0)$, $\ln(\boldsymbol{a}(i, j, k)) = \boldsymbol{c}_i(j) - \boldsymbol{c}_i(j-1)$.

Put

$$X_i = \{a_i{}^\frown a(i, 1, k_1){}^\frown a(i, 2, k_2){}^\frown \dots {}^\frown a(i, i, k_i);$$
$$1 \leqslant i \leqslant k_1 < k_2 < \dots < k_i < \omega\},$$
$$X = \bigcup\{X_i; 1 \leqslant i < \omega\},$$

so $X \subseteq W$. For fixed i, all the sequences in X_i have the same length, namely

$$c_i(0) + (c_i(1) - c_i(0)) + \dots + (c_i(i) - c_i(i-1)) = c_i(i).$$

Further, the choice of $a(i, j, k)$ ensures that if $k_1 < k_2 < \dots < k_i$ then

$$a(i, 1, k_1) < a(i, 2, k_2) < \dots < a(i, i, k_i),$$

and hence the order type of X_i, in the lexicographic ordering on W, is ω^i. Moreover, if $i < j$ then $c_i < c_j$ so in particular $c_i(i) < c_j(j)$; thus all sequences in X_i precede those in X_j. Hence X has order type ω^ω.

Take a, b from X with $\ln(a) \neq \ln(b)$. Since all sequences in any X_i have the same length, we may suppose that $a \in X_i$ and $b \in X_j$ where $i < j$. No sequence from X_i has any value in common with a sequence from X_j. Thus it follows that $\{a, b\}$ has form l for some positive integer l. And if $\{c, d\}$ is the shape of $\{a, b\}$ then c is a subsequence of c_i and d is a subsequence of c_j; thus $c, d \subseteq H$. Finally, $c_i(0) > h_{2i+1}$; this ensures that $c(0) > h_l$. Thus X has the properties required, and the lemma is proved.

We are now set to prove Theorem 7.3.2, namely that $\omega^\omega \to (\omega^\omega, m)^2$ for all finite m.

Proof of Theorem 7.3.2. Take a partition $[W]^2 = \Delta_0 \cup \Delta_1$ for which there is no set H in $[W]^m$ such that $[H]^2 \subseteq \Delta_1$. As noted before, we may suppose that whenever $a, b \in W$ with $\ln(a) = \ln(b)$ then $\{a, b\} \in \Delta_0$. Take an infinite subset H of ω with the property given by Lemma 7.3.4. Thus if $H = \{h_1, h_2, h_3, \dots\}$, listed in increasing order, then for every positive l there is $i(l) = 0, 1$ so that for all pairs $\{a, b\}$ of form l, if $\{a, b\}$ has shape $\{c, d\}$ and $a, b, c, d \subseteq \{h \in H; h > h_l\}$ then $\{a, b\} \in \Delta_{i(l)}$. Given l, apply Lemma 7.3.5 to $\{h \in H; h > h_l\}$ to find M in $[W]^m$ such that all $\{a, b\}$ from $[M]^2$ have form l, and if the shape is $\{c, d\}$ then $a, b, c, d \subseteq \{h \in H; h > h_l\}$. Thus $i(l) = 0$, and this is true for all l. Now use Lemma 7.3.6 to find a subset X of W with $\mathrm{tp}(X) = \omega^\omega$ such that all $\{a, b\}$ from X with $\ln(a) \neq \ln(b)$ have form l for some l, and moreover if $\{c, d\}$ is the shape of $\{a, b\}$ then $a, b, c, d \subseteq \{h \in H; h > h_l\}$. Thus $\{a, b\} \in \Delta_{i(l)}$, that is $\{a, b\} \in \Delta_0$, for all such $\{a, b\}$. Hence $[X]^2 \subseteq \Delta_0$, and the proof is complete.

§4. Partitions of $[\omega_1]^2$

A complete discussion is possible for the relation $\omega_1 \to (\alpha_k; k < \gamma)^2$. Some results are immediate from Chapter 2. From Corollary 2.5.2 we know $\omega_1 \nrightarrow (3)^2_{\aleph_0}$, so only partitions of $[\omega_1]^2$ into finitely many classes are relevant. From Theorem 2.5.8 comes the further negative relation $\omega_1 \nrightarrow (\omega_1)^2_2$.

Theorem 7.4.1. *For s finite,* $\omega_1 \to (\omega_1, (\omega + 1)_s)^2$.

Proof. Applying Theorem 2.2.6 to the two relations $\aleph_1 \to (\aleph_1)^2_1$ and $\aleph_0 \to (\aleph_0)^1_s$ gives the relation $\aleph_1 \to (\aleph_1, (\aleph_0)_s)^2$. Only slight changes to the proof of Theorem 2.2.6 yields the marginally stronger result $\omega_1 \to (\omega_1, (\omega + 1)_s)^2$. Referring to that proof, with the notation in use there, the changes are as follows. Start by taking a well ordering $\prec$ of the set S of order type κ. When one comes to define the set $F(\mathbf{v})$ for $\mathbf{v}$ in $N \cap SEQ_\sigma$, put

$$G(\mathbf{v}) = \{x \in S(\mathbf{v}); x \prec y \text{ for some } y \text{ in } \bigcup\{F(\mathbf{v} \upharpoonright \tau); \tau < \sigma\}\},$$

and take for $F(\mathbf{v})$ the union of $G(\mathbf{v})$ and the original $F(\mathbf{v})$. Since κ is regular, $|G(\mathbf{v})| < \kappa$ and so still $|F(\mathbf{v})| < \kappa$.

The elements x_σ are defined as before, using a sequence $\mathbf{v}$ in $N \cap SEQ_{\lambda_2}$ for which $S(\mathbf{v}) \neq \emptyset$. Choose x in $S(\mathbf{v})$. Then for each σ it follws that $x \notin F(\mathbf{v} \upharpoonright \sigma + 1)$, so in particular $x \notin G(\mathbf{v} \upharpoonright \sigma + 1)$. Since $x \in S(\mathbf{v} \upharpoonright \sigma + 1)$ and $x_\sigma \in F(\mathbf{v} \upharpoonright \sigma)$ it follows that $x \not\prec x_\sigma$; hence $x_\sigma \prec x$. From $x \in S(\mathbf{v} \upharpoonright \sigma + 1)$ we know $\{x_\sigma, x\} \in \Gamma_{l(\mathbf{v} \upharpoonright \sigma+1)}$. It follows that $H_0' = H_0 \cup \{x\}$ is a subset of κ of order type at least $\eta_{l_0} + 1$ with $[H_0']^2 \subseteq \Gamma_{l_0}$.

Theorem 7.4.1 is the strongest positive result of its type. The relation $\omega_1 \nrightarrow (\omega_1, \omega + 2)^2$ was proved by Hajnal [52] assuming the continuum hypothesis. Whether the relation can be proved without this assumption is an open problem. See Erdös and Hajnal [24], Problem 8, and their discussion of this problem in [25].

Theorem 7.4.2 (CH). $\omega_1 \nrightarrow (\omega_1, \omega + 2)^2$.

Proof. Take any set S ordered with order type ω_1 by a relation $\prec$. By Corollary 3.2.8 there is a set mapping $f: S \to \mathfrak{P}S$ of order $\aleph_1$ with $|f(x) \cap f(y)| < \aleph_0$ for any pair x, y from S, such that f has no free set of power $\aleph_1$. Since always $|f(x)| \leqslant \aleph_0$, we can choose inductively elements x_α of S for α with $\alpha < \omega_1$ so that

$$\alpha < \beta < \omega_1 \Rightarrow x_\alpha < x_\beta \text{ and } f(x_\alpha) < \{x_\beta\}.$$

Put $T = \{x_\alpha; \alpha < \omega_1\}$ so $\mathrm{tp}(T) = \omega_1$. Define a partition $[T]^2 = \Delta_0 \cup \Delta_1$ by, if $\alpha < \beta$ then

$$\{x_\alpha, x_\beta\} \in \Delta_0 \Leftrightarrow x_\alpha \notin f(x_\beta) ,$$

$$\{x_\alpha, x_\beta\} \in \Delta_1 \Leftrightarrow x_\alpha \in f(x_\beta) .$$

Any subset H_0 of T with $[H_0]^2 \subseteq \Delta_0$ is free for f, so $\mathrm{tp}(H_0) < \omega_1$. Suppose there is H_1 in $[T]^{\omega+2}$ with $|H_1|^2 \subseteq \Delta_1$. If x and y are the ω-th and $(\omega + 1)$-st elements of H_1 then $H_1 - \{x, y\} \subseteq f(x) \cap f(y)$ so $|f(x) \cap f(y)| = \aleph_0$, contrary to the choice of f. So there is no such set H_1, and the theorem is proved.

We have left to consider relations of the form $\omega_1 \to (\alpha_k; k < s)^2$ where s is finite and the α_k are all countable. Trivially Ramsey's theorem gives $\omega_1 \to (\omega)^2_s$. Erdös and Rado [29] showed that $\omega_1 \to (\omega + n)^2_2$ for any finite n. In [60], Hajnal strengthened this to $\omega_1 \to (\omega 2, \omega n)^2$ for any finite n. The next result was due to Galvin (1970, unpublished) that $\omega_1 \to (\omega 3)^2_2$. Later still Prikry [79] obtained $\omega_1 \to (\omega^2 + 1, \alpha)^2$ for every countable α. Finally Baumgartner and Hajnal [4] settled the problem by proving the best possible result, namely $\omega_1 \to (\alpha)^2_s$ for every countable α and finite s. In fact they proved somewhat more. They showed that if φ is an order type such that $\varphi \to (\omega)^1_\omega$ then $\varphi \to (\alpha)^2_s$ (for all countable α and finite s). Their proof uses deep methods from mathematical logic. They show first that the relation $\varphi \to (\alpha)^2_s$ holds in a particular model of set theory, and then use absoluteness criteria to conclude that the relation is true in the real world. Galvin [44] has since given a combinatorial proof of the Baumgartner-Hajnal theorem. We shall devote the remainder of this section to Galvin's proof for the relation $\omega_1 \to (\alpha)^2_s$.

Theorem 7.4.3. *Let α be a countable ordinal and let s be finite. Then $\omega_1 \to (\alpha)^2_s$.*

For the rest of this section, we adopt the convention that the letters $\alpha, \beta, \gamma, \ldots$ (with maybe subscripts, superscripts or the like) range over only the ordinals less than ω_1.

Let the partition $\mathbf{\Delta}$ of $[\omega_1]^2$ be given,

$$[\omega_1]^2 = \Delta_0 \cup \ldots \cup \Delta_{s-1} .$$

For any α, we shall show that there is H in $[\omega_1]^{\omega^\alpha}$ such that H is homogeneous for $\mathbf{\Delta}$; from this Theorem 7.4.3 follows.

We introduce the following property, somewhat weaker than being a homogeneous set for $\mathbf{\Delta}$.

Definition 7.4.4. A subset X of ω_1 is *almost homogeneous* (for $\boldsymbol{\Delta}$) if whenever $A \in [X]^{\omega^\alpha}$ (for any α) then for all β with $\beta < \alpha$ there are B in $[A]^{\omega^\beta}$, A' in $[A]^{\omega^\alpha}$ and i with $i < s$ such that $B < A'$ and $B \otimes A' \subseteq \Delta_i$, where $B \otimes A' = \{\{x, y\}; x \in B, y \in A' \text{ and } x \neq y\}$.

Definition 7.4.5. The symbol $\beta \to AH(\alpha(0), ..., \alpha(s-1))$ indicates that whenever X is an almost homogeneous subset of ω_1 with $\text{tp}(X) = \omega^\beta$, then there are i with $i < s$ and A in $[X]^{\omega^{\alpha(i)}}$ such that $[A]^2 \subseteq \Delta_i$.

The next lemma reduces the problem of proving Theorem 7.4.3 to the problem of constructing almost homogeneous subsets of arbitrarily large (countable) order type.

Lemma 7.4.6. *For all α there is β such that $\beta \to AH(\alpha, ..., \alpha)$.*

Proof. We shall show the following: given ordinals $\alpha(0), ..., \alpha(s-1)$ such that for all i with $i < s$ and all γ with $\gamma < \alpha(i)$ there is $\beta(i, \gamma)$ such that $\beta(i, \gamma) \to AH(\alpha(0), ..., \alpha(i-1), \gamma, \alpha(i+1), ..., \alpha(s-1))$, then there is β such that $\beta \to AH(\alpha(0), ..., \alpha(s-1))$. It follows by induction on $\alpha_0, ..., \alpha_{s-1}$ that for all $\alpha_0, ..., \alpha_{s-1}$ there is β such that $\beta \to AH(\alpha_0, ..., \alpha_{s-1})$; in particular the lemma is true.

If any $\alpha(i)$ is zero, the claim above is trivially true (with β arbitrary), so suppose that always $\alpha(i) \geq 1$. Choose sequences $\langle \alpha(i, n); n < \omega \rangle$ such that $\alpha(i, 0) \leq \alpha(i, 1) \leq ... < \alpha(i)$ and $\omega^{\alpha(i)} = \Sigma_0(\omega^{\alpha(i,n)}; n < \omega)$. Put

$$\beta(n) = \max \{\beta(i, \alpha(i, n)); i < s\}$$

and $\beta = \Sigma_0(\beta(n); n < \omega)$. We shall show that

$$\beta \to AH(\alpha(0), ..., \alpha(s-1)) \,. \tag{1}$$

So take X from $[\omega_1]^{\omega^\beta}$ such that X is almost homogeneous. Then there are X_0 in $[X]^{\omega^{\beta(0)}}$, A_0 in $[X]^{\omega^\beta}$ and $i(0)$ with $i(0) < s$ such that $X_0 < A_0$ and $X_0 \otimes A_0 \subseteq \Delta_{i(0)}$. Inductively we can continue and find X_n in $[A_{n-1}]^{\omega^{\beta(n)}}$, A_n in $[A_{n-1}]^{\omega^\beta}$ and $i(n)$ such that $X_n < A_n$ and $X_n \otimes A_n \subseteq \Delta_{i(n)}$. Then each X_n is almost homogeneous, being a subset of the almost homogeneous set X, and further

$$\text{if } n < m, \text{ then } X_n < X_m \text{ and } X_n \otimes X_m \subseteq \Delta_{i(n)} \,. \tag{2}$$

There is N in $[\omega]^{\aleph_0}$ such that $i(n)$ is constant for n in N, say with value j. For each n, the choice of $\beta(n)$ ensures that $\beta(n) \to AH(\alpha(0), ..., \alpha(j-1), \alpha(j, n), \alpha(j+1), ..., \alpha(s-1))$. Thus for each n, either there are i with $i \neq j$ and C_n in

$[X_n]^{\omega^{\alpha(i)}}$ such that $[C_n]^2 \subseteq \Delta_i$, or there is B_n in $[X_n]^{\omega^{\alpha(j,n)}}$ such that $[B_n]^2 \subseteq \Delta_j$. If for any n the first of these alternatives is true this gives i and C in $[X]^{\omega^{\alpha(i)}}$ with $[C]^2 \subseteq \Delta_i$. On the other hand, if for all n the second alternative holds, consider B where $B = \bigcup\{B_n; n \in N\}$. By (2),

$$\text{tp}(B) = \Sigma_0(\omega^{\alpha(j,n)}; n \in N) = \omega^{\alpha(j)} .$$

From (2), $B_n \otimes B_m \subseteq \Delta_j$ if $\{n, m\} \in [N]^2$, and also $[B_n]^2 \subseteq \Delta_j$ by the choice of B_n. Thus $[B]^2 \subseteq \Delta_j$, so also in this case there is B in $[X]^{\omega^{\alpha(j)}}$ with $[B]^2 \subseteq \Delta_j$. Thus the relation (1) holds. This completes the proof.

We shall later make use of the following lemma to construct large almost homogeneous sets from smaller ones.

Lemma 7.4.7. *For each n with $n < \omega$ let X_n be an almost homogeneous subset of ω_1. Suppose $X_m < X_n$ whenever $m < n$, and suppose there are integers $i(m)$ (with $i(m) < s$) such that*

$$m < n \Rightarrow X_m \otimes X_n \subseteq \Delta_{i(m)}.$$

Then $\bigcup\{X_n; n < \omega\}$ is almost homogeneous.

Proof. Take any subset A of $\bigcup\{X_n; n < \omega\}$ with $\text{tp}(A) = \omega^\alpha$. Take any β with $\beta < \alpha$. If $A \subseteq X_0 \cup \ldots \cup X_m$ for some m, then by Theorem 7.1.3 there must be k with $k \leqslant m$ such that $\text{tp}(A \cap X_k) = \omega^\alpha$. Since X_k is almost homogeneous this gives B in $[A \cap X_k]^{\omega^\beta}$ and A' in $[A \cap X_k]^{\omega^\alpha}$ such that $B < A'$ and $B \otimes A' \subseteq \Delta_i$ for some i. On the other hand, if A is cofinal in $\bigcup\{X_n; n < \omega\}$ then $\text{tp}(A \cap X_m) \geqslant \omega^\beta$ for some m. Then if $B \in [A \cap X_m]^{\omega^\beta}$ and $A' = A - (X_0 \cup \ldots \cup X_m)$ (so $\text{tp}(A') = \omega^\alpha$) it follows that $B \otimes A' \subseteq \Delta_{i(m)}$. Thus $\bigcup\{X_n; n < \omega\}$ is indeed almost homogeneous.

The construction of almost homogeneous sets will depend on refining pairs $\langle A, X \rangle$ of subsets of ω_1 (with X uncountable) such that $A \otimes X$ is "almost contained" in Δ_i for some i. We shall write $\boldsymbol{G}(i)$ for the set of pairs $\langle A, X \rangle$ where A is in $[\omega_1]^{\omega^\alpha}$ for some α, X is in $[\omega_1]^{\aleph_1}$, $A < X$ and $A \otimes X$ is "almost contained" in Δ_i. "Almost contained" is to have the sense that no matter what subset A' of A of the same order type as A or what uncountable subset X' of X are chosen, for every β smaller than α there is a subset A'' in $[A']^{\omega^\beta}$, not cofinal in A', and an uncountable subset X'' of X' such that $A'' \otimes X''$ is "almost contained" in Δ_i. This leads to the following inductive definition of $\boldsymbol{G}(i)$.

The relation $\langle A, X \rangle \in \boldsymbol{G}(i)$ where $i < s$, $A \in [\omega_1]^{\omega^\alpha}$ and $X \in [\omega_1]^{\aleph_1}$ is defined by induction on α. If $\alpha = 0$ (so $|A| = 1$) then

$$\langle A, X\rangle \in G(i) \text{ if and only if } A < X \text{ and } A \otimes X \subseteq \Delta_i .$$

If $\alpha > 0$, supposing the relation $\langle A', X'\rangle \in G(i)$ has already been defined whenever $\text{tp}(A') = \omega^\beta$ where $\beta < \alpha$, then

> $\langle A, X\rangle \in G(i)$ if and only if $A < X$ and whenever $A' \in [A]^{\omega^\alpha}$ and $X' \in [X]^{\aleph_1}$, for every β with $\beta < \alpha$ there are A'' in $[A']^{\omega^\beta}$ not cofinal in A' and X'' in $[X']^{\aleph_1}$ such that $\langle A'', X''\rangle \in G(i)$.

It is clear from the definition that if $\langle A, X\rangle \in G(i)$ and $X' \in [X]^{\aleph_1}$, $A' \subseteq A$ with $\text{tp}(A') = \text{tp}(A)$, then $\langle A', X'\rangle \in G(i)$.

Lemma 7.4.8. *Given A in $[\omega_1]^{\omega^\alpha}$ and X in $[\omega_1]^{\aleph_1}$ then there are A' in $[A]^{\omega^\alpha}$ and X' in $[X]^{\aleph_1}$ such that $\langle A', X'\rangle \in G(i)$, for some i with $i < s$.*

Proof. Since $|A| \leqslant \aleph_0$ and $|X| = \aleph_1$, we may suppose $A < X$. If $\alpha = 0$, for i with $i < s$ put $X_i = \{x \in X; A \otimes \{x\} \subseteq \Delta_i\}$. For some i we have $|X_i| = \aleph_1$; then $\langle A, X_i\rangle \in G(i)$.

Now suppose $\alpha > 0$, and suppose inductively that the lemma is true whenever $\text{tp}(A) = \omega^\beta$ with $\beta < \alpha$. Suppose the lemma is false for particular A_0 in $[\omega_1]^{\omega^\alpha}$ and X_0 in $[\omega_1]^{\aleph_1}$; thus

$$\forall i < s \ \forall A' \in [A_0]^{\omega^\alpha} \ \forall X' \in [X_0]^{\aleph_1} (\langle A', X'\rangle \notin G(i)) . \tag{1}$$

Inductively define chains $A_0 \supseteq A_1 \supseteq \ldots \supseteq A_s$ from $[A_0]^{\omega^\alpha}$ and $X_0 \supseteq X_1 \supseteq \ldots \supseteq X_s$ from $[X_0]^{\aleph_1}$, so by (1) $\langle A_j, X_j\rangle \notin G(j)$ whenever $j < s$, as follows. Since $\langle A_j, X_j\rangle \notin G(j)$, by the definition of $G(j)$ there are A_{j+1} in $[A_j]^{\omega^\alpha}$, X_{j+1} in $[X_j]^{\aleph_1}$ and $\beta(j) < \alpha$ such that

$$\forall A'' \in [A_{j+1}]^{\omega^{\beta(j)}} \ \forall X'' \in [X_{j+1}]^{\aleph_1} (A'' \text{ not cofinal in } A_{j+1} \Rightarrow \langle A'', X''\rangle \notin G(j)) . \tag{2}$$

Let $\beta = \max\{\beta(j); j < s\}$. Choose B from $[A_s]^{\omega^\beta}$ not cofinal in A_s. By the inductive hypothesis applied to B and X_s, there are B' in $[B]^{\omega^\beta}$ and X' in $[X_s]^{\aleph_1}$ such that $\langle B', X'\rangle \in G(j)$ for some j. If $\beta(j) = \beta$, since $X' \subseteq X_s \subseteq X_{j+1}$ and $B' \subseteq B \subseteq A_s \subseteq A_{j+1}$ with B' not cofinal in A_{j+1}, this contradicts (2). If $\beta(j) < \beta$ by definition of $G(j)$ there are B'' in $[B']^{\omega^{\beta(j)}}$ not cofinal in B' and X'' in $[X']^{\aleph_1}$ such that $\langle B'', X''\rangle \in G(j)$. Now $X'' \subseteq X_{j+1}$, $B'' \subseteq A_{j+1}$ and B'' is not cofinal in A_{j+1} so again (2) is violated. In either event a contradiction has been reached, and thus the lemma cannot be false for A_0 and X_0. This completes the proof.

Lemma 7.4.9. *Let* $\langle A, X\rangle \in G(i)$. *Then there are* x *in* X *and a subset* A' *of* A *with* $\mathrm{tp}(A') = \mathrm{tp}(A)$ *such that* $A' \otimes \{x\} \subseteq \Delta_i$.

Proof. Let A and X be given with $\langle A, X\rangle \in G(i)$, and suppose $\mathrm{tp}(A) = \omega^\alpha$. If $\alpha = 0$, the result in trivial, so suppose $\alpha > 0$. Make the inductive assumption that the lemma is true for all pairs $\langle A_0, X_0\rangle$ where $\mathrm{tp}(A_0) < \mathrm{tp}(A)$. For a in A and β with $\beta < \alpha$ define

$$X(a, \beta) = \{x \in X;\ \nexists\ B \in [A]^{\omega\beta}\ (a < B, B \text{ is not cofinal in } A, \text{ and } B \otimes \{x\} \subseteq \Delta_i)\}\ .$$

Put $X^* = X - \bigcup\{X(a, \beta); a \in A \text{ and } \beta < \alpha\}$.

We shall show that $X^* \neq \emptyset$. Suppose on the contrary that $X^* = \emptyset$. Then $|X(a, \beta)| = \aleph_1$ for some a and β. Let $A' = \{a' \in A; a < a'\}$, then it follows from Theorem 7.1.3 that $\mathrm{tp}(A') = \omega^\alpha$. Since $\langle A, X\rangle \in G(i)$, there must be X'' in $[X(a, \beta)]^{\aleph_1}$ and A'' in $[A']^{\omega\beta}$ not cofinal in A' such that $A'' \otimes X'' \in G(i)$. Thus by the inductive hypothesis applied to A'' and X'', there are x in X'' and B in $[A'']^{\omega\beta}$ such that $B \otimes \{x\} \subseteq \Delta_i$. However then $a < B$ and B is not cofinal in A, so $x \notin X(a, \beta)$. This contradicts $x \in X'' \subseteq X(a, \beta)$. Hence $X^* \neq \emptyset$

Now choose $x \in X^*$. Take ordinals $\alpha(n)$ where $n < \omega$ such that always $\alpha(n) < \alpha$ and $\omega^\alpha = \Sigma_0(\omega^{\alpha(n)}; n < \omega)$. Then $x \notin X(a, \alpha(n))$ for all a in A and all n. Thus we can choose inductively sets B_n in $[A]^{\omega\alpha(n)}$ but not cofinal in A, for n with $n < \omega$, so that $\sup B_{n-1} < B_n$ and $B_n \otimes \{x\} \subseteq \Delta_i$. Put $A' = \bigcup\{B_n; n < \omega\}$; then $A' \in [A]^{\omega\alpha}$ and $A' \otimes \{x\} \subseteq \Delta_i$. This proves the lemma.

The next lemma will serve as the starting point for an inductive construction of an almost homogeneous set.

Lemma 7.4.10. *For* n *with* $n < \omega$ *let* $\langle A_n, X\rangle \in G(i_n)$. *Then there are* x *in* X *and subsets* A'_n *of* A_n *with* $\mathrm{tp}(A'_n) = \mathrm{tp}(A_n)$ *such that* $A'_n \otimes \{x\} \subseteq \Delta_{i_n}$ *for each* n.

Proof. Suppose given X and the A_n such that $\langle A_n, X\rangle \in G(i_n)$. Put $X_n = \{x \in X; \nexists\ A'_n \subseteq A_n (\mathrm{tp}(A'_n) = \mathrm{tp}(A_n) \text{ and } A'_n \otimes \{x\} \subseteq \Delta_{i_n})\}$. If $X = \bigcup\{X_n; n < \omega\}$ then there is some X_n with $|X_n| = \aleph_1$. Thus $\langle A_n, X_n\rangle \in G(i_n)$, so by Lemma 7.4.9 there is x in X_n and A' contained in A_n with $\mathrm{tp}(A') = \mathrm{tp}(A_n)$ such that $A'_n \otimes \{x\} \subseteq \Delta_{i_n}$; this contradicts $x \in X_n$. Thus $X \neq \bigcup\{X_n; n < \omega\}$, and so the lemma holds..

For the remainder of this section, for each non-zero α fix upon a sequence of ordinals $\langle \alpha(n); n < \omega\rangle$ such that always $\alpha(n) \leqslant \alpha(n+1) < \alpha$ and $\omega^\alpha =$

$\Sigma_0(\omega^{\alpha(n)}; n < \omega)$. We shall eventually construct almost homogeneous sets of order type ω^β for arbitrarily large β by induction on β. We shall do this by finding for each n almost homogeneous sets D_n with $\mathrm{tp}(D_n) = \omega^{\beta(n)}$ such that there are subsets B_n in $[D_n]^{\omega^{\beta(n)}}$ which fit together to give an almost homogeneous set of type ω^β. In order to do this, we shall need a way of choosing, ω times, elements in each D_n in such a way that the set of those elements which are chosen at every step constitutes a subset of D_n of type $\omega^{\beta(n)}$.

We now look at a method of making such choices. In general, let a set A in $[\omega_1]^{\omega^\alpha}$ be given. Let also a finite family $\mathcal{A}$ of subsets of ω_1 each of order type an ordinal power of ω be given. For each integer k, define operations S_k on such sets A and such finite families $\mathcal{A}$ as follows: if $\alpha = 0$, then $S_k(A) = \{\{A\}\}$, if $\alpha > 0$, then

$$S_k(A) = \{\{B, C\}; B \in [A]^{\omega^{\alpha(k)}}, C \in [A]^{\omega^\alpha} \text{ and } B < C\} ;$$

if $\mathcal{A} = \{A_0, \ldots A_m\}$ then

$$S_k(\mathcal{A}) = \{\{B_0, C_0, \ldots, B_m, C_m\}; \{B_i, C_i\} \in S_k(A_i) \text{ for } i = 0, \ldots, m\}$$

with the understanding that if $\mathrm{tp}(A_i) = \omega^0$ then $B_i = C_i = A_i$. Thus $S_k(\{A\}) = S_k(A)$. If $\mathcal{A}$ is a finite pairwise disjoint family then so is any member of $S_k(\mathcal{A})$.

Given A from $[\omega_1]^{\omega^\alpha}$, suppose we take a sequence of families $\langle \mathcal{A}_n; n < \omega \rangle$ such that $\mathcal{A}_0 = \{A\}$ and there is some integer k so that always

$$\mathcal{A}_{n+1} \in S_{k+n}(\mathcal{A}_n) \text{ when } n \geq 0 .$$

Thus $\mathcal{A}_1$ provides two disjoint subsets of A, one small, one large; $\mathcal{A}_2$ provides similarly two subsets of each of these; and so on. If we put $A_n = \bigcup \mathcal{A}_n$, then A_n is to be the set of elements of A chosen at step n. Thus $\bigcap \{A_n; n < \omega\}$ is the set of elements chosen at every step.

Lemma 7.4.11. *In the above situation,* $\bigcap \{A_n; n < \omega\} \in [A]^{\omega^\alpha}$.

Proof. By induction on α, where $\mathrm{tp}(A) = \omega^\alpha$. If $\alpha = 0$ then for every k and n we have $S_{k+n}(A) = \{\{A\}\}$ so $\bigcap \{A_n; n < \omega\} = A \in [A]^{\omega^0}$. So now suppose that $\alpha > 0$ and make the inductive assumption that for any β with $\beta < \alpha$, given a sequence $\langle \mathcal{B}_n; n < \omega \rangle$ where $\mathcal{B}_0 = \{B\}$ and for all n, $\mathcal{B}_{n+1} \in S_{l+n}(\mathcal{B}_n)$ for some finite l, where $\mathrm{tp}(B) = \omega^\beta$, then $\mathrm{tp}(\bigcap \{\bigcup \mathcal{B}_n; n < \omega\}) = \omega^\beta$.

Now take the sequence $\langle \mathcal{A}_n; n < \omega \rangle$ described above, with $\mathcal{A} = \{A\}$ where $\mathrm{tp}(A) = \omega^\alpha$. Inductively choose members B_n and C_n of $\mathcal{A}_{n+1}$ with $\mathrm{tp}(C_n) = \omega^\alpha$ as follows. Since $\mathcal{A}_1 \in S_k(A)$ there are B_0 in $[A]^{\omega^{\alpha(k)}}$ and C_0 in $[A]^{\omega^\alpha}$ such that $B_0 < C_0$ and $\{B_0, C_0\} \subseteq \mathcal{A}_1$. Supposing $C_n \in \mathcal{A}_{n+1}$, since $\mathcal{A}_{n+2} \in S_{k+n+1}(\mathcal{A}_{n+1})$ there are subsets B_{n+1} of C_n with $\mathrm{tp}(B_{n+1}) = \omega^{\alpha(k+n+1)}$ and $\mathrm{tp}(C_{n+1}) = \omega^{\alpha}$ such that $B_{n+1} < C_{n+1}$ and $\{B_{n+1}, C_{n+1}\} \subseteq \mathcal{A}_{n+2}$. Thus always $B_n < B_{n+1}$.

Consider now the set B_0. There is one subset of B_0 in $\mathcal{A}_1$, namely B_0 itself. In $\mathcal{A}_2$ there are two subsets of B_0 and together they give a member, say $\mathcal{B}_0^*$, of $S_{k+1}(\mathcal{B}_0)$. In $\mathcal{A}_3$, these get divided to give a member of $S_{k+2}(\mathcal{B}_0^*)$, and so on. In general, if we put $\mathcal{B}_{0n} = \{X \in \mathcal{A}_{n+1}; X \subseteq B_0\}$ then

$$\mathcal{B}_{00} = \{B_0\}; \text{if } n \geqslant 0 \text{ then } \mathcal{B}_{0n+1} \in S_{(k+1)+n}(\mathcal{B}_{0n}) .$$

Since $\mathrm{tp}(B_0) = \omega^{\alpha(k)}$, by the inductive hypothesis, if

$$B_0' = \bigcap \{\bigcup \mathcal{B}_{0n}; n < \omega\}$$

then $B_0' \in [B_0]^{\omega^{\alpha(k)}}$. We can repeat this with $B_1, B_2, \ldots$. For any m, put $\mathcal{B}_{mn} = \{X \in \mathcal{A}_{n+m+1}; X \subseteq B_m\}$. Then

$$\mathcal{B}_{mn} = \{B_m\}; B_{mn+1} \in S_{(k+m+1)+n}(\mathcal{B}_{mn}) .$$

So by the inductive hypothesis, if

$$B_m' = \bigcap \{\bigcup \mathcal{B}_{mn}; n < \omega\}$$

then $B_m' \in [B_m]^{\omega^{\alpha(k+m)}}$. Now for any n we have

$$B_m' \subseteq \bigcup \mathcal{B}_{mn} \subseteq \bigcup \mathcal{A}_{m+n+1} = A_{m+n+1} ,$$

and if $l \leqslant m$ then

$$B_m' \subseteq B_m \subseteq C_{m-1} \subseteq C_{l-1} \subseteq \bigcup \mathcal{A}_l = A_l .$$

Hence always $B_m' \subseteq \bigcap \{A_l; l < \omega\}$. Thus if $B = \bigcup \{B_m'; m < \omega\}$ then $B \subseteq \bigcap \{A_l; l < \omega\}$ and $\mathrm{tp}(B) = \Sigma_0(\omega^{\alpha(k+m)}; m < \omega) = \omega^{\alpha}$. This proves Lemma 7.4.11.

Lemma 7.4.12. *Take any finite pairwise disjoint family* $\mathcal{A}$ *and any uncountable set* X *such that for all* A *in* $\mathcal{A}$ *there is* i_A *with* $i_A < s$ *such that* $\langle A, X\rangle \in G(i_A)$. *Then for any integer* k *there are a family* $\mathcal{B}$ *in* $S_k(\mathcal{A})$ *and a set* X' *in* $[X]^{\aleph_1}$ *such that for all* B *in* $\mathcal{B}$,

$$B \subseteq A \in \mathcal{A} \Rightarrow \langle B, X'\rangle \in G(i_A) .$$

Proof. Write $\mathcal{A} = \{A_0, \ldots, A_m\}$. Suppose $\mathrm{tp}(A_j) = \omega^{\alpha_j}$ and $\langle A_j, X\rangle \in G(i_j)$. Take subsets A_j' and A_j^* of A_j with $A_j' < A_j^*$ such that $\mathrm{tp}(A_j') = \omega^{\alpha_j(k)}$ and $\mathrm{tp}(A^*) = \omega^{\alpha_j}$ (but if $\alpha_j = 0$, put $A_j' = A_j^* = A_j$). Choose sets X_j from $[X]^{\aleph_1}$

for j with $j \leqslant m+1$ so that $X_0 = X$ and X_{j+1} is in $[X_j]^{\aleph_1}$ such that there is a subset A_j'' of A_j' with $\mathrm{tp}(A_j'') = \omega^{\alpha_j(k)}$ for which $\langle A_j'', X_{j+1}\rangle \in G(i_j)$. This choice is possible since $\langle A_j, X\rangle \in G(i_j)$. Put $X' = \bigcap\{X_j; j \leqslant m+1\}$ so $X' \in [X]^{\aleph_1}$ and $\langle A_j'', X'\rangle \in G(i_j)$ for each j. Also $\langle A_j^*, X'\rangle \in G(i_j)$ since $\mathrm{tp}(A_j^*) = \mathrm{tp}(A_j)$. Now $\{A_j'', A_j^*\} \in S_k(A_j)$ so if we put $\mathcal{B} = \{A_0'', A_0^*, ..., A_m'', A_m^*\}$ then $\mathcal{B} \in S_k(\mathcal{A})$ and $\mathcal{B}$ with X' has the required property.

In the next lemma we shall finally construct the almost homogeneous set of order type ω^β for arbitrary β. The more complicated statement in the lemma is needed for the inductive construction.

Lemma 7.4.13. *Let β be given. Take any set X in $[\omega_1]^{\aleph_1}$ and any finite pairwise disjoint family $\mathcal{A} = \{A_0, ..., A_k\}$ with $\langle A_j, X\rangle \in G(i_j)$ (where $j = 0, ..., k$). Then there are B in $[X]^{\omega^\beta}$ and subsets A_j' of A_j with $\mathrm{tp}(A_j') = \mathrm{tp}(A_j)$ such that B is almost homogeneous and $A_j' \otimes B \subseteq \Delta_{i_j}$ (where $j = 0, ..., k$).*

Proof. The plan of the proof is this. We shall construct the almost homogeneous set B by finding almost homogeneous sets B_n with $\mathrm{tp}(B_n) = \omega^{\beta(n)}$ and $B_m < B_n$ whenever $m < n$ such that there is $i(m)$ with $i(m) < s$ for which $B_m \otimes B_n \subseteq \Delta_{i(m)}$ whenever $m < n$; by Lemma 7.4.7 this will suffice. At stage n in the contruction we shall introduce a set D_n of order type $\omega^{\beta(n)}$ from which the elements of B_n will be drawn. When finding D_n, we shall choose elements of all the earlier D_m and of the A_j, say in sets D_m^* and A_j^*, such that $D_m^* \otimes D_n \subseteq \Delta_{i(m)}$ and $A_j^* \otimes D_n \subseteq \Delta_{i_j}$. Those elements that are chosen at every stage will form the B_m and A_j'.

The proof is by induction on β. The case $\beta = 0$ is given by Lemma 7.4.10. So suppose that $\beta > 0$, and make the inductive assumption that the statement of the lemma is true for any γ with $\gamma < \beta$.

Put $\mathcal{A}_0 = \mathcal{A}$ and $X_0 = X$. By Lemma 7.4.12 we have a family $\mathcal{C}_0$ in $S_0(\mathcal{A}_0)$ and a set X_0' in $[X_0]^{\aleph_1}$ such that for all C in $\mathcal{C}_0$ if $C \subseteq A_j$ then $\langle C, X_0'\rangle \in G(i_j)$. Since $\beta(0) < \beta$ by the inductive hypothesis applied to $\mathcal{C}_0$ and X_0' there are an almost homogeneous set D_0' in $[X_0']^{\omega^{\beta(0)}}$ and subsets C' of C for C in $\mathcal{C}_0$ with $\mathrm{tp}(C') = \mathrm{tp}(C)$ such that $C' \otimes D_0' \subseteq \Delta_{i_j}$ for that i_j for which $\langle C, X_0'\rangle \in G_{i_j}$. Note that also $\{C'; C \in \mathcal{C}_0\} \in S_0(\mathcal{A}_0)$. By Lemma 7.4.8 there are D_0 in $[D_0']^{\omega^{\beta(0)}}$ and X_1 in $[X_0]^{\aleph_1}$ such that $\langle D_0, X_1\rangle \in G(i(0))$ for some $i(0)$ with $i(0) < s$. Then D_0 is also almost homogeneous. Put $\mathcal{A}_1 = \{C'; C \in \mathcal{B}_0\} \cup \{D_0\}$. Since $D_0 \subseteq X_0$ and necessarily $A_j < X_0$ for each j, it follows that $\mathcal{A}_1$ is a finite pairwise disjoint family. And further, for each A in $\mathcal{A}_1$ there is $i(A)$ such that $\langle A, X_1\rangle \in G(i(A))$. Thus we can repeat the process.

This leads to the following inductive construction of finite pairwise disjoint families $\mathcal{A}_n$, uncountable sets X_n, and almost homogeneous sets D_n in $[X_n]^{\omega^{\beta(n)}}$. Suppose that $\mathcal{A}_n$ and X_n have already been constructed, such that for all A in $\mathcal{A}_n$ there is $i(A)$ with $i(A) < s$ for which $\langle A, X_n \rangle \in G(i(A))$. By Lemma 7.4.12 there are a family $\mathcal{C}_n$ in $S_n(A_n)$ and a set X'_n in $[X_n]^{\aleph_1}$ such that for each C in $\mathcal{C}_n$, if $C \subseteq A$ with $A \in \mathcal{A}_n$ then $\langle C, X'_n \rangle \in G(i(A))$. Noting that $\beta(n) < \beta$, by the inductive hypothesis applied to $\mathcal{C}_n$ and X'_n, we can find an almost homogeneous set D'_n in $[X'_n]^{\omega\beta(n)}$ and subsets C' of each C in $\mathcal{C}_n$ with $\mathrm{tp}(C') = \mathrm{tp}(C)$ and $C' \otimes D'_n \subseteq \Delta_{i(A)}$ for that A in $\mathcal{A}_n$ where $C' \subseteq C \subseteq A$. We mention that $\{C'; C \in \mathcal{C}_n\} \in S_n(\mathcal{A}_n)$. Using Lemma 7.4.8 gives D_n in $[D'_n]^{\omega\beta(n)}$ and X_{n+1} in $[X_n]^{\aleph_1}$ such that $\langle D_n, X_{n+1} \rangle \in G(i(n))$ for some $i(n)$. Then D_n, being a subset of D'_n, is almost homogeneous. Also $C' < D_n$ for every C', since $C' \subseteq C$, $D_n \subseteq X_n$ and $C < X_n$ (for each C in $\mathcal{C}_n$). Put $\mathcal{A}_{n+1} = \{C'; C \in \mathcal{C}_n\} \cup \{D_n\}$, so $\mathcal{A}_{n+1}$ is a finite pairwise disjoint family. And for each A in $\mathcal{A}_{n+1}$, there is $i(A)$ such that $\langle A, X_{n+1} \rangle \in G(i(A))$.

Now take any one of the original sets A_j (so $j = 0, \ldots, k$). In each of the families $\mathcal{A}_n$ consider those members which are subsets of A_j. As in Lemma 7.4.11, this gives a sequence of subfamilies $\langle \mathcal{A}_{jn}; n < \omega \rangle$ where

$$A_{j0} = \{A_j\}; A_{jn+1} \in S_n(\mathcal{A}_{jn}) .$$

Hence by Lemma 7.4.11, if $A'_j = \bigcap\{\bigcup \mathcal{A}_{jn}; n < \omega\}$ then $A'_j \subseteq A_j$ and $\mathrm{tp}(A'_j) = \mathrm{tp}(A_j)$.

Take any D_m (where $m < \omega$), and in each $\mathcal{A}_n$ consider the subsets of D_m. There are none in $\mathcal{A}_0, \ldots, \mathcal{A}_m$; in $\mathcal{A}_{m+1}$ there is D_m alone; in $\mathcal{A}_{m+2}$ there are two giving a member of $S_{m+2}(D_m)$; in A_{m+3} these are further divided, and so on. Thus there is a sequence $\langle \mathcal{D}_{mn}; n < \omega \rangle$ with $\mathcal{D}_{mn} \subseteq \mathcal{A}_{m+n+1}$ such that

$$\mathcal{D}_{m0} = \{D_m\}; \mathcal{D}_{mn+1} \in S_{(m+1)+n}(\mathcal{D}_{mn}) .$$

Thus by Lemma 7.4.11, if $B_m = \bigcap\{\bigcup \mathcal{D}_{mn}; n < \omega\}$ then $B_m \subseteq D_m$ and $\mathrm{tp}(B_m) = \mathrm{tp}(D_m) = \omega^{\beta(m)}$.

Then each B_m is an almost homogeneous subset of X. And if $m < n$ then $B_m < B_n$ since $B_m \subseteq D_m$, $B_n \subseteq D_n \subseteq X_n$ and $D_m < X_n$. Further, if $m < n$ then $B_m \otimes B_n \subseteq \Delta_{i(m)}$ since $D_m \otimes X_{m+1} \subseteq \Delta_{i(m)}$ by choice of D_m and X_{m+1}, and $B_m \subseteq D_m$, $X_n \subseteq X_{m+1}$. Put $B = \bigcup\{B_m; m < \omega\}$. By Lemma 7.4.7, B is almost homogeneous. And $\mathrm{tp}(B) = \Sigma_0(\omega^{\beta(m)}; m < \omega) = \omega^{\beta}$.

Finally, we show $A'_j \otimes B \subseteq \Delta_{i_j}$ if $j = 0, \ldots, k$. Given A_j, for any integer n let $\mathcal{A}_{jn} = \{A \in \mathcal{A}_n; A \subseteq A_j\}$ so $\bigcup \mathcal{A}_{jn}$ gives the set of members of A_j chosen at the n-th step. An easy induction on n shows that $\langle A, X'_n \rangle \in G(i_j)$ for every A in $\mathcal{A}_{jn}$. Then the choice of D'_n and $\mathcal{A}_{n+1}$ ensures that $A' \otimes D'_n \subseteq \Delta_{i_j}$ for every A' in $\mathcal{A}_{jn+1}$, so $(\bigcup \mathcal{A}_{jn+1}) \otimes D'_n \subseteq \Delta_{i_j}$. Since $A'_j \subseteq \bigcup \mathcal{A}_{jn+1}$ and

$B_n \subseteq D'_n$, thus $A'_j \otimes B_n \subseteq \Delta_{i_j}$; hence $A'_j \otimes B \subseteq \Delta_{i_j}$. This completes the proof of the lemma.

Lemma 7.4.13 contains the last step necessary for the proof of Theorem 7.4.3.

Proof of Theorem 7.4.3. Let the countable ordinal α be given, and take the partition $[\omega_1]^2 = \Delta_0 \cup ... \cup \Delta_s$. By Lemma 7.4.6 there is β such that $\beta \to AH(\alpha, ..., \alpha)$. From Lemma 7.4.13 there is B in $[\omega_1]^{\omega^\beta}$ such that B is almost homogeneous for this partition. So by Definition 7.4.5 of the symbol $\beta \to AH(\alpha, ..., \alpha)$, there is A in $[B]^{\omega^\alpha}$ such that A is homogeneous for the partition. Hence there is surely a subset of ω_1 of order type α homogeneous for the partition. Thus indeed $\omega_1 \to (\alpha)^2_s$.

APPENDIX

CARDINAL AND ORDINAL NUMBERS

§1. Set Theory

In this appendix we shall list a set of axioms for set theory and develop briefly the properties of ordinal and cardinal numbers that have been used in this book. For a more detailed treatment and the proofs that are not given here, see for example Bachmann [1] or Sierpinski [89].

Any one of the standard systems of set theory provides an adequate foundation for the material in this book (for example, the systems of Zermelo-Fraenkel (Fraenkel [43]), Bernays-Gödel (Gödel [49]) or Morse-Kelly (Kelly [60])). We shall list the axioms for the Zermelo-Fraenkel system (with the Axiom of Choice), ZFC. The system is formalized in first order predicate calculus with identity = and one other binary predicate symbol $\in$ for set membership. The axioms are as follows.

Axiom I (Extensionality).

$$\forall a \, \forall b \, [a = b \Leftrightarrow \forall x \, (x \in a \Leftrightarrow x \in b)] \, .$$

Informally, $a = b$ if and only a and b have the same members.

Axiom II (Pair set).

$$\forall a \, \forall b \, \exists \, c \, \forall x \, (x \in c \Leftrightarrow x = a \text{ or } x = b) \, .$$

Intuitively, for any two sets a, b the pair set $\{a, b\}$ exists.

Axiom III (Union set).

$$\forall a \, \exists \, c \, \forall x (x \in c \Leftrightarrow \exists y (x \in y \text{ and } y \in a)) \, .$$

If a is a set, so is $\bigcup a$, the union of the members of a.

Axiom IV (Power set).

$$\forall a\ \exists\ c\ \forall x(x \in c \Leftrightarrow \forall y(y \in x \Rightarrow y \in c)) .$$

For any set a, the power set $\mathfrak{P}a$ of a exists.

Axiom V (Replacement). For each formula $\varphi(x, y)$ (may be with further free variables) in which z, a, b do not occur:

$$\forall x\ \forall y\ \forall z\,(\varphi(x, y) \text{ and } \varphi(x, z) \Rightarrow y = z) \Rightarrow$$
$$\forall a\ \exists\ b\ \forall y(y \in b \Leftrightarrow \exists\ x(x \in a \text{ and } \varphi(x, y))) .$$

Informally, if $\varphi(x, y)$ is a function-like formula then for any set a there exists the set $\{y;\ \exists\ x(x \in a \text{ and } \varphi(x, y))\}$.

Axiom VI (Infinity).

$$\exists\ b(\exists\ x(x \in b) \text{ and } \forall y(y \in b \Rightarrow \exists\ x(x \in b \text{ and } y \in x))) .$$

Ensures the existence of an infinite set.

Axiom VII (Foundation (or Regularity)).

$$\forall a(\exists\ x(x \in a) \Rightarrow \exists\ x(x \in a \text{ and } \forall y(y \in a \Rightarrow y \notin x))) .$$

Intuitively, every non-empty set a has a *foundation member,* that is, a member disjoint from a.

Although Axiom VII is one of the standard axioms of ZFC, the results in the text do not require its use.

Axiom VIII (Choice).

$$\forall a\ \exists\ f(f \text{ is a function and } \mathrm{dom}(f) = a) \text{ and}$$
$$\forall x(x \in a \text{ and } \exists y(y \in x) \Rightarrow f(x) \in x) .$$

Every set has a choice function.

§2. Ordinal numbers

A variety of definitions of the ordinal numbers can be given, all producing the same family of sets. (Bachmann [1, p24] proves the equivalence of five such definitions.) A convenient definition is the following. A set x is said to be *transitive* if members of members of x are members of x, that is, if

$$y \in x \Rightarrow y \subseteq x .$$

A set x is said to be *well-founded* if every non-empty subset of x has a foundation member. Then we pose the following.

Definition A.2.1. A set x in an *ordinal* if x is a transitive, well-founded set all the members of which are transitive.

(If the Axiom of Foundation is assumed, it is unnecessary to require specifically that an ordinal be well-founded, since the axiom asserts that every set is well-founded.)

It is easily seen that every member of an ordinal is again an ordinal. Define the ordering $<$ between ordinals by, for ordinals α, β

$$\beta < \alpha \Leftrightarrow \beta \in \alpha .$$

Then each ordinal α is the set of all smaller ordinals,

$$\alpha = \{\beta ; \beta \text{ is an ordinal and } \beta < \alpha\} .$$

Each ordinal α is well ordered by $<$ (that is, by $<$ restricted to the members of α). Any descending sequence of ordinals must be finite, for if $\ldots < \alpha_2 < \alpha_1 < \alpha_0$ is an infinite descending sequence, by transitivity $\{\alpha_n; 1 \leqslant n < \omega\}$ is a subset of α_0, but yet it has no foundation member, contrary to the well-foundedness of α_0. If A is any non-empty set of ordinals then both $\bigcap A$ and $\bigcup A$ are ordinals; $\bigcap A$ is the least member of A and $\bigcup A$ is the supremum of A (the least ordinal α with $\beta \leqslant \alpha$ for every β in A).

If α is a non-zero ordinal such that $\alpha = \bigcup \alpha$ then α is said to be a *limit* ordinal. If α (with $\alpha \neq \emptyset$) is not a limit ordinal then there is an ordinal β such that $\alpha = \beta \cup \{\beta\}$. In this case α is said to be the *successor* of β, and one writes $\alpha = \beta + 1$.

The *finite* ordinals are defined to be those ordinals which are well ordered by $>$, and one puts $0 = \emptyset$, $1 = 0 + 1$, $2 = 1 + 1$, $3 = 2 + 1$, etc. Let ω be the set of all finite ordinals; then ω is an ordinal and in fact ω is the smallest limit ordinal.

There are the following methods for giving a proof by transfinite induction over the ordinals. Let $\varphi(\alpha)$ be a formula (maybe with other free variables).

Theorem A.2.2 (Transfinite induction, first form). *Suppose that for all ordinals β, if $\forall \gamma < \beta (\varphi(\gamma))$ then $\varphi(\beta)$. Then $\varphi(\alpha)$ holds for all ordinals α.*

Theorem A.2.3 (Transfinite induction, second form). *Suppose that*

(i) $\varphi(0)$,

(ii) $\varphi(\alpha) \Rightarrow \varphi(\alpha + 1)$,

(iii) *if* α *is a limit ordinal then* $\forall \beta < \alpha (\varphi(\beta)) \Rightarrow \varphi(\alpha)$.

Then $\varphi(\alpha)$ *holds for all ordinals* α.

Both theorems are proved by considering the least counter-example to $\varphi(\alpha)$, should there be one.

Corresponding to the two forms of proof by induction, there are the following methods of giving a definition by transfinite recursion over the ordinals. (There are more general forms of definition by transfinite recursion; however the two given below suffice to justify the uses of transfinite recursion made in the text.) Let $\varphi(x, y)$ be a function-like formula (maybe with further free variables).

Theorem A.2.4 (Transfinite recursion, first form). *Given the ordinal* α, *there is a unique function* f *with domain* α *such that for all* β *with* $\beta < \alpha$,

$$f(\beta) = y \text{ for that } y \text{ such that } \varphi(\{f(\gamma); \gamma < \beta\}, y) .$$

Theorem A.2.5 (Transfinite recursion, second form). *Given the ordinal* α *and a set* u, *there is a unique function* f *with domain* α *such that for all* β *with* $\beta + 1 < \alpha$ *and all limit ordinals* γ *with* $\gamma < \alpha$,

(i) $f(0) = u$,

(ii) $f(\beta + 1) = y$ *for that* y *such that* $\varphi(f(\beta), y)$,

(iii) $f(\gamma) = \bigcup\{f(\beta); \beta < \gamma\}$.

The operations of ordinal addition, multiplication and exponentiation are defined by recursion as follows:

$$\alpha + 0 = \alpha, \alpha + (\beta + 1) = (\alpha + \beta) + 1 ,$$

$$\alpha + \gamma = \bigcup\{\alpha + \beta; \beta < \gamma\} \text{ if } \gamma \text{ is a limit ordinal;}$$

$$\alpha 0 = 0 , \alpha(\beta + 1) = (\alpha\beta) + \alpha ,$$

$$\alpha\gamma = \bigcup\{\alpha\beta; \beta < \gamma\} \text{ if } \gamma \text{ is a limit ordinal;}$$

$$\alpha^0 = 1 , \alpha^{\beta+1} = (\alpha^\beta)\, \alpha ,$$

$$\alpha^\gamma = \bigcup\{\alpha^\beta; \beta < \gamma\} \text{ if } \gamma \text{ is a limit ordinal.}$$

Sets with the order types $\alpha + \beta$, $\alpha\beta$, α^β can be realized as follows. Take sets a, b ordered by $<_a$, $<_b$ with order types α, β respectively. For addition, suppose $a \cap b = \emptyset$ and order $a \cup b$ as follows:

$$x <_1 y \Leftrightarrow (x, y \in a \text{ and } x <_a y) \text{ or } (x, y \in b \text{ and } x <_b y)$$
$$\text{or } (x \in a \text{ and } y \in b)\,.$$

Then $\mathrm{tp}(a \cup b, <_1) = \alpha + \beta$. For multiplication, order $a \times b$ by the reverse lexicographic ordering, that is

$$\langle u, x\rangle <_2 \langle v, y\rangle \Leftrightarrow (x <_b y) \text{ or } (x = y \text{ and } u <_a v)\,.$$

Then $\mathrm{tp}(a \times b, <_2) = \alpha\beta$. For exponentiation, let $Z(a, b)$ be the set of all functions $f : b \to a$ such that $f(x)$ is non-zero for only finitely many values of x. Order $Z(a, b)$ by last differences, that is, for f, g from $Z(a, b)$ if y is the greatest element in b such that $f(y) \neq g(y)$ then

$$f <_3 g \Leftrightarrow f(y) <_a g(y)\,.$$

Then $\mathrm{tp}(Z(a, b), <_3) = \alpha^\beta$.

The basic properties of the ordinal operations will not be developed here. We mention however the following result.

Theorem A.2.6. *Every ordinal ξ has a representation in the form*

$$\xi = \omega^{\beta_1} n_1 + \omega^{\beta_2} n_2 + \ldots + \omega^{\beta_m} n_m$$

where $\beta_1 > \beta_2 > \ldots > \beta_m$ and $m, n_1, n_2, \ldots, n_m$ are finite.

Proof. Let ξ be given. There is a least ordinal α such that $\xi < \omega^\alpha$; then α is a successor ordinal, say $\alpha = \beta + 1$. Thus

$$\omega^\beta \leqslant \xi < \omega^{\beta+1} = \omega^\beta\omega = \bigcup\{\omega^\beta n; n < \omega\}\,,$$

and so there is n with $n < \omega$ for which $\omega^\beta n \leqslant \xi < \omega^\beta(n + 1)$. Thus

$$\omega^\beta n \leqslant \xi < \omega^\beta(n + 1) = \omega^\beta n + \omega^\beta = \bigcup\{\omega^\beta n + \gamma; \gamma < \omega^\beta\}\,,$$

and so there is γ with $\gamma < \omega^\beta$ for which

$$\xi = \omega^\beta n + \gamma\,.$$

Now repeat the process on γ; there are β_2, n_2 with $n_2 < \omega$ and γ_2 with $\gamma_2 < \omega^{\beta_2}$ for which

$$\gamma = \omega^{\beta_2} n_2 + \gamma_2\,,$$

and since $\gamma < \omega^\beta$ and $\omega^{\beta_2} \leqslant \gamma$ necessarily $\beta > \beta_2$. Repeat with γ_2, and continue in this way. For some finite m we must have $\gamma_m = 0$, for otherwise

$\ldots < \beta_3 < \beta_2 < \beta$ would be infinite descending sequence of ordinals. Hence

$$\xi = \omega^{\beta} n + \omega^{\beta_2} n_2 + \ldots + \omega^{\beta_m} n_m \;.$$

In fact the representation of the ordinal ξ given by Theorem A.2.6. is unique; this is called the *Cantor normal form* for the ordinal ξ.

The operations of addition and multiplication can be extended without difficulty to infinitary operations acting on a well ordered sequence of ordinals. The infinite sum $\Sigma_0(\xi_\alpha; \alpha < \beta)$ and infinite product $\Pi_0(\xi_\alpha; \alpha < \beta)$ of the sequence $\langle \xi_\alpha; \alpha < \beta \rangle$ are defined by recursion on the length β of the sequence as follows:

$$\Sigma_0(\xi_\alpha; \alpha < 0) = 0 \;,\; \Sigma_0(\xi_\alpha; \alpha < \beta + 1) = (\Sigma_0(\xi_\alpha; \alpha < \beta)) + \xi_\beta \;,$$

$$\Sigma_0(\xi_\alpha; \alpha < \gamma) = \bigcup\{\Sigma_0(\xi_\alpha; \alpha < \beta); \beta < \gamma\} \text{ if } \gamma \text{ is a limit ordinal;}$$

$$\Pi_0(\xi_\alpha; \alpha < 0) = 1 \;,\; \Pi_0(\xi_\alpha; \alpha < \beta + 1) = (\Pi_0(\xi_\alpha; \alpha < \beta))\, \xi_\beta \;,$$

$$\Pi_0(\xi_\alpha; \alpha < \gamma) = \bigcup\{\Pi_0(\xi_\alpha; \alpha < \beta); \beta < \gamma\} \text{ if } \gamma \text{ is a limit ordinal.}$$

It follows that if $\langle \alpha; \gamma < \beta \rangle$ is the sequence of length β with constant value α then $\Sigma_0(\alpha; \gamma < \beta) = \alpha\beta$ and $\Pi_0(\alpha; \gamma < \beta) = \alpha^{\beta}$.

§3. Cardinal arithmetic

Two sets a and b are said to be *similar*, written $a \approx b$, if there is a one-to-one and onto function $f : a \to b$. Define $a \preccurlyeq b$ if a is similar to some subset of b, and $a \prec b$ if $a \preccurlyeq b$ and $a \not\approx b$.

There are a number of classical results concerning the relation of similarity, provable without using the Axiom of Choice.

Theorem A.3.1 (i) (Schröder–Bernstein). *If $a \preccurlyeq b$ and $b \preccurlyeq a$ then $a \approx b$.*

(ii) (Cantor) *If a is any set then $a \prec \mathfrak{P}a$.*

(iii) (Hartogs) *For every set a there is an ordinal α such that $\alpha \not\preccurlyeq a$.*

However, the Axiom of Choice is needed if a reasonable concept of "size" is to be based on the relation $\preccurlyeq$. With Choice, it follows that $\preccurlyeq$ is a total ordering, and that there are ordinals of every possible size.

Theorem A.3.2. *The Axiom of Choice is equivalent to each of the following:*

(i) *the relation $\preccurlyeq$ is total, that is, $\forall a\, \forall b\, (a \preccurlyeq b$ or $b \preccurlyeq a)$.*

(ii) *every set is similar to some ordinal, that is, $\forall a\, \exists\, \alpha\, (a \approx \alpha)$.*

An ordinal α is said to be an *initial ordinal* if α if not similar to any smaller ordinal, that is $\beta < \alpha \Rightarrow \alpha \not\approx \beta$. The sequence $\omega_0, \omega_1, \omega_2, ..., \omega_\alpha, ...$ of the infinite initial ordinals is defined by recursion: $\omega_0 = \omega$, and if $\alpha > 0$ then

$$\omega_\alpha = \text{least initial ordinal greater than } \omega_\beta \text{ for all } \beta \text{ with } \beta < \alpha .$$

For this to be a valid definition, one must check that given the ω_β for all β with $\beta < \alpha$, there is a larger initial ordinal. Put $a = \bigcup\{\omega_\beta; \beta < \alpha\}$; by Hartog's Theorem there is an ordinal γ such that $\gamma \not\preccurlyeq a$. Whenever $\beta < \alpha$ then $\omega_\beta \subseteq a$ so $\omega_\beta \preccurlyeq a$; thus $\gamma \not\preccurlyeq \omega_\beta$ and hence $\omega_\beta \prec \gamma$. Then γ is an ordinal greater than all the ω_β and similar to none of them. The least such ordinal γ will be an initial ordinal.

The similarity relation has the properties of an equivalence relation (although the equivalence classes are "proper classes"; they do not exist as sets in the system ZFC). One seeks a representative from each equivalence class, to serve as a "number" representing that size of set – the representatives so found will be called *cardinal numbers.* Since the Axiom of Choice is assumed, by Theorem A.3.2(ii) there are ordinals in each similarity class and hence exactly one initial ordinal, so this initial ordinal is a good choice. Thus we make the definition: κ is a *cardinal number* if and only if κ is an initial ordinal. When the infinite initial ordinal ω_α is used as the representative of the similarity class and ignoring its structure as an ordinal, it is usual to denote it by $\aleph_\alpha$ rather than ω_α, so we define: $\aleph_\alpha = \omega_\alpha$. Then the sequence of cardinal numbers is

$$0, 1, 2, ..., \aleph_0, \aleph_1, \aleph_2, ..., \aleph_\alpha, ...$$

Each set is similar to exactly one cardinal number; thus we define the *cardinality,* $|x|$, of a set x by:

$$|x| = \text{that cardinal number } \kappa \text{ such that } x \approx \kappa .$$

The operations of cardinal addition, multiplication and exponentiation are defined as follows. Given cardinal numbers κ, λ take any sets a, b with $|a| = \kappa$, $|b| = \lambda$ and $a \cap b = \emptyset$ for (i), and define:

(i) $\kappa \dot{+} \lambda = |a \cup b|$,

(ii) $\kappa \cdot \lambda = |a \times b|$,

(iii) $\kappa^\lambda = |{}^b a|$,

where ${}^b a = \{f; f : b \to a\}$, the set of all functions mapping from b into a. These definitions do not depend on the choice of the sets a and b.

Theorem A.3.3. *If κ is an infinite cardinal then $\kappa^2 = \kappa$.*

Proof. Suppose $\kappa = \aleph_\alpha$, and use transfinite induction on α. The case $\alpha = 0$ is easy, so we assume $\alpha > 0$. Make the inductive assumption that $\aleph_\beta^2 = \aleph_\beta$ whenever $\beta < \alpha$. We shall show that then $\aleph_\alpha \times \aleph_\alpha \approx \aleph_\alpha$; from this the result follows. Since clearly $\aleph_\alpha \preccurlyeq \aleph_\alpha \times \aleph_\alpha$, we have only to show that $\aleph_\alpha \times \aleph_\alpha \preccurlyeq \aleph_\alpha$.

Note that for ordinals σ, τ with $\sigma, \tau < \aleph_\alpha$ also $\sigma + \tau < \aleph_\alpha$ (the ordinal sum of σ and τ). For since $\sigma < \aleph_\alpha$ either $|\sigma| < \aleph_0$ or $|\sigma| = \aleph_\gamma$ for some γ with $\gamma < \alpha$, and likewise for τ. Hence there is β with $\beta < \alpha$ such that $|\sigma|, |\tau| \leqslant \aleph_\beta$. But

$$|\sigma + \tau| = |\sigma| \dot{+} |\tau| \leqslant \aleph_\beta + \aleph_\beta = 2 \cdot \aleph_\beta \leqslant \aleph_\beta^2 ,$$

and by the inductive hypothesis $\aleph_\beta^2 = \aleph_\beta$. Hence $|\sigma + \tau| \leqslant \aleph_\beta < \aleph_\alpha$, and so $\sigma + \tau < \aleph_\alpha$.

Define an ordering $<<$ on $\aleph_\alpha \times \aleph_\alpha$ as follows:

$$\langle \sigma, \tau \rangle << \langle \sigma', \tau' \rangle \Leftrightarrow (\sigma + \tau < \sigma' + \tau') \text{ or } (\sigma + \tau = \sigma' + \tau' \text{ and } \sigma < \sigma') .$$

It is easy to see that $<<$ is a well ordering. Let $\mathrm{tp}(\aleph_\alpha \times \aleph_\alpha, <<) = \gamma$. Clearly $\aleph_\alpha \leqslant \gamma$. In fact $\gamma = \aleph_\alpha$. For suppose, on the contrary, that $\aleph_\alpha < \gamma$. There is than an order isomorphism $f : \gamma \to \aleph_\alpha \times \aleph_\alpha$, and let $f(\aleph_\alpha) = \langle \sigma, \tau \rangle$. Consider

$$S = \{\langle \mu, \nu \rangle \in \aleph_\alpha \times \aleph_\alpha ; \langle \mu, \nu \rangle << \langle \sigma, \tau \rangle\} .$$

Put $\sigma + \tau = \xi$, so $\xi < \aleph_\alpha$. If $\langle \mu, \nu \rangle << \langle \sigma, \tau \rangle$ then $\mu + \nu \leqslant \sigma + \tau = \xi$, so $\mu, \nu \leqslant \xi$. Thus $S \subseteq (\xi + 1) \times (\xi + 1)$, so $|S| \leqslant |\xi + 1|^2$. Since $\xi < \aleph_\alpha$, also $|\xi + 1| < \aleph_\alpha$ so by the inductive hypothesis $|\xi + 1|^2 < \aleph_\alpha$. Thus $|S| < \aleph_\alpha$. Yet f restricted to $\aleph_\alpha$ is one-to-one and onto S, so $|S| = \aleph_\alpha$. This contradiction shows that $\gamma = \aleph_\alpha$. Hence $\aleph_\alpha \times \aleph_\alpha \preccurlyeq \aleph_\alpha$, as required.

Corollary A.3.4. *If κ, λ are infinite cardinals then*

$$\kappa \dot{+} \lambda = \kappa \cdot \lambda = \max(\kappa, \lambda) .$$

If $\kappa \leqslant \lambda$ then $\kappa^\lambda = 2^\lambda$.

Proof. Suppose $\kappa \leqslant \lambda$. Then $\lambda \leqslant \kappa \dot{+} \lambda \leqslant \lambda \dot{+} \lambda = 2.\ \lambda \leqslant \lambda^2 = \lambda$, so $\kappa \dot{+} \lambda = \lambda$. Also $\lambda \leqslant \kappa \cdot \lambda \leqslant \lambda^2 = \lambda$ so $\kappa \cdot \lambda = \lambda$. And $2^\lambda \leqslant \kappa^\lambda \leqslant \lambda^\lambda \leqslant (2^\lambda)^\lambda = 2^{\lambda^2} = 2^\lambda$, so $\kappa^\lambda = 2^\lambda$.

The definitions of sum and product of two cardinals generalize to the sum and product of an arbitrary family of cardinals as follows. Given a family

$(\kappa_i; i \in I)$ of cardinals, define $\Sigma(\kappa_i; i \in I)$ to be the cardinality of $\bigcup\{A_i; i \in I\}$ where the A_i are pairwise disjoint sets with $|A_i| = \kappa_i$, and define $\Pi(\kappa_i; i \in I)$ to be the cardinality of the cartesian product $\times(A_i; i \in I)$ of sets A_i where $|A_i| = \kappa_i$. There is the classical theorem of König connecting these two notions.

Theorem A.3.5 (König). *Let $(\kappa_i; i \in I)$ and $(\lambda_i; i \in I)$ be two families of cardinals with always $1 \leqslant \kappa_i < \lambda_i$. Then*

$$\Sigma(\kappa_i; i \in I) < \Pi(\lambda_i; i \in I) .$$

Proof. For each i in I, choose two sets A_i, B_i such that $A_i \subseteq B_i$, $|A_i| = \kappa_i$, $|B_i| = \lambda_i$ where the A_i are pairwise disjoint. Choose b_i from $B_i - A_i$. Put $A = \bigcup\{A_i; i \in I\}$ and $B = \times(B_i; i \in I)$. Define a function $F : A \to B$ by, for a in A with $a \in A_i$,

$$F(a)(j) = \begin{cases} a & \text{if } j = i , \\ b_j & \text{if } j \neq i . \end{cases}$$

Then F is one-to-one and so $|A| \leqslant |B|$.

Suppose in fact $|A| = |B|$, so there is a one-to-one and onto map $G : A \to B$. Then $B = \bigcup\{G[A_i]; i \in I\}$. For each i in I, put $C_i = \{g(i); g \in G[A_i]\}$; then $C_i \subseteq B_i$. And $C_i \neq B_i$ since $|C_i| \leqslant |G[A_i]| = |A_i| < |B_i|$. Choose x_i from $B_i - C_i$ and define f in B by $f(i) = x_i$ for all i. Then $f \notin G[A]$, contradicting that G maps onto B. Hence $|A| \neq |B|$, so $|A| < |B|$. Thus $\Sigma(\kappa_i; i \in I) < \Pi(\lambda_i; i \in I)$ as claimed.

We shall now review the concept of cofinality, particularly as applied to cardinal numbers. For any ordinal number α, the subset C of α is said to be *cofinal* in α if for all β with $\beta < \alpha$ there is γ in C with $\gamma \geqslant \beta$.

Definition A.3.6. The *cofinality* of the ordinal α (denoted by α' or $\mathrm{cf}(\alpha)$) is the least ordinal β such that there is a mapping of β onto a cofinal subset of α. The ordinal α is *regular* if $\alpha' = \alpha$; otherwise α is *singular*.

It is clear from the definition that α' is always a cardinal, and $\alpha' \leqslant \alpha$. If α is a successor ordinal then $\alpha' = 1$, if α is a limit ordinal then $\alpha' \geqslant \omega$.

Lemma A.3.7. *If α is any ordinal then α' is the least ordinal β such that there is a strictly increasing map $f : \beta \to \alpha$ with $f[\beta]$ cofinal in α.*

Proof. This is trivial if $\alpha' = 0, 1$ so we may suppose $\alpha' \geqslant \omega$. By the definition of α', there is a map $g : \alpha' \to \alpha$ with $g[\alpha']$ cofinal in α. To prove the lemma, it will suffice to find a strictly increasing map $f : \alpha' \to \alpha$ with $f[\alpha']$ cofinal in α. Define the values $f(\gamma)$ for γ with $\gamma < \alpha'$ by induction as follows: $f(\gamma)$ is the least ordinal ξ with $\xi < \alpha$ such that $\xi > g(\delta)$ and $\xi > f(\delta)$ for all δ with $\delta < \gamma$. Since $\gamma < \alpha'$, neither $\{g(\delta); \delta < \gamma\}$ nor $\{f(\delta); \delta < \gamma\}$ is cofinal in α so there is such a ξ. Then f is strictly increasing, and $f[\alpha']$ is cofinal in α since $g[\alpha']$ is.

Theorem A.3.8. *For all α, the cofinality α' is regular.*

Proof. We need to show that $\alpha'' = \alpha'$. It suffices to show that there is a mapping of α'' onto a cofinal subset of α, for then $\alpha' \leqslant \alpha''$. Since anyway $\alpha'' \leqslant \alpha'$, this would give the result.

By Lemma A.3.7 there are increasing maps $f : \alpha'' \to \alpha'$ and $g : \alpha' \to \alpha$ with $f[\alpha'']$ cofinal in α' and $g[\alpha']$ cofinal in α. Then the composition gf is increasing, $gf : \alpha'' \to \alpha$. Also $gf[\alpha'']$ is cofinal in α. For given β with $\beta < \alpha$, there is γ in α' with $g(\gamma) \geqslant \beta$ and δ in α'' with $f(\delta) \geqslant \gamma$. But then $gf(\delta) \geqslant g(\gamma) \geqslant \beta$, so $gf[\alpha'']$ is indeed cofinal in α.

Theorem A.3.9. *Let κ be an infinite cardinal. Then κ' is the least cardinal λ such that κ can be expressed as a sum of λ cardinals all less than κ. If κ is singular then κ can be expressed as the sum of a strictly increasing sequence of κ' cardinals all less than κ.*

Proof. Given κ, by Lemma A.3.7 there is a strictly increasing function $f : \kappa' \to \kappa$ with $f[\kappa']$ cofinal in κ. For ξ with $\xi < \kappa'$, put $A_\xi = f(\xi + 1) - f(\xi)$. Then $\kappa = \bigcup\{A_\xi; \xi < \kappa'\}$ and the A_ξ are pairwise disjoint, so $\kappa = \Sigma(|A_\xi|; \xi < \kappa')$. And $|A_\xi| \leqslant |f(\xi + 1)| \leqslant f(\xi + 1) < \kappa$, so certainly κ is the sum of κ' cardinals all less than κ.

Now let λ be the least cardinal such that $\kappa = \Sigma(\kappa_\xi; \xi < \lambda)$ for some cardinals κ_ξ with always $\kappa_\xi < \kappa$, so $\aleph_0 \leqslant \lambda \leqslant \kappa'$. If $\alpha < \lambda$ then $|\Sigma_0(\kappa_\xi; \xi < \alpha)| = \Sigma(\kappa_\xi; \xi < \alpha) < \kappa$, since $|\alpha| < \lambda$. Hence if $A = \{\Sigma_0(\kappa_\xi; \xi < \alpha); \alpha < \lambda\}$ then $A \subseteq \kappa$. Because λ is a limit ordinal, $\Sigma_0(\kappa_\xi; \xi < \lambda) = \bigcup\{\Sigma_0(\kappa_\xi; \xi < \alpha); \alpha < \lambda\} \leqslant \kappa$, and since $|\Sigma_0(\kappa_\xi; \xi < \lambda)| = \Sigma(\kappa_\xi; \xi < \lambda) = \kappa$ in fact $\Sigma_0(\kappa_\xi; \xi < \lambda) = \kappa$. It follows that A is cofinal in κ. Hence $\kappa' \leqslant |A| = \lambda$, and thus $\lambda = \kappa'$.

Finally, suppose κ is singular, and write $\kappa = \Sigma(\kappa_\sigma; \sigma < \kappa')$ where always $\kappa_\sigma < \kappa$. Then $\{\kappa_\sigma; \sigma < \kappa'\}$ is cofinal in κ, for if not there would be β with $\beta < \kappa$ such that $\kappa_\sigma < \beta$ for all σ, which would give the contradiction

$$\kappa = \Sigma(\kappa_\sigma; \sigma < \kappa') \leqslant |\beta| \cdot \kappa' < \kappa\ .$$

Thus inductively we may choose cardinals λ_τ for τ with $\tau < \kappa'$ such that

$\lambda_\tau \in \{\kappa_\sigma; \tau \leqslant \sigma < \kappa'\}$ and $\lambda_\tau \geqslant \bigcup\{\lambda_\nu + 1; \nu < \tau\}$, since always $\bigcup\{\lambda_\nu + 1; \nu < \tau\} < \kappa$ when $\tau < \kappa'$. Then $\langle \lambda_\tau; \tau < \kappa' \rangle$ is a strictly increasing sequence of cardinals below κ, and $\kappa = \Sigma(\lambda_\tau; \tau < \kappa')$.

It follows from Theorem A.3.9 that if κ is any infinite successor cardinal then κ is regular. For suppose $\kappa = \lambda^+$. By Theorem A.3.9, write $\kappa = \Sigma(\kappa_\sigma; \sigma < \kappa')$ where always $\kappa_\sigma < \kappa$, so $\kappa_\sigma \leqslant \lambda$. If $\kappa' < \kappa$ then $\kappa' \leqslant \lambda$ and there would be the contradiction

$$\kappa = \Sigma(\kappa_\sigma; \sigma < \kappa') \leqslant \lambda \cdot \lambda = \lambda^2 = \lambda < \kappa ;$$

hence $\kappa' = \kappa$ so that κ is regular. Thus for the infinite cardinal κ to be singular, it must be that $\kappa = \aleph_\alpha$ for some limit ordinal α. It is easy to see that then $\kappa' = \alpha'$. A regular cardinal $\aleph_\alpha$ with α a limit ordinal is said to be *weakly inaccessible*; the existence of a weakly inaccessible cardinal cannot be proved from the axioms of ZFC set theory.

We shall conclude this section with a few remarks about cardinal exponentiation. Cantor's theorem showed that always $2^\kappa > \kappa$. This result can be strengthened as follows.

Theorem A.3.10. *If κ is infinite then* $(2^\kappa)' > \kappa$.

Proof. Write $2^\kappa = \Sigma(\lambda_\sigma; \sigma < (2^\kappa)')$ where always $1 \leqslant \lambda_\sigma < 2^\kappa$. If $(2^\kappa)' \leqslant \kappa$, from König's Lemma (Theorem A.3.5) we would obtain the contradiction

$$2^\kappa = \Sigma(\lambda_\sigma; \sigma < (2^\kappa)') < \Pi(2^\kappa; \sigma < (2^\kappa)') \leqslant (2^\kappa)^\kappa = 2^{\kappa \cdot \kappa} = 2^\kappa ;$$

and hence $(2^\kappa)' > \kappa$.

The Continuum Hypothesis is the statement $2^{\aleph_0} = \aleph_1$; the Generalized Continuum Hypothesis (GCH) is the statement $\forall\alpha(2^{\aleph_\alpha} = \aleph_{\alpha+1})$. Both statements are consistent with and independent from the axioms of ZFC. Assuming the GCH, or at least particular instances of it, enables one to evaluate κ^λ for infinite κ and λ, as in Theorem A.3.11 below. In ZFC alone apart from Theorem A.3.10 very little can be proved about the value of κ^λ.

Theorem A.3.11. *Let κ, λ be infinite cardinals. Suppose $2^\theta = \theta^+$ for all infinite cardinals θ with $\theta \leqslant \max(\kappa, \lambda)$. Then*

$$\kappa^\lambda = \begin{cases} \kappa & \text{if } \lambda < \lambda' , \\ \kappa^+ & \text{if } \kappa' \leqslant \lambda < \kappa , \\ \lambda^+ & \text{if } \kappa \leqslant \lambda . \end{cases}$$

Proof. Suppose $\lambda < \kappa'$. If f is a function $f: \lambda \to \kappa$, then $f[\lambda]$ is not cofinal in κ and so $f[\lambda] \subseteq \alpha$ for some α with $\alpha < \kappa$. Hence ${}^{\lambda}\kappa = \bigcup\{{}^{\lambda}\alpha; \alpha < \kappa\}$. And if $\alpha < \kappa$ then

$$|{}^{\lambda}\alpha| = |\alpha|^{\lambda} \leqslant (2^{|\alpha|})^{\lambda} = 2^{|\alpha| \cdot \lambda} = (|\alpha| \cdot \lambda)^{+} \leqslant \kappa.$$

Hence $\kappa^{\lambda} = |{}^{\lambda}\kappa| \leqslant \Sigma(|{}^{\lambda}\alpha|; \alpha < \kappa) \leqslant \kappa \cdot \kappa = \kappa$ and so $\kappa^{\lambda} = \kappa$.

Now suppose $\kappa' \leqslant \lambda < \kappa$, so κ is singular. By Theorem A.3.9 there is a strictly increasing sequence $\langle \kappa_{\sigma}; \sigma < \kappa' \rangle$ of cardinals below κ with $\kappa = \Sigma(\kappa_{\sigma}; \sigma < \kappa')$. By König's Lemma,

$$\kappa = \Sigma(\kappa_{\sigma}; \sigma < \kappa') < \Pi(\kappa_{\sigma+1}; \sigma < \kappa') \leqslant \kappa^{\kappa'} \leqslant \kappa^{\lambda}.$$

Also $\kappa^{\lambda} \leqslant (2^{\kappa})^{\lambda} = 2^{\kappa \cdot \lambda} = 2^{\kappa} = \kappa^{+}$, so $\kappa < \kappa^{\lambda} \leqslant \kappa^{+}$ and hence $\kappa^{\lambda} = \kappa^{+}$.

Finally, if $\kappa \leqslant \lambda$ by Corollary A.3.4, $\kappa^{\lambda} = 2^{\lambda}$ and so $\kappa^{\lambda} = \lambda^{+}$.

REFERENCES

[1] H. Bachmann, *Transfinite Zahlen*, Ergebnisse der Mathematik und ihrer Grenzgebiete, Vol. 1 (2nd ed.) (Springer, Berlin, 1967).

[2] F. Bagemihl, The existence of an everywhere dense independent set, *Michigan Math. J.* **20** (1973) 1–2.

[3] J.E. Baumgartner, Improvement of a partition theorem of Erdös and Rado, *J. Combinatorial Theory Ser. A* **17** (1974) 134–137.

[4] J.E. Baumgartner and A. Hajnal, A proof (involving Martin's axiom) of a partition relation, *Fund. Math.* 78 (1973) 193–203.

[5] J.E. Baumgartner and K. Prikry, On a theorem of Silver, *Discrete Math.* **14** (1976) 17–21.

[6] N.G. de Bruijn and P. Erdös, A colour problem for infinite graphs and a problem in the theory of relations, *Nederl. Akad. Wetensch. Proc. Ser. A* **54** = *Indag. Math.* **15** (1951) 371–373.

[7] C.C. Chang, A partition theorem for the complete graph on ω^ω, *J. Combinatorial Theory Ser. A* **12** (1972) 396–452.

[8] R.O. Davies, An intersection theorem of Erdös and Rado, *Proc. Cambridge Philos. Soc.* **63** (1967) 995–996.

[9] B. Descartes, Solution to Advanced Problem No. 4526, *Amer. Math. Monthly* **61** (1954) 352–353.

[10] F.R. Drake, *Set Theory*. Studies in Logic and the Foundations of Mathematics, Vol. 76. (North-Holland, Amsterdam, 1974).

[11] B. Dushnik and E.W. Miller, Partially ordered sets, *Amer. J. Math.* **63** (1941) 600–610.

[12] W.B. Easton, Powers of regular cardinals, *Ann. Math. Logic* **1** (1970) 139–178.

[13] P. Erdös, Some set-theoretical properties of graphs, *Univ. Nac. Tucumán. Revista A* **3** (1942) 363–367.

[14] P. Erdös, Some remarks on set theory, *Proc. Amer. Math. Soc.* **1** (1950) 127–141.

[15] P. Erdös, Graph theory and probability, *Canad. J. Math.* **11** (1959) 34–38.

[16] P. Erdös and G. Fodor, Some remarks on set theory – V, *Acta Sci. Math. Szeged* **17** (1957) 250–260.

[17] P. Erdös and G. Fodor, Some remarks on set theory – VI, *Acta Sci. Math. Szeged* **18** (1958) 243–260.

[18] P. Erdös and A. Hajnal, On the structure of set-mappings, *Acta Math. Acad. Sci. Hungar.* 9 (1958) 111–131.

[19] P. Erdös and A. Hajnal, On a property of families of sets, *Acta Math. Acad. Sci. Hungar.* **12** (1961) 87–123.
[20] P. Erdös and A. Hajnal, Some remarks on set theory – IX. Combinatorial problems in measure theory and set theory, *Michigan Math. J.* **11** (1964) 107–127.
[21] P. Erdös and A. Hajnal, On chromatic number of graphs and set-systems, *Acta Math. Acad. Sci. Hungar.* **17** (1966) 61–99.
[22] P. Erdös and A. Hajnal, On decomposition of graphs, *Acta Math. Acad. Sci. Hungar.* **18** (1967) 359–377.
[23] P. Erdös and A. Hajnal, On chromatic number of infinite graphs, in: P. Erdös et al., eds., *Theory of Graphs* (Proc. Colloq., Tihany, 1966) (Academic Press, New York, 1968) pp. 83–98.
[24] P. Erdös and A. Hajnal, Unsolved problems in set theory, in. D.S. Scott, ed., *Axiomatic Set Theory* (Proc. Sympos. Pure Math., Vol. 13, Part I, Univ. California, Los Angeles, Ca., 1967) (Amer. Math. Soc., Providence, R.I., 1971) pp. 17–48.
[25] P. Erdös and A. Hajnal, Unsolved and solved problems in set theory, in: L. Henkin et al., eds., *Proceedings of the Tarski Symposium* (Proc. Sympos. Pure Math., Vol. 25, Univ. of California, Berkeley, CA, 1971) (Amer. Math. Soc., Providence, R.I., 1974) pp. 267–287.
[26] P. Erdös and E.C. Milner, A theorem in the partition calculus, *Canad. Math. Bull.* **15** (1972) 501–505; Corrigendum, *ibid.* 17 (1974) 305.
[27] P. Erdös and R. Rado, Combinatorial theorems on classifications of subsets of a given set, *Proc. London Math. Soc. Ser. 3,* **2** (1952) 417–439.
[28] P. Erdös and R. Rado, A problem on ordered sets, *J. London Math. Soc.* **28** (1953) 426–438.
[29] P. Erdös and R. Rado, A partition calculus in set theory, *Bull. Amer. Math. Soc.* **62** (1956) 427–489.
[30] P. Erdös and R. Rado, A construction of graphs without triangles having pre-assigned order and chromatic number, *J. London Math. Soc.* **35** (1960) 445–448.
[31] P. Erdös and R. Rado, Intersection theorems for systems of sets, *J. London Math. Soc.* **35** (1960) 85–90.
[32] P. Erdös and R. Rado, Partition relations and transitivity domains of binary relations, *J. London Math. Soc.* **42** (1967) 624–633.
[33] P. Erdös and R. Rado, Intersection theorems for systems of sets (II), *J. London Math. Soc.* **44** (1969) 467–479.
[34] P. Erdös, L. Gillman and M. Henriksen, An isomorphism theorem for real-closed fields, *Ann. of Math. Ser. 2,* **61** (1955) 542–554.
[35] P. Erdös, A. Hajnal and E.C. Milner, On the complete subgraphs of graphs defined by systems of sets, *Acta Math. Acad. Sci. Hungar.* **17** (1966) 159–229.
[36] P. Erdös, A. Hajnal and E.C. Milner, On sets of almost disjoint subsets of a set, *Acta Math. Acad. Sci. Hungar.* **19** (1968) 209–218.
[37] P. Erdös, A. Hajnal and E.C. Milner, Set mappings and polarized partition relations, in: P. Erdös et al., eds., *Combinatorial Theory and its Applications I* (Proc. Colloq., Balatonfüred, 1969) (North-Holland, Amsterdam, 1970) pp. 327–363.
[38] P. Erdös, A. Hajnal and R. Rado, Partition relations for cardinal numbers, *Acta Math. Acad. Sci. Hungar.* **16** (1965) 93–196.
[39] P. Erdös, E.C. Milner and R. Rado, Intersection theorems for systems of sets (III), *J. Austral. Math. Soc.* **18** (1974) 22–40.

[40] G. Fodor, Proof of a conjecture of P. Erdös, *Acta Sci. Math. Szeged* **14** (1952) 219–227.

[41] G. Fodor, Some results concerning a problem in set theory, *Acta Sci. Math. Szeged* **16** (1955) 232–240.

[42] G. Fodor, Eine Bemerkung zur Theorie der Regressiven Funktionen, *Acta Sci. Math. Szeged* 17 (1956) 139–202.

[43] A.A. Fraenkel, Einleitung in die Mengenlehre. Die Grundlehren der Mathematischen Wissenschaften in Einzeldarstellungen, Vol. 9 (3rd. ed.) (Springer, Berlin, 1928).

[44] F. Galvin, On a partition theorem of Baumgartner and Hajnal, in: A. Hajnal, ed., *Infinite and Finite Sets*, Vol. II, (Colloq., Keszthely, 1973. Colloq. Math. Janos Bolyai, Vol. 10) (North-Holland, Amsterdam, 1975) pp. 711–729.

[45] F. Galvin and A. Hajnal, Inequalities for cardinal powers, *Ann. of Math. Ser. 2,* **101** (1975) 491–498.

[46] F. Galvin and J. Larson, Pinning countable ordinals, *Fund. Math.* **82** (1974) 357–361.

[47] F. Galvin and S. Shelah, Negation of partition relations without CH, *Notices Amer. Math. Soc.* **18** (1971) 666.

[48] F. Galvin and S. Shelah, Some counterexamples in the partition calculus, *J. Combinatorial Theory Ser. A,* **15** (1973) 167–174.

[49] K. Gödel, *The Consistency of the Axiom of Choice and the Generalized Continuum Hypothesis with the Axioms of Set Theory.* Annals of Mathematics Studies No. 3 (Princeton University Press, Princeton, N.J., 1940).

[50] L. Haddad and G. Sabbagh, Calcul de certains nombres de Ramsey généralisés, *C.R. Acad. Sci. Paris Sér. A–B,* **268** (1969) A1233–A1234.

[51] L. Haddad and G. Sabbagh, Nouveaux résultats sur les nombres de Ramsay généralisés, *C.R. Acad. Sci. Paris Sér. A–B,* **268** (1969) A1516–A1518.

[52] A. Hajnal, Some results and problems on set theory, *Acta Math. Acad. Sci. Hungar.* **11** (1960) 277–298.

[53] A. Hajnal, Proof of a conjecture of S. Ruziewicz, *Fund. Math.* **50** (1961/62) 123–128.

[54] A. Hajnal, Remarks on the theorem of W.P. Hanf, *Fund. Math.* **54** (1964) 109–113.

[55] A. Hajnal, On some combinatorial problems involving large cardinals, *Fund. Math.* **69** (1970) 39–53.

[56] A. Hajnal, A negative partition relation, *Proc. Nat. Acad. Sci. U.S.A.* **68** (1971) 142–144.

[57] W. Hanf, Incompactness in languages with infinitely long expressions, *Fund. Math.* **53** (1963/64) 309–324.

[58] J.M. Henle and E.M. Kleinberg, A combinatorial proof of a combinatorial theorem, *Acta Math. Acad. Sci. Hungar.* **26** (1975) 3–7.

[59] I. Juhász, *Cardinal Functions in Topology.* Mathematical Centre Tracts Vol. 34 (Mathematisch Centrum, Amsterdam, 1971).

[60] J.L. Kelly, *General Topology* (van Nostrand, New York, 1955).

[61] E.M. Kleinberg, Infinitary combinatorics, in: A.R.D. Mathias et al., eds., *Cambridge Summer School in Mathematical Logic* (Cambridge, England, 1971. Lecture Notes in Mathematics Vol. 337) (Springer, Berlin, 1973) pp. 361–418.

[62] A.H. Kruse, A note on the partition calculus of P. Erdös and R. Rado, *J. London Math. Soc.* **40** (1965) 137–148.

[63] C. Kuratowski, Sur une caractérisation des alephs, *Fund. Math.* 38 (1951) 14–17.
[64] D. Kurepa, On the cardinal number of ordered sets and of symmetrical structures in dependence of the cardinal numbers of its chains and antichains, *Glasnik Mat.-Fiz. Astronom. Drustvo Math. Fiz. Hrvatski Ser. II* **14** (1959) 183–203.
[65] J.A. Larson, A short proof of a partition theorem for the ordinal ω^ω, *Ann. Math. Logic* **6** (1973) 129–145.
[66] D. Lázár, On a problem in the theory of aggregates, *Compositio Math.* **3** (1936) 304.
[67] S. Mazur, On continuous mappings on cartesian products, *Fund. Math.* **39** (1952) 229–238.
[68] E. Michael, A note on intersections, *Proc. Amer. Math. Soc.* **13** (1962) 281–283.
[69] E.C. Milner, Some combinatorial problems in set theory, Thesis, London, 1962.
[70] E.C. Milner, Transversals of disjoint sets, *J. London Math. Soc.* **43** (1968) 495–500.
[71] E.C. Milner, Partition relations for ordinal numbers, *Canad. J. Math.* **21** (1969) 317–334.
[72] E.C. Milner, A finite algorithm for the partition calculus, in: W.R. Eames et al., eds., *Proceedings of the Twenty-Fifth Summer Meeting of the Canadian Mathematical Congress* (Lakehead Univ., Thunder Bay, Ont., 1971) (Lakehead Univ., Thunder Bay, Ont., 1971) pp. 117–128.
[73] E.C. Milner and R. Rado, The pigeon-hole principle for ordinal numbers, *Proc. London Math. Soc. Ser. 3,* **15** (1965) 750–768.
[74] M. Morley, Partitions and models, in: M.H. Löb, ed., *Proceedings of the Summer School in Logic* (Leeds, 1967, Lecture Notes in Mathematics Vol. 70) (Springer, Berlin, 1968) pp. 109–158.
[75] E. Nosal, On a partition relation for ordinal numbers, *J. London Math. Soc. Ser. 2* **8** (1974) 306–310.
[76] S. Picard, Solution d'un problème de la théorie des relations, *Fund. Math.* 28 (1937) 197–202.
[77] S. Piccard, Sur un problème de M. Ruziewicz de la théorie des relations, *Fund. Math.* **29** (1937) 5–8.
[78] K. Prikry, On a problem of Erdös, Hajnal and Rado, *Discrete Math.* **2** (1972) 51–59.
[79] K. Prikry, On a set-theoretic partition problem, *Duke Math. J.* **39** (1972) 77–83.
[80] F.P. Ramsey, On a problem of formal logic, *Proc. London Math. Soc. Ser. 2,* **30** (1930) 264–286.
[81] F. Rowbottom, Some strong axioms of infinity incompatible with the axiom of constructibility, *Ann. Math. Logic* **3** (1971) 1–44.
[82] S. Ruziewicz, Une généralisation d'un théorème de M. Sierpinski, *Publ. Math. Univ. Belgrade* **5** (1936) 23–27.
[83] R.A. Shore, Square bracket partition relations in L, *Fund. Math.* **84** (1974) 101–106.
[84] W. Sierpinski, Sur une décomposition d'ensembles, *Monatsch. Math. Phys.* **35** (1928) 239–242.
[85] W. Sierpinski, Sur un problème de la théorie des relations, *Ann. Scuola Norm. Super. Pisa Ser. 2,* **2** (1933) 285–287.
[86] W. Sierpinski, Sur un problème de la théorie des relations, *Fund. Math.* **28** (1937) 71–74.
[87] W. Sierpinski, Sur les suites transfinies finalement disjointes, *Fund. Math.* **28** (1937) 115–119.

[88] W. Sierpinski, Sur quelques propositions concernant la puissance du continu, *Fund. Math.* **38** (1951) 1–13.

[89] W. Sierpinski, *Cardinal and Ordinal Numbers.* Polska Akademia Nauk, Monografie Matematyczne Vol. **34** (2nd ed.) (P.W.N.-Polish Scientific Publishers, Warsaw, 1965).

[90] J.H. Silver, Some applications of model theory in set theory, Thesis, Berkeley, 1966. See also *Ann. Math. Logic* **3** (1971) 45–110.

[91] J.H. Silver, The independence of Kurepa's conjecture and two-cardinal conjectures in model theory, in: D.S. Scott, ed., *Axiomatic Set Theory* (Proc. Sympos. Pure Math., Vol. 13, Part I, Univ. California, Los Angeles, CA, 1967) (Amer. Math. Soc., Providence, R.I., 1971) pp. 383–390.

[92] J.H. Silver, On the singular cardinals problem, in: *Proceedings of the International Congress of Mathematicians* (Vancouver 1974) Vol. 1 (Canadian Mathematical Congress, Montréal, 1975) pp. 265–268.

[93] E. Specker, Teilmengen von Mengen mit Relationen, *Comment. Math. Helv.* **31** (1957) 302–314.

[94] A. Tarski, Sur la décomposition des ensembles en sous-ensembles presque disjoints, *Fund. Math.* **12** (1928) 188–205 and **14** (1929) 205–215.

[95] N.H. Williams, A partition property of cardinal numbers, *Dissertationes Math. (Rozprawy Mat.)* **90** (1972) 1–26.

[96] A.A. Zykov, On some properties of linear complexes, *Mat. Sbornik N.S.* **24 (66)** (1949) (Russian) 163–188.

INDEX OF AUTHORS

INDEX OF DEFINITIONS

INDEX OF NOTATION